UNITED STATES ARMY IN THE KOREAN WAR

THE KOREAN WAR

POLICY AND DIRECTION: THE FIRST YEAR

by

James F. Schnabel

REPRINTED BY ST. JOHN'S PRESS LLC
ALEXANDRIA, VA 22301
ISBN 978-1944961374
CENTER OF MILITARY HISTORY
UNITED STATES ARMY WASHINGTON, D.C.,
1992

Library of Congress Catalog Card Number: 70-609930

First Printed 1972—CMH Pub 20-1-1

For sale by the U.S. Government Printing Office
Superintendent of Documents, Mail Stop: SSOP, Washington, DC 20402-9328
ISBN 0-16-035955-4

UNITED STATES ARMY IN THE KOREAN WAR

Maurice Matloff, General Editor

. . . to Those Who Served

Foreword

This volume describes the initial direction and strategy of the first major though limited war that the United States was to fight on the continent of Asia in the era of global tension that followed World War II. There are marked similarities as well as some basic differences between the war in Korea and the war that would follow a decade later in Southeast Asia, and certainly the study of both is necessary to understand the limitations on armed conflict under the shadow of nuclear holocaust. One can also discern in this volume the importance of individuals in altering the course of human events and the fate of nations, the wider concerns that preclude the massing by a world power of its military strength in one direction, and many other facets of the nation's recent military history it behooves all thoughtful Americans to ponder.

Colonel Schnabel's work is the third to appear in a planned 5-volume history of the United States Army in the Korean War. It complements the detailed account of operations from June to November 1950, *South to the Naktong, North to the Yalu,* published in 1961, and the sequel to it still in preparation that will cover tactical operations through June 1951. The volume entitled *Truce Tent and Fighting Front,* published in 1966, covers the last two years of the war, and a logistical history of the Korean conflict is also scheduled to appear.

Both military and civilian students and the scholarly reading public should find in this book much that is illuminating and provocative of reflection, and not only about events that happened more than two decades ago.

Washington, D.C.
15 May 1971

JAMES L. COLLINS, JR.
Brigadier General, USA
Chief of Military History

The Author

James F. Schnabel enlisted in the U.S. Army in July 1942 and, after graduation from officer candidate school, the Field Artillery School, Fort Sill, Oklahoma, was commissioned as a second lieutenant in January 1943. Remaining in the service after World War II, he was assigned, as a captain, to the Military History Section, GHQ, FEC, in late 1949. He served as a historian in Tokyo until 1953, transferring in May 1953 to the Office of the Chief of Military History, U.S. Army, in Washington, D.C., where he remained until 1956. From 1956 until 1960 as a lieutenant colonel he held the post of Chief Historian, Supreme Headquarters, Allied Powers, Europe, and Allied Command Europe stationed in Paris, France. In 1960 he returned to the Office, Chief of Military History, and became Historical Liaison Officer to the Office of the Chief of Staff, U.S. Army. He retired from the Army in 1964 and is presently a historian with the Office of the Joint Chiefs of Staff.

A graduate of North Idaho Teachers College, Colonel Schnabel has published several articles on the Korean War and has appeared as a guest lecturer at the Army War College.

Preface

This book is intended to elucidate United States policy during the Korean War and to describe the strategies and command methods by which that policy was carried out. The major decisions that determined the United States course in Korea and continued to influence the nation's responses to Communist aggression during the two decades that followed were taken during the first twelve months of the Korean fighting. Although the war continued for another two years, no significant change was made in the policy developed between President Truman's decision to intervene in June 1950 and the beginning of Armistice negotiations at Kaesong in July 1951. This book concentrates on that initial period.

One of the unique aspects of the Korean War was the close control which Washington maintained at all times over operations in the field. Routine transactions and problems which during World War II would have been handled by a theater commander became, during Korea, matters of great concern to the nation's highest officials in Washington. These exceptional practices were owing in large part to the scarcity of United States military resources when the war began and to the real danger that a miscalculation in Korea might result in a full-scale war with the Soviet Union and/or Communist China. The vast distance between Washington and the Far East served to hinder effective, timely communication, further complicating the problems of directing the war.

I was not aware of it at the time, but work on this book began three days after the North Korean invasion when I, as an Army captain of artillery assigned to the Historical Branch, G–2, GHQ, FEC, in Tokyo, was called to General MacArthur's personal file room in the Dai Ichi Building to examine copies of first teleconferences between CINCFE and Washington. Notes taken that day marked the beginning of nearly three years of research in Tokyo and, briefly, in Korea. Upon my return to Washington in mid-1953 I was designated to prepare the present volume. The first draft of this work was submitted to the Office, Chief of Military History, in June 1956 concurrently with my transfer to Paris, where I served until June 1960 as Chief Historian, SHAPE and Allied Command Europe. Returning to the Office of the Chief of Military History in July 1960 I was able, although assigned additional duties, to make revisions indicated as a result of the intensive review and criticism of the manuscript that had taken place in my four-year absence. Following my retirement from the Army in August 1964 further revision was performed by Mr. Billy C. Mossman and by Dr. Stetson Conn, then Chief Historian.

So many individuals have contributed to the present volume that it would be impossible for me to thank all of them publicly. There are those, however, to whom I owe special debts of gratitude. During the early years of my work on this history I received particular encouragement and very wise counsel from Col. Allison R. Hartman, then Chief, Historical Branch, G–2, GHQ, FEC. Among those outside critics who have reviewed all or part of the manuscript in its various stages and to whom I am indebted for valuable comments and ideas are General J. Lawton Collins, General Matthew B. Ridgway, Lt. Gen. Edward M. Almond, Mr. Robert Amory, formerly Deputy for Intelligence, Central Intelligence Agency, and Mr. Wilber W. Hoare, Jr., Chief, Historical Division, Joint Chiefs of Staff.

Other individuals within the Office of the Chief of Military History whose help and advice have been exceptionally valuable to me include the Editor in Chief, Mr. Joseph R. Friedman, whose literary guidance has contributed greatly in the final revision of the manuscript; Mr. Charles V. P. von Luttichau, who is responsible for the fine maps which accompany the text; Dr. Louis Morton and Dr. John Miller, jr., for their suggestions and criticism while they were with the Office of the Chief of Military History; Mr. David Jaffé, whose diligence and skill as an editor I have good reason to appreciate; and Mrs. Stephanie B. Demma, who rendered most able assistance to Mr. Jaffé. The index was prepared by Mr. Nicholas J. Anthony.

Finally, I would be remiss were I not to recognize gratefully the support I received from the several Chiefs of Military History under whom I served during the preparation of this volume: Maj. Gen. Orlando Ward, Maj. Gen. Albert C. Smith, Brig. Gen. James A. Norell, Brig. Gen. William H. Harris, and Brig. Gen. Hal C. Pattison.

That I have acknowledged the contributions made by those persons named above in no way implies that they share responsibility for the interpretations of this book or for any deficiencies that it may have. Responsibility for them is mine alone.

Washington, D.C.
15 May 1971

JAMES F. SCHNABEL

Contents

Color Maps

Illustrations

Illustrations are from Department of Defense files.

CHAPTER I

Korea, Case History of a Pawn

The Soviet-sponsored government of North Korea, having failed to conquer its southern neighbor by less violent means, invaded the Republic of Korea on 25 June 1950. When the United States, with other United Nations, came to the aid of the South Koreans, a 3-year war resulted that cost more than 142,000 American battle casualties.

The campaigns set in motion by the invasion of South Korea later were characterized as a "limited war." The fighting was deliberately confined in geographic terms, political decisions placed restrictions upon military strategy, and none of the belligerents, with the exception of the two Korean governments, used its full military potential. But there was nothing limited about the ferocity of the battles.

Erupting from the rivalries of great nations, the Korean War was greatly influenced by domestic conditions rooted deep in the history of Korea, and by the topography of the peninsula where it took place.

The Land

Korea is a harsh Asian peninsula inhabited by a hardy, harassed people who rarely if ever had been completely free. War and tragedy form the main theme of Korea's history. Suppression and ill-use have been the heritage of its long-suffering people. Few habitable areas of the earth are more unsuited to large-scale, modern military operations. The rugged landscape, a lack of adequate roads, rail lines, and military harbors, the narrow peninsula, and, not least, climatic extremes restrict and hamper maneuver, severely limit logistic support, and intensify the normal hardships of war.

Jutting from the central Asian mainland, the Korean peninsula has an outline resembling Florida's. In the north, a river-mountain complex separates Korea from Manchuria and the maritime provinces of the USSR. Eastward, across the Sea of Japan, the Japanese islands flank the peninsula. To the west, the Yellow Sea stands between Korea and China. The Korean peninsula stretches south for more than 500 miles, while east and west, it spans only 220 miles at its widest. Thousands of islets, some scarcely more than large rocks, rim its 5,400-mile coastline.

In area, Korea equals the combined states of Tennessee and Kentucky, covering about 85,000 square miles. The facetious claim that Korea, ironed flat, would cover the whole world has an element of truth, for the terrain throughout the peninsula is mountainous. Roads and railways wind through tortuous valleys. Ice-free ports exist on Korea's southern and western coasts, but the latter shore is distinguished by some of the most extreme tidal variations in the world. On the eastern shore, there are only a few adequate harbors. Although geographers place Korea in a temperate zone, the classification hardly mitigates the harsh winters, particularly in the wind-swept northern mountains, or the sweltering, dusty, and no less harsh summers in the south.

Korea's Past

The forces shaping Korea into a nation arose from its unfortunate proximity to three powers, China, Japan, and Russia. The periodic surges of ambition in each of these neighbors turned Korea into a battleground and a spoil. Sometimes described as a "dagger pointed at the heart of Japan," Korea became instead Japan's steppingstone to the Asian mainland. For China and, later, Russia, Korea was a back gate both to be locked against intruders and to be opened during any opportunity for expansion. Korea's ice-free ports fronting the Sea of Japan were especially coveted by the Russians. Korea therefore has seldom been completely free of domination by one of its stronger neighbors.[1]

China reached the Korean scene first, making its impact felt on northern Korea several centuries before the beginning of the Christian era. By the 7th century, A.D., the Chinese had forced their thought, customs, and manners into the Korean culture and had turned Korea into a virtual satellite. Late in that century, a native dynasty, Chinese-controlled, unified the peninsula. Before then Japan had occasionally invaded southern Korea, but with little lasting effect. Badly defeated by the Koreans in 663 A.D., Japan retired for nearly a thousand years.

Like China, Korea endured the Mongol armies in the 13th century. For nearly a hundred years the savages from the steppes ruled and ravaged Korea. Kublai Khan launched two abortive invasions of Japan from Korea, ruthlessly squandering Korean lives and property in his depredations.

With the gradual dissipation of Mongol power by the mid-14th century, Korea again basked in the reflected glory of a revitalized China. Adapting Chinese culture to their own talents, the Koreans

[1] Unless otherwise cited, material on Korea's history is based on the following: H. Frederick Nelson, *Korea and the Old Orders in Eastern Asia* (Baton Rouge: Louisiana State University Press, 1946); Yoshi Kuno, *Japanese Expansion on the Asiatic Continent*, 2 vols. (Berkeley: University of California Press, 1937), vol. I; Ernest W. Clement, *A Short History of Japan* (Tokyo: Christian Literature Society, 1926); Andrew Grajdanzev, *Modern Korea* (New York: The John Day Company, 1944); Cornelius Osgood, *The Koreans and Their Culture* (New York: Ronald Press, 1951); Harold M. Vinacke, *A History of the Far East in Modern Times* (New York: Appleton-Century-Crofts, 1941), pp. 123–24; A. Whitney Griswold, *Far Eastern Policy of the United States* (New York: Harcourt, Brace, 1938); George M. McCune and John A. Harrison, *Korean-American Relations*, 3 vols. (Berkeley: University of California Press, 1951), vol. I; Tyler Dennett, *Americans in Eastern Asia* (New York: The Macmillan Company, 1922).

flourished. Skilled artisans, craftsmen, and inventors, as well as philosophers and scholars, brought Korea a level of civilization rivaling that of China. But the Japanese violently disrupted this happy era. In a brutal expedition beginning in 1592, Japanese samurai under the brilliant Hideyoshi pillaged the peninsula for seven years. Aided by China, the Koreans eventually expelled the Japanese, but their home had become a wasteland. Their best artisans and scholars, along with the greater part of their portable treasure, were taken home by the Japanese.

In the following centuries, Korea kept loose cultural and political ties with China but withdrew from contact with the rest of the world. It never again reached the level of civilization the Japanese had destroyed. When Western influence spread to Asia in the 19th century, China's peculiar relationship with Korea baffled the West. Western efforts to trade with Korea were thwarted by this misunderstanding. The Koreans received Western overtures coldly. They impartially murdered French missionaries and American and Dutch seamen. Several punitive expeditions by these Western nations against Korea failed to improve relations.

Unfortunately for Korea's privacy, in 1860 Russia reached Korea's borders and later in the century westernization again whetted Japan's appetite for territorial expansion. With China, Japan, and Russia fighting for control of Korea throughout the rest of the 19th century, the Korean people had little chance to learn self-government. They remained separate from the modern world emerging around them.

Japan won Korea by defeating China and Russia, in turn, in short but decisive wars. In the Sino-Japanese War of 1894–95 Japan used Western military techniques to beat its larger but tradition-bound enemy. Ten years later, Japan astounded the world by defeating Russia. Having occupied Korea to fight Russia, Japan left its troops there. Ignoring Korean objections, Japan disbanded the Korean Army and abolished the Korean Department of Post and Communications. It allowed a semblance of self-rule in Korea for several years, but remained the real master. Japanese seizure of governmental functions, the forced abdication of Korea's Emperor, and encroachment in all aspects of Korean society culminated in an agreement in July 1907 placing Korea completely under Japanese control. The annexation of Korea by Japan in August 1910 was simply a formality.[2]

The United States and Korea

In the quarter century before the Japanese take-over, the United States showed a mild interest in Korea and made some effort to support Korean independence, at least in principle. In 1882, an American naval officer, Commodore Robert W. Shufeldt, negotiated a commercial treaty with the Korean Emperor. The result of four years' effort, this treaty was achieved through the reluctant good offices of the Chinese

[2] An account of Korean life under the Japanese can be found in History of the Occupation of Korea, August 1945–May 1948, 3 vols. (hereafter cited as History of Occupation of Korea), prepared in 1948 by historians of the XXIV Corps, vol. I, ch. 2, copy in OCMH.

Government. It provided for exchange of diplomatic representatives, protection of navigation and of United States citizens, extraterritoriality, and trade under a most-favored nation clause. The treaty could have given the United States overriding influence in Korea. But when the Emperor sought an American foreign affairs adviser and Army military advisers, the United States moved slowly. The matter dragged on for several years. The American representative in Korea repeatedly appealed to Washington for action. Although requested in 1884, military advisers reached Korea only in 1888.

The United States treated Korea casually in the late 19th century. Its only significance lay in the effect it had upon relationships with other major powers in the Far East. According to one authority, "The Korean Government was in the position of an incompetent defective not yet committed to guardianship. The United States was her only disinterested friend—but had no intention of becoming her guardian." [3]

When the Japanese took over Korea, the United States made no objection. President Theodore Roosevelt remarked, "We cannot possibly interfere for the Koreans against Japan. . . . They could not strike one blow in their own defense." On 29 July 1905, Secretary of War William H. Taft negotiated a secret "agreed memorandum" with the Japanese Prime Minister. The United States approved Japan's "suzerainty over" Korea in return for its pledge not to interfere with American interests in the Philippine Islands. The Korean Emperor's appeal to the United States for help under the "good offices" clauses of the Shufeldt Treaty fell on deaf ears.[4]

Between 1905 and 1910, uprisings and rebellions erupted frequently throughout Korea. Japan crushed them with efficient savagery. The Koreans had few weapons, and Japan was a powerful and merciless nation. According to Japanese statistics, 14,566 Korean "rebels" were killed between July 1907 and December 1908. By 1910, when Japan formally annexed Korea, little open resistance remained in the land; and no Western nation spoke out against Japan's seizure of the peninsula.

Complete suppression marked the ensuing thirty-five years of Japanese rule. The Japanese exploited the people and the land. But they also modernized Korea, building highways, railroads, dams, and factories. Much of this development was designed for military use. The port of Pusan, for example, was built for military, rather than commercial, reasons; and the rail line running from Pusan north to the Manchurian border had much more military than commercial value.

The Japanese integrated Korean industry into their own economy. Korea became completely dependent upon Japan for semimanufactured commodities, for repair parts, and for markets. Many key Korean plants produced only parts used in the final assembly of products in Japan. As Japan embarked on its program of conquest in Asia in the 1930's, the Japanese turned Korean in-

[3] Dennett, *Americans in Eastern Asia*, p. 495.

[4] (1) Griswold, *Far Eastern Policy of the United States*, p. 125. (2) Robert T. Oliver, *Verdict in Korea* (State College, Pa.: Bald Eagle Press, 1952), p. 37.

dustry almost exclusively to military use. The heavy, sustained use of machinery without adequate maintenance during World War II ruined Korean factories and equipment. The use of almost all chemical production, especially of nitrogen, in behalf of Japan's war effort caused severe soil depletion in Korea.[5]

Banning Koreans from responsible positions and from educational opportunities, the Japanese controlled key governmental and economic functions. Comprising only 3 percent of the population of Korea, the 750,000 Japanese residents were absolute masters of the country. Nearly 80 percent of the Korean people could neither read nor write.[6]

The Koreans deeply resented Japanese exploitation. Judged in Japanese courts under Japanese laws, they received severe sentences for minor offenses, more severe than those given Japanese for similar infractions. The Japanese-controlled Bank of Chosen charged Koreans interest rates 25 percent higher than those assessed Japanese competitors. The Korean national debt increased thirtyfold between 1910 and 1945, and the taxation of Koreans was oppressive. In most industries, Japanese received twice as much as Koreans doing the same work. Large numbers of farms were transferred from Korean to Japanese owners.[7]

Despite iron-handed Japanese rule that sought to crush Korean national aspirations, the flame of patriotism and independence remained alive in Korea. Revolutionary groups and movements sustained the Korean hope for freedom, defying the Japanese whenever possible. One strong group working to free Korea from alien rule called itself the "Provisional Government of the Republic of Great Korea." It originated on 1 March 1919 when a declaration of independence, signed by Korean students, was read before a student gathering in Seoul. The Japanese ruthlessly hunted down the instigators of this declaration, and many patriots fled Korea to escape torture and death. On 10 April 1919 some of these refugees met in Shanghai and established the Provisional Government. Dr. Syngman Rhee headed the group as Premier. After the Manchurian incident in 1931, the Provisional Government moved to Nanking and, later, to Chungking.

This group sought to achieve complete independence for Korea and to establish itself as the Korean Government. Differences on how these goals should be reached brought frequent clashes in the leadership of the Korean Provisional Government. Two men, Rhee and Kim Koo, emerged at the top. When Kim Koo became Premier in the mid-1930's, Rhee served as unofficial representative of the Provisional Government in the United States. The group acquired a considerable following among Koreans in the United States and China and attracted widespread passive support within Korea. Both Rhee and Kim were revered by the Korean people.[8]

A strong Korean Communist party also

[5] Testimony of Hon. Paul G. Hoffman, Administrator of the Economic Cooperation Administration (ECA) before the House Committee on Foreign Affairs, 81st Congress, 1st Session, 8 June 1949, in House Report No. 962, *Korean Aid*, H.R. 5330, June 1949, p. 9.

[6] *Ibid.*

[7] History of Occupation of Korea, vol. I, ch. 2.

[8] *Ibid.*, pp. 46–48.

SYNGMAN RHEE

KIM KOO

sprang up in Korea. Organized in 1925, it pushed the underground movement against Japan. Communist power in Korea grew under the well-organized leadership of the anti-Japanese underground. The Korean Communists were in contact with the Russian Communists through the Far Eastern Division of the Comintern. It is believed, however, that, owing to a secret agreement with Japan, the Russians abstained from encouraging too greatly the Communists in Korea during Japanese occupation. Many Communist Koreans took refuge in Manchuria, China, and Russia.[9]

In this setting of turbulent and long-suppressed patriotic emotions, it was inevitable that the political void caused by the fall of the Japanese Empire at the end of World War II should touch off a struggle for power.

[9] *Ibid.*, vol. II, ch. 2, pp. 7–20.

Korea 1945

When World War II began, Korea was regarded by the Allies as a victim of, not a party to, Japanese aggression. One of the earliest signs that the Allied Powers were concerned about Korea appeared in a joint statement by the United States, China, and Great Britain in December 1943, after the Cairo Conference, which said: "The aforesaid three great powers, mindful of the enslavement of the people of Korea, are determined that in due course Korea shall become free and independent." [10]

Divergencies between American and Russian policies appearing in the latter stages of World War II affected Korea. The destruction of the Axis in 1945 left

[10] Department of State, *Foreign Relations of the United States: The Conferences at Cairo and Tehran, 1943*, Dept. of State Publication 7187 (Washington, 1961), p. 448.

power vacuums in many areas of the world and brought the differences between the United States and the Soviet Union into sharp focus. Countries newly freed from German or Japanese domination assumed significance as possible targets of clashing American-Soviet interests.

Unlike the Soviet Union, the United States attached little importance to Korea as a strategic area. Korea supported a relatively small population, and had neither important industrial facilities nor many natural resources. If at some future date Korea fell into hands unfriendly to the United States, the United States recognized that the occupation of Japan might be hampered and American freedom of movement might be restricted in the general area. But with China in 1945 under control of a friendly government, such a situation appeared unlikely.

Russia, on the other hand, maintained its traditional regard for Korea as a strategic area. As later events demonstrated, the Soviet Union would not countenance control of Korea by another power and sought to control Korea itself.

President Franklin D. Roosevelt and Premier Josef V. Stalin at the Yalta Conference in 1945 touched upon Korea's future. Roosevelt advocated a trusteeship for Korea administered by the United States, the Soviet Union, and China. Looking at American experience in the Philippines, he surmised that such a trusteeship might last for twenty or thirty years. Stalin said he believed that Great Britain should also be a trustee. No actual mention of Korea was made in the document recording the agreements at Yalta. The secret protocol developed by Roosevelt and Stalin and agreed to by Prime Minister Winston S. Churchill only provided territorial and other concessions to the USSR in the Far East as conditions for Russian entrance into the war against Japan after the defeat of Germany. Later, soon after Roosevelt's death, Stalin told Harry Hopkins, President Harry S. Truman's representative in Moscow, that Russia was committed to the policy of a 4-power trusteeship for Korea.[11]

Though American military planners ostensibly paid little attention to Korea, they had Korea in mind. On 25 July 1945, the Army Chief of Staff, General of the Army George C. Marshall, sent a note to President Truman at Potsdam, advising him that some guidance on handling Korea would assist the Joint Chiefs of Staff. General of the Army Douglas MacArthur, Commander in Chief of the United States Army Forces, Pacific, had already received instructions to prepare for occupying Japan, and shortly before Potsdam these orders were broadened to include Korea. In response to the additional directive, General MacArthur suggested that Tokyo and Seoul have first priority for occupation, Pusan second priority, and the Kunsan area on Korea's west coast, third priority. General Marshall then informed the President that MacArthur should be able to land a division at Pusan within a short time of the end of the war. The other strategic areas in Korea, Mar-

[11] (1) Department of State, *Foreign Relations of the United States: The Conference at Malta and Yalta, 1945*, Dept of State Publication 6199 (Washington, 1955), pp. 770, 984. (2) Harry S. Truman, *Memoirs*, 2 vols., vol. II, *Years of Trial and Hope* (New York: Doubleday and Co., Inc., 1956), pp. 316–17.

shall added, were Seoul, near the west coast, and Ch'ongjin, in the north on the Sea of Japan. Marshall expected that the Russians, if they participated in the occupation, would occupy Ch'ongjin and would undoubtedly move into Manchuria and perhaps into north China. He considered it desirable, therefore, to establish early control over any areas to be held by the United States.[12]

Korea was only briefly considered at the Potsdam conference. Among the questions discussed were the Soviet timetable for entering the war in the Pacific and the Allied proclamation demanding Japan's unconditional surrender. Looking ahead to the surrender of the Japanese on the Asiatic mainland, the Allied military representatives drew a tentative line across the map of Manchuria, above which the Soviet Union was to accept surrender of Japanese forces. No mention was at first made of Korea. But since thousands of Japanese troops were stationed in Korea, there was a later discussion of Allied operations in that area.[13]

At Potsdam, the chief of the Russian General Staff told General Marshall that Russia would attack Korea after declaring war on Japan. He asked whether the Americans could operate against Korean shores in co-ordination with this offensive. General Marshall told him that the United States planned no amphibious operation against Korea until Japan had been brought under control and Japanese strength in South Korea was destroyed. Although the Chiefs of Staff developed ideas concerning the partition of Korea, Manchuria, and the Sea of Japan into U.S. and USSR zones, these had no connection with the later decisions that partitioned Korea into northern and southern areas.[14]

Russian entry into the war against Japan on 9 August, and signs of imminent Japanese collapse on 10 August 1945 changed U.S. Army planning from defeating Japan to accepting its surrender. Military planners in the War Department Operations Division began to outline surrender procedures in General Order No. 1, which General MacArthur would transmit to the Japanese Government after its surrender. The first paragraph of the order specified the nations and commands that were to accept the surrender of Japanese forces throughout the Far East.[15]

The Policy Section of the Strategy and Policy Group in the Operations Division drafted the initial version of the order.

---

[12] (1) Lt. Paul C. McGrath, U.S. Army in the Korean Conflict, n.d., pp. 26–27, OCMH draft MS. (2) Memo, Marshall for President (delivered at Potsdam), 25 Jul 45, file OPD 370.9, Case 17/8.

[13] (1) Interv, 1st Lt Paul C. McGrath with Vice Adm M. B. Gardner, 28 Jan 53, the Pentagon. (2) Interv, McGrath with Lt Gen Charles P. Cabell, Dir of the Joint Staff, JCS, OSD, 27 Jan 53. Both in OCMH.

[14] (1) McGrath, U.S. Army in the Korean Conflict, pp. 24–25. (2) History of Occupation of Korea, vol. II, ch. 3, p. 6. (3) Roy E. Appleman, *South to the Naktong, North to the Yalu*, UNITED STATES ARMY IN THE KOREAN WAR (Washington, 1961), pp. 2–3. (4) See also discussions of 24 and 26 July in Department of State, *Foreign Relations of the United States: The Conference at Berlin (The Potsdam Conference), 1945*, 2 vols., Dept of State Publications 7015, 7163 (Washington, 1960), II, 345–52, 408–15. (5) There was widespread misconception that the division of Korea had been agreed upon at the high-level conference of the Big Three. In June 1946, the Institute of Pacific Relations published a categorical statement that this agreement had been made at Yalta. The New York *Times* in October 1946 named Potsdam as the place where the agreement had been made.

[15] McGrath, U.S. Army in the Korean Conflict, p. 42.

Under pressure to produce a paper as quickly as possible, members of the Policy Section began work late at night on 10 August. They discussed possible surrender zones, the allocation of American, British, Chinese, and Russian occupation troops to accept the surrender in the zone most convenient to them, the means of actually taking the surrender of the widely scattered Japanese military forces, and the position of Russia in the Far East. They quickly decided to include both provisions for splitting up the entire Far East for the surrender and definitions of the geographical limits of those zones.[16]

The Chief of the Policy Section, Col. Charles H. Bonesteel, had thirty minutes in which to dictate Paragraph 1 to a secretary, for the Joint Staff Planners and the State-War-Navy Coordinating Committee were impatiently awaiting the result of his work. Colonel Bonesteel thus somewhat hastily decided who would accept the Japanese surrender. His thoughts, with very slight revision, were incorporated into the final directive.[17]

Bonesteel's prime consideration was to establish a surrender line as far north as he thought the Soviets would accept. He knew that Russian troops could reach the southern tip of Korea before American troops could arrive. He knew also that the Russians were on the verge of moving into Korea, or were already there. The nearest American troops to Korea were on Okinawa, 600 miles away. His problem therefore was to compose a surrender arrangement which, while acceptable to the Russians, would at the same time prevent them from seizing all of Korea. If they refused to confine their advance to North Korea, the United States would be unable to stop them.

At first Bonesteel had thought of surrender zones conforming to the provincial boundary lines. But the only map he had in his office was hardly adequate for this sort of distinction. The 38th Parallel, he noted, cut Korea approximately through the middle. If this line was agreeable to President Truman and to Generalissimo Stalin, it would place Seoul and a nearby prisoner of war camp in American hands. It would also leave enough land to be apportioned to the Chinese and British if some sort of quadripartite administration became necessary. Thus he decided to use the 38th Parallel as a hypothetical line dividing the zones within which Japanese forces in Korea would surrender to appointed American and Russian authorities.

The determination of the surrender zones for the Pacific involved other countries besides Korea. Since the job had to be done in a hurry, Colonel Bonesteel had the paragraphs of the general order rushed through the Chief of the Strategy and Policy Group, Brig. Gen. George A. Lincoln, to the Joint Staff Planners who were meeting in an all-night session. This channel was the same as for all important military policy papers in 1945. Drafts were routed in turn through General Lincoln, the Joint Planners, the State-War-Navy Coordinating Committee, the Joint Chiefs of Staff, and the Secretaries of State, War, and Navy, until they finally reached the President.

---

16 (1) *Ibid.* (2) See also Truman, *Memoirs*, II, 317.

17 The remainder of this subsection is based on McGrath, U.S. Army in the Korean Conflict, pp. 40–53.

When Bonesteel's draft paper reached the Joint Planners in the predawn hours of 11 August, Admiral M. B. Gardner suggested moving the surrender line north to the 39th Parallel, a recommendation that the planners believed the Navy Secretary, James C. Forrestal, favored. Gardner pointed out that the 39th Parallel would place Dairen in the military zone to be occupied by the Americans. General Lincoln, however, felt that the Russians would hardly accept a surrender line that barred them from Dairen and other parts of the Liaotung Peninsula; besides, American units would have great difficulty reaching the Manchurian port ahead of the Russians. Calling Assistant Secretary of State James Dunn, Lincoln ascertained that his opinion was shared. Mr. Dunn believed that Korea was more important politically to the United States than Dairen, and he felt this to be the view of Secretary of State James F. Byrnes. As a result, the 38th Parallel remained in the draft when the Joint Planners handed the general order to the State-War-Navy Coordinating Committee.

While General Lincoln was shepherding the document through the State-War-Navy Coordinating Committee on 11 and 12 August, the Russians invaded Korea, landing on the northeast coast near Rashin. Russian troops then poured out of the maritime provinces of Siberia, down the Korean peninsula, and into the Kaesong-Ch'unch'on area above Seoul, where they looted much equipment, including locomotives and rolling stock. Reports of the Russian troop movements reaching Washington underscored the need for concurrence in the proposed general order. Otherwise, the Russian advance would render academic the American acceptance of the Japanese surrender in southern Korea. At the same time, swift Russian troop movements into key areas of southern Manchuria eliminated the possibility of including Dairen in the American surrender zone.

Between 11 and 14 August, the State-War-Navy Coordinating Committee and the Joint Chiefs of Staff discussed the wording of the surrender instrument. Meanwhile, General MacArthur informed the Joint Chiefs of Staff that he would adhere to three priorities for the use of the forces under his command. After the Japanese surrender, the occupation of Japan would come first, Korea second, China third.

In Washington, the War Department Operations Division rephrased General Order No. 1 to the satisfaction of the Joint Chiefs of Staff and the heads of the State, War, and Navy Departments. On 15 August 1945, clean copies of the draft order were sent to Fleet Admiral William D. Leahy's White House office. Within a few hours President Truman gave his approval, directing at the same time that General Order No. 1 be sent also to the capitals of Great Britain and the USSR with requests for concurrence by the heads of those states. The Joint Chiefs of Staff telegraphed the general order to General MacArthur and directed that he furnish an estimated time schedule for the occupation of a port in Korea.

Among the items it specified, General Order No. 1 stated that Japanese forces north of the 38th Parallel in Korea would surrender to the Russian commander, while those south of the parallel would surrender to the commanding general of the U.S. expeditionary forces. As Washington waited for the Moscow reaction to President Truman's message, there was a short period of suspense. Russian troops had entered Korea three days before the President accepted the draft of General Order No. 1. If the Russians failed to accept the proposal, and if Russian troops occupied Seoul, General Lincoln suggested that American occupation forces move into Pusan.

Stalin replied to President Truman on 16 August 1945. He said nothing specifically about the 38th Parallel but offered no objection to the substance of the President's message. He asked that the general order be "corrected" to authorize Russian forces to accept the surrender of the Japanese in the northern half of Hokkaido. Stalin also reminded the President that the Liaotung Peninsula, upon which Dairen and Port Arthur are located, was part of Manchuria and thus within the USSR military zone. Though President Truman parried Stalin's proposal to place Russian forces on Hokkaido, Stalin's message settled the surrender zones in Korea and canceled American plans to land troops at Dairen.

The New Zones

The new dividing line, about 190 miles across the peninsula, sliced across Korea without regard for political boundaries, geographical features, waterways, or paths of commerce. The 38th Parallel cut more than 75 streams and 12 rivers, intersected many high ridges at variant angles, severed 181 small cart roads, 104 country roads, 15 provincial all-weather roads, 8 better-class highways, and 6 north-south rail lines.[18] It was, in fact, an arbitrary separation.

South of the 38th Parallel, the American zone covered 37,000 square miles and held an estimated 21,000,000 persons. North of the line of latitude, the USSR zone totaled 48,000 square miles and had about 9,000,000 people.[19] Of the 20 principal Korean cities, 12 lay within the American zone, including Seoul, the largest, with a population of nearly 2,000,000. The American zone included 6 of Korea's 13 provinces in their entirety, the major part of 2 more, and a small part of another.

The two areas, North and South Korea, complemented each other both agriculturally and industrially. South Korea was mainly a farming area, where fully two-thirds of the inhabitants worked the land. It possessed three times as much irrigated rice land as the northern area, and furnished food for the north. But North Korea furnished the fertilizer for the southern rice fields, and the largest nitrogenous fertilizer plant in the Far East was in Hungnam. Although North Korea also had a high level of agricultural production, it was deficient in some crops. The barrier imposed serious adverse effects on both zones.[20]

[18] Shannon C. McCune, "Physical Basis for Korean Boundaries," *Far Eastern Quarterly*, No. 5 (May 1946), pp. 286–87.

[19] (1) Andrew Grajdanzev, "Korean Divided," *Far Eastern Survey*, XIV (October 1945), 282. (2) History of Occupation of Korea, vol. I, ch. 4, p. 16.

[20] The closing paragraphs of this chapter are based on information in (1) Testimony of Hoffman, 8 June 1949, House Report 962, June 1949, and (2) George A. McCune, *Korea Today* (Cambridge: Harvard University Press, 1950), pp. 52–56.

South Korea had in 1940 turned out about 74 percent of Korea's light consumer goods and processed products. Its industry consisted of some large and many small plants producing textiles, rubber products, hardware, and ceramics. Many of these plants had been built to process raw materials from North Korea.

North Korea, a largely mountainous region, held valuable mineral deposits, especially coal. Excellent hydroelectric plants, constructed during the last ten years of Japanese domination, ranked with the largest and best in the world. Because of its power resources, North Korea housed almost all of Korea's heavy industry, including several rolling mills and a highly developed chemical industry. In 1940, North Korea produced 86 percent of Korea's heavy manufactured goods. The only petroleum processing plant in the country, a major installation designed to serve all of Korea, was located in the north, as were seven of eight cement plants. Almost all the electrical power used by South Korea came from the north, as did iron, steel, wood pulp, and industrial chemicals needed by South Korea's light industry.

Sharp differences between north and south had traditionally been part of the Korean scene. South Koreans considered their northern neighbors crude and culturally backward. North Koreans viewed southerners as lazy schemers. During the Japanese occupation Koreans in the north had been much less tractable than those in the south. Differences in farming accounted for some of the social differences in the two zones. A dry-field type of farming in the north opposed a rice-culture area in the south to produce marked variations in points of view. In the south were more small farms and a high tenancy rate, while in the north larger farms and more owner-farmers prevailed. Those differences the 38th Parallel promised to exacerbate.

CHAPTER II

The House Divided

The Americans Occupy South Korea

On 13 August 1945 the Joint Chiefs of Staff designated General Douglas MacArthur to receive the surrender of Japanese forces in those areas for which the United States was responsible, including the southern half of Korea.[1]

General MacArthur stood at the pinnacle of a distinguished military career. A man of outstanding intellect and physical stamina, son of a Civil War hero, he was marked early for posts of high responsibility. Graduated from West Point with the highest scholastic rating ever recorded there, he rose swiftly in World War I to the rank of brigadier general and displayed great courage in combat. He later served as superintendent of the Military Academy, Chief of Staff of the Army, and thereafter became the military adviser to the Philippines which gave him the rank of field marshal in 1936. General MacArthur after retiring from the United States Army in 1937 was recalled to active duty in July 1941 and led Allied forces to victory over the Japanese in the Southwest Pacific Area, planning and directing a series of brilliant campaigns.

MacArthur received little guidance at the outset on how to handle Korea. He designated the XXIV Corps, commanded by Lt. Gen. John R. Hodge, to carry out the terms of surrender in Korea and to occupy and administer South Korea on behalf of the United States. General Hodge became commander of the United States Army Forces in Korea (USAFIK) on 27 August 1945.[2]

As already noted, the possibility of establishing a 4-power trusteeship over Korea had been discussed between President Roosevelt and Marshal Stalin at Yalta in February 1945, and in conferences with Mr. Harry Hopkins in May 1945 Stalin had agreed to such a 4-power trusteeship. In June the Chinese Government had also agreed. The British Government, although informed of plans for trusteeship, had made no commitment.[3]

The paucity of specific guidance in

---

[1] (1) Hq, United States Army Military Government in Korea, Statistical Research Division, History of the United States Army Military Government in Korea, Period of September 1945–30 June 1946, 3 vols. (hereafter cited as History of USAMGIK), I, 22–23, copy in OCMH. (2) WD GO No. 1, 13 Aug 45.

[2] (1) History of USAMGIK, I, 22–23. (2) USAFIK GO No. 1, 27 Aug 45.

[3] Department of State, *Foreign Relations of the United States: The British Commonwealth and the Far East, 1945* (hereafter cited as *Foreign Relations: The British Commonwealth and the Far East, 1945*), Dept of State Publication 8451 (Washington, 1969), vol. VI, pp. 1021, 1095.

advance of occupation reflected an assumption that fairly simple solutions could be found for Korea's problems in close co-operation with Great Britain, the Soviet Union, and the Chinese Government of Generalissimo Chiang Kai-shek. But neither in Washington nor in the Far East were serious preparations made for carrying out an American program in South Korea that would take into account the entirely different Soviet outlook with respect to Korea's future.

That General MacArthur expected problems with the Russians became evident on 29 August when he warned Hodge that the Russians might already be in Seoul when he arrived. He told Hodge to take over Seoul nevertheless, to make friendly contact with the Russian commander, and to act with caution to avoid troublesome incidents. MacArthur believed that Korea would be occupied on a quadripartite basis, with British, Chinese, Russian, and American participation, although he had no exact knowledge of the areas the four powers would occupy. General Hodge continued to believe that guidance from Washington was inadequate. He contended that unless he were provided more specific and positive policy instructions the United States would fail in Korea.[4]

[4] (1) Rad, MacArthur to Hodge, 29 Aug 45, quoted in History of Occupation of Korea, vol. I, ch. 1, pp. 60–61. (2) The history written on this period by officers of Hodge's headquarters and approved by him states: "General Hodge had been given little or no practical guidance by his instructions on such thorny questions as the eventuality of Korean independence, methods of handling various political factions or the severance of Korea from Japanese influence, economic or otherwise. If Washington or GHQ had given much constructive thought to Korean problems, it had not been reflected in orders issued the Corps Commander." History of Occupation of Korea, vol. I, ch. 1, p. 63.

In the years before World War II the President of the Korean Provisional Government in Chungking, Kim Koo, and its representative in the United States, Dr. Syngman Rhee, who styled himself Chairman of the Korean Commission in the United States, had sought United States recognition and support but without success. In the closing months of the war, these men increased their efforts, seeking not only recognition by the United States and other governments, but membership in the United Nations. American national policy forbade such actions however and Rhee was told in June 1945, after an appeal to President Truman, that "It is

> the policy of this Government in dealing with groups such as the 'Korean Provisional Government' to avoid taking action which might, when the victory of the United Nations is achieved, tend to compromise the right of the Korean people to choose the ultimate form and personnel of the government which they may wish to establish."[5]

On 17 August, with the approach of allied victory over Japan, Kim Koo petitioned President Truman, through the United States Ambassador to China, for permission to send representatives of his Provisional Government to Korea and sought to participate in "all Councils affecting the present and future destiny of Korea and Koreans." No immediate action was taken on this request, but General Hodge, a few days after arriving in Korea, suggested to General MacArthur that leaders of the Chungking government in exile be returned to Korea under allied sponsorship to act as "figureheads" until the political

[5] *Foreign Relations: The British Commonwealth and the Far East, 1945*, vol. VI, pp. 1023, 1027, 1030–32.

General of the Army MacArthur

General Hodge

situation stabilized and elections could be held.[6]

While this action was not taken in the manner Hodge had suggested, the return of individual members of the Korean Provisional Government was approved in late September and transportation and support provided them by the United States. Each individual returning to Korea was required to sign a statement agreeing to abide by the laws and regulations of the Military Government.[7]

On 28 August the commander of Japanese forces in Seoul had appealed for a quick entry into Korea by American troops to preserve order and to maintain government functions. He charged that Korean communists were creating trouble as an excuse to bring Russian troops into the area below the 38th Parallel.

A small advance party from the XXIV Corps landed at Kimp'o Airfield near Seoul at noon on 4 September. Four days later, the bulk of the corps landed at Inch'on and entered Seoul. Contrary to prior fears, the Russians had not taken over the Korean capital. A few Soviet soldiers had entered smaller towns in the American sector close to the 38th Parallel, but no organized units appeared to be south of the line.[8]

General MacArthur issued a proclamation to the people of Korea on 7 September establishing American military control over all Korea south of the 38th Parallel. "Having in mind the long enslavement of the people of Korea and the determination that in due course Korea shall become free and independent," he declared, "the Korean people are assured that the purpose of the occupation is to enforce the Instrument of Surrender and to protect them in their personal and religious rights. In giving effect to these purposes, your active aid and compliance are required. . . . All persons will obey promptly all my orders and orders issued under my authority. Acts of resistance to the occupying forces or any acts which

[6] *Ibid.*, pp. 1036–37, 1053.

[7] *Ibid.*, pp. 1053–60.

[8] (1) History of Occupation of Korea, vol. II, ch. 3, p. 19. (2) History of USAMGIK, I, 24–26. (3) For details of arrival of U.S. forces in Korea, including preparation for movement from Okinawa and events in Korea between the announcement of surrender and the actual landing, see History of Occupation of Korea, vol. I, chs. 1–4.

PARADE HELD BY KOREANS *for the American Advance Party.*

may disturb public peace and safety will be punished severely." [9]

General Hodge appointed Maj. Gen. Archibald V. Arnold, commander of the U.S. 7th Division—the initial occupation force—Military Governor of South Korea on 12 September 1945,[10] and a Department of State official had, at Hodge's request, been assigned as his Political Adviser. The latter, Mr. H. Merrell Benninghoff, described for the Secretary of State a disturbed and chaotic situation in South Korea on 15 September. "USAFIK," he commented,

> is operating under two great difficulties, neither of which can be corrected at this end. The first is that this headquarters has no information in regard to the future policy of the United States or its allies as to the future of Korea. What is going to happen to the nation and what will be the solution of the now almost complete division of the country into two parts? What will be our general policies beyond immediate military necessity? The second difficulty is that USAFIK is in small strength, and has too few competent military government and

[9] *Foreign Relations: The British Commonwealth and the Far East, 1945*, vol. VI, pp. 1043–44.

[10] USAFIK GO No. 7, 12 Sep 45.

other officers that it can operate only in a limited area and with little overall effect.[11]

American forces did indeed face urgent problems in Korea. Industry and commerce had virtually ceased. Public utilities and services hardly existed. The Korean economy was in a perilous state. Complicating all these problems, the political atmosphere was turbulent and tense. There was little prospect of an early stabilization of this political situation and even less chance that the Koreans themselves could assume orderly control of their own affairs. Although more than seventy political parties had been formed in South Korea in the brief period between Japanese capitulation and the arrival of American troops, none appeared competent to govern. Not only were these parties at odds with one another, but their leaders had little if any political experience.[12]

The Japanese heritage had left very few Koreans qualified for responsible posts either in government or in industry. Railway jobs, for example, even as yardmen, much less as engineers, were beyond the experience and skills of most Koreans. No trained public administrators existed. Faced with these facts, General Hodges decided to keep some Japanese officials in responsible posts during a transition period. On the day after he reached Korea, Hodge appointed General Nobuyuki Abe, wartime governor-general of Korea, temporary head of the Korean Government, to serve under American supervision. Hodge promised that Americans would replace the Japanese officials as soon as possible, and Koreans would, in turn, replace the Americans. His assurances proved to be a mistake. Deeply offended at seeing their old rulers apparently still in control, the Koreans reacted violently, forcing Hodge to dismiss the Japanese and to place many less able Koreans in governmental offices. By December 1945, almost 75,000 Koreans, many of them of dubious qualification, were holding governmental positions.[13]

The course of events indicates all too clearly that the United States had not foreseen what its role might be in Korea and had made no effective plans for military government. The first instructions sent from Washington to General Hodge were vague. Subsequent instructions were, according to his reports, incomplete.

Without benefit of specific guidance, General Hodge tried to keep order, to restore public utilities, and to shore up the sagging economy. His efforts were hampered by the fact that his XXIV Corps had been organized to fight the Japanese, not to occupy Korea. Keeping experienced men and officers in Korea was next to impossible. A steady and considerable rotation brought unqualified people into positions at all levels of responsibility in the Korean occupation.[14]

Another indication of how little prepared the United States was to occupy Korea in 1945 was the almost total ab-

[11] *Foreign Relations: The British Commonwealth and the Far East, 1945*, vol. VI, p. 1052.

[12] (1) History of Occupation of Korea, vol. II, ch. 1. (2) E. Grant Meade, *American Military Government in Korea* (New York: King's Crown Press, Columbia University, 1951), pp. 54–58.

[13] (1) History of USAMGIK, I, 26–27, 29–32. (2) History of Occupation of Korea, vol. I, ch. 4, pp. 16–18.

[14] Meade, *American Military Government in Korea*, p. 48.

sence in the country of Americans who could speak or understand Korean. Americans were forced, in dealing with Korean officials and the general public, to rely upon English-speaking Koreans. General Hodge used a Korean to interpret his first press conference. A Korean translated his first address to the Korean public. The U.S. military government became known among the people as a "government by interpreters." A survey in October 1945 showed that Koreans distrusted native interpreters and rated their influence on American officials among the biggest problems disturbing them. South Koreans strongly suspected that interpreters were dishonest and were trying, in many cases successfully, to influence occupation policy. The situation improved as trained military government officers began arriving in Korea in increasing numbers late in October.[15]

The Soviets had not been idle meanwhile. An inkling of their intentions existed within the Department of State even before Japanese surrender. In a policy paper prepared in June 1945, State planners had predicted, "The Soviet Government will, no doubt, establish military government in the portion of Korea under its control and may subsequently wish to establish a Korean regime friendly to the Soviet Union composed at least partially of Korean leaders groomed in the Soviet Union." [16]

Dr. Rhee, whose prestige with the Korean people was believed by Washington officials and Generals MacArthur and Hodge to be strong enough to instill a sense of purpose into the politics of his native land, reached Korea on 16 October 1945. Kim Koo arrived in Korea slightly later from Chungking, China.[17]

Their arrival coincided with the issuance to General MacArthur of specific guidance from Washington. This guidance, which had been under preparation within the SWNCC since 1 September, was sent MacArthur on 17 October. The basic initial directive stated that the United States "ultimate objective" in Korea was "to foster conditions which will bring about the establishment of a free and independent nation capable of taking her place as a responsible and peaceful member of the family of nations." MacArthur was further instructed, "In all your activities you will bear in mind the policy of the United States in regard to Korea, which contemplates a progressive development from this initial interim period of civil affairs administration by the United States and the U.S.S.R., to a period of trusteeship under the United States, the United Kingdom, China, and the U.S.S.R., and finally to the eventual independence of Korea with membership in the United Nations organization." [18]

The presence of Syngman Rhee and Kim Koo coincided, perhaps accidentally, with a noticeable rise in communist activity in southern Korea, all of it directed against the American occupation. Other antioccupation groups, not necessarily communist, stirred up increasing trouble. General Hodge criticized in-

[15] (1) History of Occupation of Korea, vol. I, ch. 4. (2) History of USAMGIK, I, 69.

[16] *Foreign Relations: The British Commonwealth and the Far East, 1945*, vol. VI, pp. 556–80.

[17] History of Occupation of Korea, vol. II, ch. 1, p. 20.

[18] *Foreign Relations: The British Commonwealth and the Far East, 1945*, vol. VI, pp. 1073–74.

RUSSIAN MEMBER ADDRESSES FIRST FORMAL MEETING OF THE U.S. - USSR COMMISSION IN SEOUL

decisiveness in Washington as a "drifting" which could only lead to an untenable position for his forces. "The Koreans want their independence more than any one thing, and they want it now," he stated. "This stems from the Allied promise of freedom and independence which is well known by every Korean without the qualifying phrase 'in due course.' I am told that there are no Korean words expressing 'in due course.'" The United States, he insisted, must either take some positive action at an international level or empower and direct him to seize the initiative in South Korea. As a drastic alternative, he proposed that both the United States and Russia withdraw forces from Korea simultaneously and leave Korea to its own devices and an inevitable internal upheaval for its self-purification.[19]

At the same time General Hodge ex-

[19] (1) Rad, CX 56045, CINCAFPAC to JCS, 16 Dec 45. (2) See also Meade, *American Military Government in Korea*, p. 48.

SOUTH KOREAN MOB PROTESTS PROPOSED TRUSTEESHIP

perienced unexpected trouble from Dr. Rhee. After forming a Central Council for the Rapid Realization of Korean Independence, Rhee found that he could not control the Korean Communist party. Invited by General Hodge to give a series of nonpartisan radio talks to the Korean people, Rhee used the opportunity to castigate the Communists. From then on, a bitter enmity grew between Rhee and the Communists, and the consequent divisiveness complicated Hodge's problems.[20]

[20] History of Occupation of Korea, vol. II, ch. 1.

Although American policy-makers pinned their hopes on trusteeship, the Korean people opposed it vehemently. Foreign control by any name was inimical to Korean national aspirations. The Koreans wanted at once the freedom about which their liberators kept talking. Nevertheless, viewed in perspective, trusteeship represented at least a step toward the eventual solution of Korea's problems.

The United States succeeded in bringing about a meeting in Moscow in late December 1945 of foreign ministers of the United States, Great Britain, and the USSR. A seemingly constructive plan of trusteeship for Korea was worked out among these officials. Under this plan a U.S.–USSR joint commission would be formed to recommend, after consulting with Korean political parties and social organizations, the creation of a provisional Korean democratic government for all of Korea. The ministers directed the commission to consult with this provisional Korean government and to draw up a program, which would be considered by their own governments. The object would be an agreement to form a 4-power trusteeship of Korea for a period of up to five years.[21]

When news of the trusteeship proposal with its "up to five years" clause reached South Korea, many of the Koreans reacted violently. Riots, which had to be quelled by U.S. troops, broke out on 29 December. In contrast, the South Korean Communists, presumably acting on instructions from their Russian mentors, announced their support of the trusteeship proposals on 3 January.

The conference of U.S. and USSR officials in Korea began on 16 January and ran for fifteen formal sessions through 5 February 1946. The Americans wanted to integrate the two zones, but the Russians wanted to keep both zones and merely to co-ordinate activities between them. Since neither side would budge on this basic issue, the sessions produced little of consequence. On 2 February, Hodge reported that there was nothing in the attitude of the Russians to show that they had any thought of unifying Korea so long as American forces were present. "So far," he said:

> all discussion includes adjustment of the flow of everything from mail to persons through central posts along the boundary. My best guess now is that north and south will never be really united until the Russians are sure that the whole will be soundly communistic. Based on current trends, I question our ability to stem the propaganda and controlled political maneuvering of the Soviets.[22]

The Russian propaganda campaign in Korea was indeed cleverly contrived and handled. Taking full advantage of Korean sentiment, the Russians presented trusteeship to the Korean people as the brainchild of the United States. *Tass* made it appear that the Russians had been trying to arrange for everything the Koreans wanted, including full and immediate independence, but that the Americans were fighting for a 10-year trusteeship. General Hodge was bitter about the Russian success in this venture. "As the significance of the Tass statement . . . sinks in, the Korean people are feeling that the U.S. has again 'sold them down the river,'" he charged, "this time to the Russians instead of the Japanese."[23]

After the military-level conferences, which resulted only in some vague agreements on an exchange of mail, an allocation of radio frequencies, and military liaison, the Joint Commission of the U.S.-USSR began deliberations at Seoul on 20 March 1946. The pattern of stalemate was repeated. The Americans

---

[21] (1) Department of State, *Korea, 1945 to 1948*, Dept of State Publication 3305, FE Series (Washington, 1948). (2) McCune, *Korea Today*, p. 61, app. A, Doc 1. (3) Truman, *Memoirs*, II, 319–20.

[22] Rad, TFGCG 272, Hodge to MacArthur, 1 Feb 46.

[23] *Ibid.*

claimed that the Russians obstinately refused to co-operate or to make any constructive attempt toward agreement. The Russians insisted that only Korean groups fully supporting the Moscow agreement were eligible for membership in a provisional government. The Joint Commission adjourned on 8 May without resolving this fundamental issue.

The Russians in North Korea

Every world area in which Soviet and American interests touched had become increasingly sensitive by the end of World War II. The establishment of the United Nations Organization in 1945 gave some reason to hope that Russian - American differences could eventually be settled by reasonable process, but it produced no immediate magic. The failure of Russian and American negotiators at the conference table in Korea was symptomatic of doctrinal differences. Unilateral Russian actions in North Korea extended these differences into a tangible form.

The Soviets had sent forces into Korea with definite objectives. From the beginning, they sealed off their zone. They stopped interzonal communication and transportation and set up a solid line of roadblocks. They emplaced machine guns with fields of fire covering the line which they chose to interpret as the 38th Parallel, for in actuality parts of the Russian line were 1,000 to 1,200 yards south of the latitude shown on American maps. In spite of the Russian guards, a daily flow of 5,000 to 6,000 destitute refugees from North Korea poured into the American zone during the first few months of dual control.

KIM IL SUNG

American attempts to set up liaison in the north proved futile during the first month of occupation. Suggestions by General Hodge to his Russian counterpart that interzonal commerce and communication be allowed, even encouraged, met flat rejection. The major contacts between the Americans and the Russians in Korea consisted of an exchange of mail trains once every two weeks, a small Russian liaison mission in Seoul, and a similar tiny American group at the Russian headquarters in P'yongyang. Telephone communications between zones were subject to Russian whims and mainly used by the Russians.[24]

[24] Statement, Mr. John M. Allison, Deputy Director, Office of Far Eastern Affairs, Department of State, 81st Congress, 1st Session, 9 June 1949, in House Report No. 962, *Korean Aid*, H.R. 5330, June 1949.

Finally, in October 1945, the Soviet commander slammed the door on any further efforts by Hodge to work out agreements. The Communist official informed Hodge by letter that there would be no negotiation of any sort at the military level until decisions were made and relationships established at the top political level. On 11 October the Russian liaison detachment was withdrawn from South Korea.[25]

American and other observers who penetrated North Korea reported some alarming developments. The Russians were molding North Korea into a model communist state. Korean political parties which fitted the Soviet design were being placed in nominal power. Behind a façade of native government the Russians were communizing North Korea without arousing the storms of critical protest that met the Americans in their efforts to democratize South Korea.

Russian policy in North Korea was aimed at creating an indigenous government which would be a replica of the Russian political system and subservient to the Soviet Union. The ready-made strong Communist organization in North Korea as well as the area's nearness to Manchuria and USSR territory made the job easy for the Russians. They brought back to Korea thousands of Korean expatriates who had lived, studied, and become completely communized in the USSR. A few had held government or party posts in Moscow.

On 3 October 1945 the Russians introduced into their new nation one of these Koreans, born Kim Sung Chu but traveling under the alias of Kim Il Sung. The Russians hailed him as the leading exponent of Korean nationalism. The original Kim Il Sung had been a famous leader of Korean resistance against the Japanese. The Russian-sponsored interloper had served as a captain in the Russian Army. After going to Manchuria in 1930, he became a small-time bandit leader, and finally disappeared into the USSR in 1941 or 1942. Backed by the Russians, Kim Il Sung assumed control of the Korean Communist party in late October 1945. At the same time other Russian-trained Koreans took over key posts in the North Korean regime. This seizure of power by the Korean Communist party in North Korea was carried out boldly with complete Russian backing.[26]

A central North Korean government—the Interim People's Committee—was created on 12 February 1946. This committee, headed by Kim Il Sung and dominated by Korean Communist party members, gave wide publicity to Communist measures and reforms. Within limits defined by the Russians and subject to their advisory control, the Korean Communists functioned with marked initiative. By mid-1946 the USSR position in North Korea had become sufficiently secure to permit withdrawal of all but 10,000 occupation troops. Thereafter, the occupiers further reduced their interference in purely administrative functions. Assured of reliable leadership, the USSR could supervise developments in North Korea

---

[25] *Foreign Relations: The British Commonwealth and the Far East, 1945*, vol. VI, p. 1071.

[26] (1) Department of State, *North Korea: A Case Study of a Soviet Satellite*, Report No. 5600 (Washington, May 1951). (2) Truman, *Memoirs*, II, 320–22.

through a relatively small number of strategically placed Russian personnel.

The United States Seeks a Solution

Communist-inspired riots throughout the southern zone marked the close of the first year of American occupation. In the fall of 1946, after Korean mobs overran several police stations and seized arms and ammunition, General Hodge declared martial law. But he would have been hard pressed had a full-scale uprising occurred, for he then had only 43,500 soldiers in Korea and the over-all combat effectiveness of his entire XXIV Corps had dropped to an estimated 10 percent. Meanwhile, reports kept filtering in from North Korea that the Russians were training hundreds of thousands of young North Koreans and forming a native army; and South Korean communists passed the word in the American zone that the North Korean Army would invade and "liberate" South Korea. Communists sentenced to prison terms after the October riots shrugged off the punishment as unimportant since they believed that the Russians would set them free in six months anyway.

The U.S.-USSR Joint Commission resumed its meetings on 22 May 1947. In mid-July, General Hodge reported pessimistically, "Based upon performance to date I feel sure that the U.S.-USSR Joint Commission will fail, with the break-up coming when the Kremlin gives the order. So far as I can determine there is no change in the Soviet stand." [27] He charged that the Russian delegation was under orders to turn Korea into a USSR satellite.

"We have wasted well over a year on South Korean rehabilitation in attempts to placate the Russians and to make the Moscow decision work," Hodge claimed, and he recommended that the United States abrogate the Moscow decision and go it alone if the Joint Commission failed again. "I have always been aware that Korea has been low on the agenda of national foreign policy," he said, "but I feel that the situation here is reaching the point where Washington must become aware that it may soon reach the point of explosion." He asked that he be given a definite long-range plan to use if the Joint Commission failed, that "all concerned" stop commenting about Korean plans in the press until some definite facts had been established, and that his command be raised to full authorized strength.[28]

In July, acting on advice from the Department of State, President Truman directed the transfer of the responsibility for civil administration in Korea from military to civilian control. On 25 July 1947 the War Department notified General MacArthur that the Department of State would gradually assume civil affairs responsibility. "In order to facilitate this transfer," he was told, "CG USAFIK will henceforth report to the Joint Chiefs of Staff on political, economic, cultural, social and nonmilitary operational aspects of the occupation, the War Department acting as the Executive Agent for the Joint Chiefs of Staff in all routine matters." The military command relationship be-

[27] Rad, CM-IN 2987, CINCFE to JCS (forwarding message from Hodge), 18 Jul 47.

[28] *Ibid.*

tween the Commander in Chief, Far East (CINCFE), and the Commanding General, USAFIK, was not to be affected, and CINCFE's military responsibilities for Korea would remain unchanged.[29]

The fanatic Korean dislike for trusteeship meanwhile continued to foment resistance to the Joint Commission, a resistance in which Dr. Rhee was a principal factor. He kept up a continual attack against communism and against General Hodge personally. But Rhee need not have concerned himself with opposing the negotiations toward trusteeship: the Joint Commission got nowhere. In a direct move to break the deadlock, the United States proposed on 26 August 1947 that the four major powers meet again to decide how the Moscow agreement could be carried out. China and Great Britain agreed, but the Soviet Union refused. Consequently, after two years of occupation, and with no arrangement for unification and independence of Korea yet in sight, the United States placed the problem before the General Assembly of the United Nations on 23 September 1947.[30]

In a draft resolution on 16 October 1947 the United States recommended that both zones of Korea hold elections before 31 March 1948 under observation of the United Nations. A United Nations temporary commission would view the elections and supervise the formation of a national government. When a unified Korean government had thus been established, foreign troops were to withdraw.[31]

During consideration of this proposal in the General Assembly, the USSR representative protested that the United Nations had no jurisdiction over Korea and that foreign troops must withdraw before creation of a unified Korean government. His counterproposal was that the occupying powers immediately withdraw their troops. This was rejected. When the General Assembly, on 14 November 1947, approved a resolution supporting the United States proposal and establishing the U.N. Temporary Commission on Korea, Russia refused to take part in the U.N. commission.[32]

The Russians did more than refuse to co-operate. The main source of hydroelectric power for South Korea was located in their zone, and in November 1947, upon the formation of the U.N. Temporary Commission, they cut in half the amount of electricity allowed South Korea.

Elections took place in South Korea on 10 May 1948. The North Koreans did not participate, nor did they recognize the results of the elections. The U.N. commission itself was barred from North Korea. But the elections brought out an estimated 80 percent of the eligible voters in the south who chose representatives for their National Assembly, and the U.N. commission reported the results to be valid.[33]

The new assembly of the Republic of Korea convened for the first time on 31 May 1948 and elected 73-year-old Dr.

[29] Rad, WARX 82849, WD to CINCFE, 25 Jul 47.
[30] Department of State, *Korea, 1945 to 1948*, p. 5.
[31] *Ibid.*, pp. 7–8.

[32] (1) *Ibid.*, pp. 8–9. (2) Testimony, Hoffman, H.R. 5330, 7 Jun 49.
[33] (1) Rpt, House Comm on Foreign Affairs, 81st Congress, 2d Session, *Background Information on Korea*, House Report 2495, 11 Jul 50, pp. 11, 12. (2) Testimony, Hoffman, H.R. 5330, 7 Jun 49.

A SOUTH KOREAN WOMAN CASTS HER BALLOT *at the United Nations supervised election.*

Syngman Rhee as its chairman. After considerable debate, the assembly produced a constitution in July 1948 and on the 20th of the month elected Rhee President of the republic. Whereupon General Hodge, because of his past differences with Rhee, recommended his own relief as commanding general, USAFIK. When Hodge left Korea in August 1948 he was succeeded by his deputy, Maj. Gen. John B. Coulter, and he left Korea in August 1948.[34]

[34] For details of Rhee's biography and his opposition to Hodge, see the following: *Current Biography Yearbook 1947* (New York: H. W. Wilson Company, 1948), pp. 534-36; Robert T. Oliver, *Why War Came in Korea* (New York: Fordham University Press, 1950), pp. 200-203; History of Occupation of Korea, vol. II, ch. 1, p. 33, and ch. 2, p. 50; McCune, *Korea Today*, p. 244; Memo, CSUSA, 13 May 48, sub: Replacement of CG USAFIK, in G-3, DA file 091 Korea, sec. V, Case 22/2.

AMBASSADOR MUCCIO WITH GENERAL COULTER

On 15 August 1948, during elaborate ceremonies at Seoul, General MacArthur proclaimed the new Republic of Korea (ROK), Rhee was formally inaugurated as President, and USAFIK's governmental authority came to an end. The United States formally recognized the Republic of Korea on 1 January 1949; and John J. Muccio, who had been special representative to the republic since August 1948, became the first U.S. ambassador on 21 March 1949.

Withdrawal From Korea

Soon after Rhee was inaugurated, he quoted General MacArthur as having promised in private conference: "Personally, I will do anything I can to help the Korean people and to protect them. I will protect them as I would protect the United States or California against aggression." [35] But in their postwar planning to meet Russian aggression American military planners were and had always been opposed to any concept that included Korea as an area of military importance.

In 1946, the United States was prepared to stay in Korea as long as necessary, that is, until agreement could be reached with Russia.[36] In September 1947, Lt. Gen. Albert C. Wedemeyer investigated the Korean situation and reported to President Truman that American troops were still needed there. He believed that Russian forces would stay until the North Korean puppet government and armed forces were sufficiently strong to carry out USSR objectives without the presence of Russian troops. He warned that Russia might withdraw when conditions were favorable, primarily to force the United States to fol-

[35] New York *Times*, October 22, 1948.

[36] Rad, WAR 87750, WARCOS to CINCAFPAC, 11 May 46.

low suit, and that after American troops withdrew the Russians had plans for North Korean forces to seize South Korea. The USSR delegation on the Joint Commission suggested on 26 September 1947 that U.S. and USSR troops be withdrawn simultaneously at the beginning of 1948, and the Russian foreign minister followed up on 9 October by making the same suggestion to Secretary of State Marshall.[37]

The USSR proposal was declined, but on 29 September U.S. officials had decided to try for a Korean settlement which would let the United States withdraw as soon as possible and with minimum ill effects. Military leaders concurred inasmuch as the United States had little strategic interest in keeping forces or bases in Korea, and because forces then in Korea were sorely needed elsewhere. President Truman, on 8 April 1948, called for every effort to create conditions which would allow a military withdrawal by the end of the year.

The Department of State held that American forces should remain in Korea until a strong South Korean military force had been established, and a strong South Korean government formed. It also desired full United Nations approval of the withdrawal. But the Army had already started to plan its retirement. Planning dates, in which the Department of State eventually concurred, set tactical withdrawals to start on 15 August 1948. The ambassador and a military mission of sixty-one men and officers would handle United States interests in Korea.[38]

A month later than planned, on 15 September, USAFIK units began to leave Korea. But new political developments in both North and South Korea soon reduced the American departures. On 9 September the North Koreans had formed a government, the Democratic People's Republic of Korea, which immediately claimed jurisdiction over all of Korea. The Soviet Union and its satellites quickly recognized this government. On 19 September the USSR notified the United States that all Russian forces in Korea would depart by the end of the year and expressed the hope that American troops would do likewise. Both the rise of the communist state in the north and the Russian eagerness for the withdrawal of all foreign troops argued against any rapid removal of American forces. Furthermore, a rebellion within the South Korean defense force in October, although short-lived, underlined the seething unrest within the republic and prompted an appeal from President Rhee to President Truman for the retention of American troops until the complete loyalty of his own forces was assured and until the latter were capable of dealing with any threat from without or within.[39]

Although a State Department repre-

---

[37] (1) Lt Gen A. C. Wedemeyer, Report to the President, Korea, Sep 47, pp. 13 and 25. (2) Department of State, *Korea, 1945 to 1948,* pp. 6–7.

[38] This Defense-State disagreement may be traced in DA file P&O XO91 Korea, sec. V.

[39] (1) Major Robert K. Sawyer, *Military Advisors in Korea: KMAG in Peace and War,* ARMY HISTORICAL SERIES (Washington, 1962) (hereafter cited as *Military Advisors in Korea*), pp. 35–37. (2) Department of State, *Korea, 1945 to 1948,* pp. 114–15. (3) Rad, ZPOL 1936, COMGENUSAFIK (Muccio) to Department of State, 20 Nov 48.

sentative in Korea observed in November 1948 that the "presence of U.S. troops would have a stabilizing effect locally," earlier withdrawals and normal attrition had destroyed the ability of American units, even if augmented by South Korean forces, to repel a serious invasion. The United States had decided in September not to match the USSR plan to withdraw all forces by the end of 1948, a plan ostensibly carried out, but to consider the removal of foreign troops as just one facet of the Korean question and to await further action on the question by the U.N. General Assembly. When that body on 12 December 1948 called for the departure of all American forces, the Joint Chiefs of Staff ordered the 16,000 troops then in Korea to be reduced to a single regimental combat team (RCT) of 7,500 men.[40]

Early in 1949 the Joint Chiefs of Staff asked General MacArthur's advice on the possible effects of withdrawing and the best time to withdraw these remaining troops. In response to the first part of the question, MacArthur told the Joint Chiefs that Russia would never agree to United States proposals on Korea. North and South Korea would, in his words, "continue quarreling." He bluntly predicted that the United States could not establish Korean forces in the south capable of stopping a full-scale invasion from the north. "The threat of invasion possibly supported by Communist Armies from Manchuria will continue in foreseeable future," he said, and he entertained a pessimistic view of Korea's chances for survival as an independent state. "It should be recognized," he said, "that in the event of any serious threat to the security of Korea, [U.S.] strategic and military considerations will force abandonment of any pretense of active [U.S.] military support." As to the best time to withdraw, he believed that 10 May would be a suitable date since it was the anniversary of the Korean elections and "Koreans are much affected by tradition."[41] Subsequently, the National Security Council recommended that all U.S. combat troops be pulled out of Korea by 30 June 1949. President Truman approved this recommendation, and on the date specified USAFIK's last tactical troops left Korea.

Despite the American appraisal of Korea as an area of little strategic value, the U.S. Government made some provision for its ward. It granted limited financial aid and laid the foundations of a self-sustaining defense force. In June 1949, in explaining to a Congressional committee the necessity for giving $150,000,000 to the South Koreans, Secretary of State Dean Acheson insisted that failure to provide this economic help portended the loss of all Korea to the communists within two or three months. He could not guarantee that the Republic would withstand all pressures. But he believed that the money and the military assistance then being given to the South Koreans would at least permit them to hold their own against the North Koreans.[42]

[40] (1) Rad, STFGCG 1888, COMGENUSAFIK to State Department, 12 Nov 48. (2) Department of State, *Korea, 1945 to 1948,* pp. 22 and 115–16. (3) Sawyer, *Military Advisors in Korea,* p. 36.

[41] Rad, CX 67198, CINCFE to DA, 19 Jan 49.

[42] (1) Testimony, Secretary of State Dean Acheson, *Korean Aid,* H.R. 5330, 23 Jun 49. (2) The House

MEMBERS OF THE KOREAN CONSTABULARY *in training at Taegu.*

Building a Native Defense Force in South Korea

In early November 1945 General Marshall instructed General MacArthur to prepare plans for raising a police-type force in Korea as the first step toward reducing the number of U.S. troops in the country. On 28 November General MacArthur reported plans for creating a Korean national police force of 25,000 by 1 January 1946, and asked permission to use surplus U.S. arms for this force. He pointed out at the same time that it might be advisable to set up a complete Korean national defense force.[43]

On 9 January 1946, the Joint Chiefs

of Representatives rejected the Korean aid bill in a close vote on 19 January 1950. President Truman was very concerned over this rejection and made a strong public statement to this effect two days after the vote. A new bill finally became law on 14 February 1950. See Dean Acheson, *Present at the Creation* (New York: W. W. Norton & Company, 1969), p. 358.

[43] (1) Rad, CM-OUT 80645, Marshall to MacArthur, 3 Nov 45. (2) Rad, CM-IN 9260, CINCAFPAC to WD, 28 Nov 45.

of Staff authorized General MacArthur to form a Korean police force equipped with surplus U.S. weapons. They said that establishing armed forces should await Korean independence. General MacArthur thereupon reported that South Korean police forces would comprise 25,000 regular police, 25,000 state police, and a coast guard for inshore patrol. They would be armed only with rifles, possibly light machine guns.

The first battalion of this police force, designated the Korean Constabulary, was activated in late January. General Hodge assigned a handful of American officers to guide it, and he equipped it with captured Japanese rifles.[44]

The development of the constabulary was hampered by a lack of equipment, the language barrier, a scarcity of advisers, and unsettled political conditions. General Hodge did not press the buildup of the constabulary because he was concerned about its political reliability. Training went on with little fanfare. The constabulary received minimum publicity. Fewer than a dozen American advisers were assigned to work with the constabulary at any one time in 1946 and 1947. By April 1946 its strength had reached only 2,000 men. By the end of November 1946 the figure had risen to 5,000. But by the close of 1947 the ranks of the constabulary had swollen to nearly 20,000 men.[45]

Meanwhile, when the Korean problem was handed to the United Nations, the question of a South Korean army, as distinguished from a police force, arose again. In October 1947, Washington told MacArthur that, in view of the probable U.S. withdrawal from Korea it might be desirable to create a South Korean army without fanfare, perhaps by expanding the constabulary. Washington authorities asked whether a South Korean army sufficiently strong to hold off North Korean communists could be produced in less than a year. They were concerned also about the optimum size of a South Korean army and how it should be equipped.[46] The War Department asked MacArthur to give his views as early as possible since the United States might be required to withdraw within the next year.

MacArthur turned to General Hodge for answers, and on 22 October 1947 sent Hodge's reply to Washington. "I believe that a South Korean force sufficiently equipped and trained to defend South Korea against the armed forces of North Korea could be formed within one year if equipment can be made available at an early date, and additional personnel become available to train it," Hodge declared. The minimum goal must be 100,000 men and officers, organized into an army headquarters, service troops, and six infantry divisions. Hodge would have recommended twice as many men, but he felt that there would be considerable defection in North Korean ranks in the event of a showdown.[47]

Equipment would be the bottleneck. Excess ordnance equipment, including small arms, 105-mm. howitzers, and

[44] For definitive coverage of the Korean national defense forces in the prewar period, see Sawyer, *Military Advisors in Korea.*

[45] Sawyer, *Military Advisors in Korea*, pp. 17, 23n, 29n.

[46] Rad, WAR 88572, WD to CINCFE, 16 Oct 47.

[47] Rad, CS 56266, CINCFE to DA, 22 Oct 47.

vehicles, was available in Korea for only about 25,000 troops. "Complete organization of 100,000 can be organized and basically trained in from 8 to 12 months if equipment is supplied," Hodge said. "It is believed that the equipping of 3 divisions and part of Army Service troops could be accomplished in 90 days from date authority is given provided equipment is available in Japan." Hodge recommended that, when his forces pulled out of Korea, equipment for a Korean army of 100,000 be left behind and that small arms for an additional 100,000 also be provided.[48]

Even though raising an adequate force before an American withdrawal might be impossible, Hodge recommended that the constabulary at least be brought at once to its full authorized strength of 25,000 and equipped with 81-mm. mortars and 105-mm. howitzers. He asked for authority to issue it U.S. equipment at once. Whatever was done, he said, must be done in secrecy, for the North Korean communists seemed eager to invade the south. Although Hodge doubted that the Russians would instigate an invasion while they still had forces in North Korea, he considered an attack on South Korea by North Korean armed forces likely if the Russians withdrew their forces unilaterally. General MacArthur threw cold water on the whole proposition. "I believe no definite decisions can be made until action is reached by the United Nations," he told Washington on forwarding Hodge's views. "Unilateral action by the United States at this time would be inconsistent with the proposal submitted by it to the United Nations. If the United Nations accepts the problem, decisions such as the one under discussion will pass to it."[49]

American planners doubted that a constabulary could be effective against Russian-sponsored aggression. The Joint Strategic Survey Committee (JSSC) told the Joint Chiefs of Staff in January 1948:

> Present information indicates that the withdrawal of U.S. forces will probably result in Communist domination, and it is extremely doubtful if it would be possible to build up the constabulary in time and with facilities available . . . to prevent Soviet encroachment. Therefore eventual domination of Korea by the USSR will have to be accepted as a probability if U.S. troops are withdrawn. However, an augmented constabulary might be a temporary deterrent to overt acts by North Korean forces.[50]

General MacArthur advised against the establishment of a South Korean army but proposed in February 1948 that the constabulary be increased to 50,000 men, equipped with heavy infantry weapons from stocks in Korea.[51] The Joint Chiefs of Staff authorized this action on 10 March 1948. General Hodge assigned more American officers to advise the constabulary and set up schools for Koreans in the use of American equipment.

Because the Department of the Army had proposed early in 1948 that the augmented U.S. diplomatic mission to South Korea include a military section, and because General MacArthur had concurred in this proposal, President Rhee formally asked for a U.S. military

[48] *Ibid.*

[49] *Ibid.*

[50] Rpt, JSSC to JCS, 1483/50, 30 Jan 48.

[51] Rad, CX 58437, CINCFE to DA, 6 Feb 48.

mission in November 1948.[52] His reasons were that a constabulary of 50,000 men was entirely inadequate to defend South Korea, that these forces were too weak to hold back a North Korean Army, and that a U.S. military mission would immediately assuage South Korean feelings of insecurity and assure the public of safety and protection.

A military mission already existed in the Provisional Military Advisory Group (PMAG) established by MacArthur's headquarters on 15 August 1948. It was headed by Brig. Gen. William L. Roberts. Little more than a grouping of 100 advisers for administrative purposes initially, PMAG had grown by the end of 1948 to 92 officers and 149 enlisted men. On 1 July 1949 PMAG was redesignated the United States Military Advisory Group to the Republic of Korea (KMAG). Authorized 472 officers and men, it was assigned to the American Mission in Korea (AMIK).

General Roberts was named chief of KMAG with headquarters at Seoul. His mission was to develop and train a South Korean force capable of preserving internal security, preventing border raids and incursions, and deterring armed attack or other aggression by North Korean forces. Although Roberts was authorized direct communication with the Department of the Army on military matters, he was instructed to keep General MacArthur informed of his activities.[53] General MacArthur became responsible for the logistical support of AMIK to the Korean water line and for the evacuation of U.S. nationals from Korea in an emergency.[54] During the year preceding the North Korean attack these were his only responsibilities in Korea.

In late November 1948, the Republic of Korea (ROK) passed the Armed Forces Organization Act, and on 15 December set up a department of national defense, which redesignated the constabulary brigades as divisions.

Although the United States had been transferring weapons and equipment to the Republic of Korea for only 50,000 men, ROK forces by 1 March 1949 totaled about 114,000, including a 65,000-man army, 45,000 police, and a coast guard of 4,000. When the United States agreed in March to support a Korean army of 65,000 men, the Republic of Korea moved forward rapidly, and within five months recruited nearly 100,000 men for the new Army. During 1949 the Army was organized into eight divisions; and KMAG furnished advisers to most battalions.[55]

General MacArthur wanted the ROK Army to be strong enough to maintain internal security within the republic, but no stronger, and he saw no need for a ROK air force or navy which had no internal security role and which could not become strong enough to defeat

[52] (1) Rad, WARX 91520, DA to CINCFE, Jan 48. (2) Rad, CX 58237, CINCFE to DA, 28 Jan 48. (3) President Rhee and General Hodge signed an interim agreement on 24 August 1948 providing for American assistance in training and equipping Republic of Korea security forces. *Background Information on Korea*, House Report 2495, 11 Jul 50, pp. 15–16. (4) Rad, No. 186, Muccio, Seoul, to Secy of State, 5 Nov 48.

[53] Rad, WX 90992, DA to CG USAFIK, 2 Jul 49.

[54] (1) JCS 1483/44, 17 Oct 47. (2) GHQ, FEC Annual Narrative Historical Rpt, 1 Jan–31 Oct 50, app. IV, pt. 1, p. 8.

[55] (1) Sawyer, *Military Advisors in Korea*, p. 41. (2) Hist Rpt, GHQ SCAP and FEC, 1 Jan–31 Dec 49, vol. II, p. 26.

North Korean air and naval forces.[56] The ROK Army, he felt, should be capable of offering "token resistance" to invasion, but should "be so organized as to indicate clearly its peaceful purpose and to provide no plausible basis for allegations of being a threat to North Korea." [57]

In June 1949, justifying an American withdrawal from Korea, Maj. Gen. Charles L. Bolté, Director, Plans and Operations Division, Department of the Army, announced that South Korean forces were better equipped than the North Korean troops. Bolté drew this conclusion from reports submitted by General Roberts, the KMAG chief. Largely on that basis, the Army, as the executive agent for the Joint Chiefs of Staff for the Far East, was not only agreeable to the withdrawal of American tactical units but was heartily in favor of it.[58]

When USAFIK withdrew from Korea in 1949, it transferred to the ROK, under the Surplus Property Act through the Office of Foreign Liquidation, military equipment that originally cost the United States approximately $56,000,000 and that had a 1949 replacement value of about $110,000,000. The ground force equipment was sufficient for a force of 50,000 men. It included 100,000 small arms, 50,000,000 rounds of small arms ammunition, more than 2,000 rocket launchers, more than 40,000 vehicles of all types, and a number of light artillery pieces and mortars with over 700,000 rounds of ammunition for them. Individual organizational equipment for an additional 15,000 men subsequently arrived in Korea from American stocks in Japan. Although the United States Government made plans for further material aid to the Republic of Korea and allotted Military Defense Assistance Program funds for that purpose, low priorities, administrative red tape, and procurement difficulties prevented this aid from reaching Korea before June 1950.[59]

President Rhee sent an almost frantic request for greater support to President Truman in August 1949. He said:

> Unless I and my government with the aid of our friends, do find solutions, the immediate future for our nation is bleak and bloody. . . . Some American advisors assure us that the Communists will never attack in force, and therefore we may rest easily defended by our brave army. We Koreans believe that the Communists, under Soviet direction intend to attack in force, that they will do so, and if they do, it is we, the Koreans, civilian and military, who will pay the price, not the good-willed American advisors. . . . American officers tell me we have sufficient ammunition for two months of combat; my own officers tell me it is only sufficient for two days.

He asked for more equipment and ammunition and for M2 howitzers to replace M3's of limited range. On 26 September 1949, President Truman assured Rhee that KMAG would continue

[56] GHQ, FEC Annual Narrative History Rpt, 1 Jan–31 Oct 50, app. IV, pt. 1, p. 9.

[57] Memo, signed Maddocks for CSUSA, 7 Mar 49, sub: Strength of SK Armed Forces, Tab A, in G–3, DA file 091 Korea, sec. I–G, Case 11.

[58] (1) Testimony, Maj Gen Charles L. Bolté, *Korean Aid*, H.R. 5330, Jun 49, p. 120. (2) J. Lawton Collins, *War in Peacetime* (Boston: Houghton Mifflin Company, 1969), p. 42.

[59] Senate Comm on Armed Services and Senate Comm on Foreign Relations, 82d Congress, 1st Session, *Hearings on Military Situation in the Far East and the Relief of General MacArthur, 1951* (hereafter cited as the *MacArthur Hearings*), pp. 1992–93.

to make recommendations for the equipping and support of the ROK Army and that, when Congress appropriated more military aid funds for Korea, Mr. Muccio would so advise him.[60] The triumph of the Communists on the mainland of China in late 1949 apparently had little effect on expediting further military aid to Korea.

In October 1949, the ROK Minister of National Defense asked for 189 M25 tanks. Col. William H. Sterling Wright, acting for General Roberts who was in Japan at the time, advised General J. Lawton Collins, Chief of Staff, U.S. Army, against fulfilling the request. The rough terrain, poor roads, and primitive bridges, he said, militated against efficient tank operations.[61]

At almost the same time that Colonel Wright was minimizing the usefulness of tanks in Korea, Col. John E. Baird, acting chief, KMAG, in the absence of both General Roberts and Colonel Wright, informed Ambassador Muccio that the type and quality of matériel available to South Korea were inadequate for war. On 26 October 1949, he warned that the South Korean Army was outnumbered in all weapons except individual arms and that the Russians had given North Korea much better armament. North Korean artillery had 122-mm. howitzers with a maximum range of 12,980 yards as against the South Korean 105-mm. howitzer M3 which could reach only 7,600 yards. During border clashes, North Koreans placed their artillery just beyond maximum range of the 105-mm. howitzer and shelled at will. They also had the 120-mm. mortar. "The presence in North Korea of high performance aircraft of fighter and bomber type, artillery of medium range and a preponderance of mortars are matters seriously affecting the spirit of the Security Forces." Colonel Baird recommended F–51 aircraft for the Republic of Korea, saying, "It is imperative that Korea be given some means of defense against air attack." But the only aircraft the Republic of Korea received were twenty liaison-type planes.[62]

The U.S. and ROK Governments signed a military assistance agreement on 26 January 1950. This authorized substantial aid to the new government and formalized the establishment of the military advisory groups. The final stipulation of this agreement came on 15 March 1950, when the United States promised the Republic of Korea a total of $10,970,000 in military aid. Of this, only a few hundred dollars' worth of signal wire reached the peninsula before 25 June, although signal equipment and spare parts worth $350,000 were en route from San Francisco.[63]

North Korea Prepares

President Rhee's fears of attack from the north were not unreasonable. The Soviet Government was developing a strong native army in North Korea.

---

[60] (1) Ltr, Rhee to Truman, 20 Aug 49. (2) Ltr, Truman to Rhee, 26 Sep 49. Both in DA file P&O 091 Korea, sec. I–E, Book I, Case 16, Incl 1.

[61] Memo, Minister of National Defense, Seoul, for Gen Collins, 20 Oct 49, in G–3, DA file P&O 091 Korea, sec. I, Case 18.

[62] Ltr, KMAG to Mr. Muccio, 26 Oct 49, sgd Col John E. Baird, CMP, Actg Chf, KMAG, in DA file P&O 091 Korea, sec. I, Case 18.

[63] *MacArthur Hearings*, pp. 1992–93.

Trained by Russian officers and equipped with material furnished by the Soviet Union, the North Korean Army grew into a powerful and efficient striking force between 1946 and 1950. The North Koreans began recruiting their Army in August 1946 and built it up by the end of that year to a force of 20,000 men. In conjunction with its political consolidation of North Korea during 1948, Russia provided weapons for 60,000 men. Total mobilization was declared in 1949, and the addition of 40,000 draftees, 20,000 Koreans who had been serving the Chinese Communist Army, and several thousand men trained for three years in the USSR as cadres for air and tank corps doubled the size of the military force. All units received additional Soviet equipment and training programs were intensified. Early in 1950, the tempo of military expansion increased sharply. The Army expanded to 135,000 men with the addition of new conscripts and 10,000 more returnees from the Chinese Communist Army. Civilians received basic military training. In April and May 1950, large shipments of arms coming from the Soviet Union re-equipped the Army and Air Force. North Korea received heavy artillery prime movers, armor, automatic weapons, and propeller-driven aircraft in considerable quantity.[64]

The organization and training of the North Korean Army remained under the close control of the Russians. Key army commands fell only to men completely amenable to Russian direction. Russian advisers accompanied North Korean Army units from the first, but gradually decreased in numbers as trusted North Korean officers were developed. In 1948, 150 Russian advisers worked with each division. The number dwindled to twenty per division in 1949 and from three to eight per division by 1950. But Russian control remained strong because of North Korea's dependence on the USSR for training in critical military skills and supplies and also for weapons. Japanese rifles were gradually replaced by Russian pieces, and gasoline was allocated to the North Korean Army on a monthly basis that was closely watched.[65]

The North Korean Communists used every conceivable means, including propaganda and armed violence, to instigate the overthrow of the South Korean Government. Agents and terrorists from Communist-dominated political groups in North Korea infiltrated the south and carried out subversive actions, for example, opposing the rice collection program instituted by the American military government to bring food into the cities.[66] General Wedemeyer reported in late 1947, "Current political and economic unrest in Southern Korea is aggravated by Communistic terrorism and by Communist-inspired riots and revolutionary activities in the occupied area." [67] The elections in South Korea were preceded by violent communist activity. Between 29 March and 10 May 1948, 589 persons were killed and 10,000

[64] (1) Dept of State, *North Korea: A Case Study of a Soviet Satellite*. (2) For detailed information of the North Korean Army prior to 1950, see Appleman, *South to the Naktong, North to the Yalu*, Chapter II.

[65] Dept of State, *North Korea: A Case Study of a Soviet Satellite*, pp. 117–18.

[66] Rad, C 54133, CINCFE to JCS, 18 Jul 47.

[67] Wedemeyer, Report to the President, pp. 13, 24.

arrested.[68] When the American troops began to withdraw from Korea, even more serious attempts to seize control developed.

Just before the last American forces left Korea, North Korean communists launched their first open attack across the 38th Parallel. On 3 May 1949, they struck across the border in the Kaesong area. ROK units repulsed them, but a mass defection of two battalions of the ROK Army resulted, the ROK battalion commanders moving their units into North Korea and surrendering their men and equipment. About half of these troops returned to South Korea later.

In July 1949, North Korean units again crossed the parallel near Kaesong, only to be thrown back. Hundreds of small-scale assaults occurred in the next year. In every case the ROK Army pushed the invaders back. While most skirmishes were confined to small-arms fire fights, some involved artillery duels and inflicted heavy casualties on both sides.[69]

A strong and effective guerrilla movement in South Korea, subsidized and directed by the North Korean Government, was also functioning under orders to overthrow the Republic of Korea. A series of uprisings on the island of Cheju-do spread to the mainland by late 1948, and keeping the guerrillas under control became a major task for the ROK Army, but, by June 1950, the ROK Army had virtually stamped them out, in some cases after full-scale battles. The ROK Government claimed that its forces had killed 5,000 guerrillas in South Korea in the period from September 1949 to April.[70]

Situation in Korea—June 1950

Korea in 1950 was quite different from the country entered by the Allies late in 1945. Two political entities with widely divergent forms of government existed on one small peninsula separated by an artificial boundary. Each government existed only through the support of opposing major powers. Indigenous industrial and economic development remained impossible for either of the two portions of Korea. Political unity seemed out of the question, and bitter hatreds had developed between them.

From the autumn of 1949, the North Korean Government had intensified its "hate" campaign against the Rhee Government. Increasing stress was placed on service in the national defense as the highest duty to the communist state. By June 1950, the North Korean military machine was ready and the populace was psychologically prepared for war. As part of this build-up, the communist regime conducted a "peaceful unification" campaign. During the spring of 1950 it made a last effort at a guerrilla-led overthrow of the Republic of Korea, but failed. At this juncture, under cover of two unification proposals to the Republic of Korea, offered on 7 June and 20 June 1950, the final steps for invasion were taken, as the main body of the

---

[68] George C. McCune, "The Korean Situation," *Far Eastern Survey*, XVII (8 September 1948), 197–202.

[69] KMAG, Semi-Annual Rpt, 31 Dec 49, sec. IV, p. 22.

[70] Statement by Ambassador Muccio, *Hearings Before Committee on Armed Services*, MDP 1950, 81st Congress, 6 Jun 50.

North Korean Army moved to positions along the parallel.[71]

The current estimates of ROK intelligence agencies on 25 June 1950 set the strength of the North Korean forces at 10 infantry divisions, 1 tank division, 1 air force division, and an antiaircraft gun regiment—120,000 infantry soldiers, 34,000 constabulary troops, 5,000 armored troops, and 2,000 air force personnel. Weapons strength, according to ROK figures, amounted to 1,600 artillery pieces, 50 T–34 tanks and SU–76 self-propelled (SP) guns, 211 YAK–9 fighters and IL–10 attack planes.[72] A State Department report from Seoul as of 11 May 1950, at some variance with these estimates, credited the North Korean Army with 103,000 soldiers and constabulary troops of all types (excluding 25,000 provincial police), 65 tanks, including some T–34's, 296 light and medium artillery pieces, 780 medium and heavy mortars, and 356 45-mm. antitank guns. Aircraft attributed to the North Korean Air Force were set at 100 YAK aircraft, 70 IL–10 attack planes, and 10 reconnaissance planes. Later reports, believed more accurate, gave the North Korean Army 135,000 men organized into 8 infantry divisions, 1 armored brigade, 2 half-strength divisions, 1 separate infantry regiment, and 1 motorcycle reconnaissance regiment. Many of these troops were veterans from the armies of the USSR and Communist China. In addition to large amounts of artillery, the North Koreans possessed 150 T–34 Russian-made tanks and 180 high-performance combat aircraft.[73]

In March 1950, General Roberts still believed that the ROK Army was stronger than its potential opponent in the north, but he feared the air capability of North Korea. Pointing out that the Russians had given their protégés about 100 combat-type high-performance aircraft, General Roberts said:

> If South Korea were attacked today by the inferior ground forces of North Korea plus their Air Corps, I feel that South Korea would take a bloody nose. Again, then, knowing these people somewhat, I feel that they would follow the apparent winner and South Korea would be gobbled up to be added to the rest of Red Asia.[74]

The United States Government received a clear warning that the ROK Army was not strong enough when Ambassador Muccio, in the same month South Korea was attacked, told the Senate Committee on Armed Services that the matériel superiority of the North Korean forces, particularly in heavy infantry support weapons, tanks, and combat aircraft which the USSR had supplied, would provide North Korea with the margin of victory in any full-scale invasion of the republic. Ambassador Muccio told the legislators that it was vital that the ROK Army be maintained on an effective defensive level of equality in manpower, equipment, and training, in relation to those forces which immediately threatened it.[75]

---

[71] Dept of State, *North Korea: A Case Study of a Soviet Satellite*, pp. 17–18.

[72] ROK Army, Military History of Korea, translated from Korean, by Hq, U.S. Army Forces, Far East, Military Intelligence Service Group, p. 9, copy in OCMH.

[73] (1) Rad, No. 683, State Dept, Seoul, to Secy of State, 11 May 50. (2) Appleman, *South to the Naktong, North to the Yalu*, p. 11.

[74] Ltr, Gen Roberts to Gen Bolté, 8 Mar 50, in G–3, DA file OPS 091 Korea, sec. I–B, Book I, Case 4.

[75] Statement, Mr. Muccio, *Hearings Before Committee on Armed Services*, MDP 1950, 81st Congress, 6 Jun 50, p. 80.

In opposition to Ambassador Muccio's testimony was that of William C. Foster, then deputy administrator of the Economic Cooperation Administration (ECA), given before the Senate Appropriations Committee one week later. Speaking about the ROK Army, Mr. Foster said:

> The rigorous training program has built up a well-disciplined force of 100,000 soldiers, one that is prepared to meet any challenge by North Korean forces, and one that has cleaned out the guerrilla bands in South Korea in one area after another.

If American legislators were somewhat confused at this point they could scarcely be blamed.[76]

By June 1950, the ROK Army reached a strength of 95,000, the bulk of which comprised eight infantry divisions and a cavalry regiment. But only four of the divisions were near full strength of 10,000 men each. In artillery, the South Koreans owned 91 105-mm. M3 howitzers, and in armor, had about two dozen armored cars and about half that many half-tracks. To oppose the 180-plane North Korean Air Force, the ROK Air Force had a dozen serviceable liaison planes and ten trainers.[77]

Meanwhile, in South Korea, elections for a new National Assembly had been conducted during May 1950. The U.N. Temporary Commission on Korea supervised the elections in which 130 seats went to Independents, 49 to parties supporting Syngman Rhee, and 44 to other parties. In the north, these elections and the presence of the U.N. commission were loudly condemned, and the campaign for a unified assembly was revived. On 20 June the "Supreme People's Assembly" passed a decree which demanded the establishment of an all-Korean legislative body to draw up a constitution and organize a government of the republic. The decree designated leading figures of the South Korean Government as national traitors, called for the unification of military and security forces, and demanded the withdrawal of the U.N. commission.

John Foster Dulles visited Korea as a special representative of the President in the middle of June 1950. After inspecting South Korean defenses, which he was assured were adequate, Mr. Dulles addressed the National Assembly of the Republic of Korea on 19 June 1950. He told the legislators that the American people granted them their support ". . . consistent with your own self-respect and primary dependence upon your own efforts." He said that the United States considered the Republic of Korea a part of the United Nations and ended saying, "You are not alone; you will never be alone, as long as you continue to play worthily your part in the great design of human freedom." [78]

[76] (1) *MacArthur Hearings*, p. 2009. (2) See also Collins, *War in Peacetime*, p. 43.

[77] (1) Appleman, *South to the Naktong, North to the Yalu*, pp. 12–17. (2) KMAG Semi-Annual Rpt, 15 Jun 50, sec. V, p. 16, and Annex X. (3) Sawyer, *Military Advisors in Korea*, p. 110.

[78] Speech, John Foster Dulles to ROK National Assembly, 19 Jun 50, quoted in *MacArthur Hearings*, p. 2020.

CHAPTER III

National Defense and the United States Army

It has become almost a truism that nations inevitably try to prepare for the war they have just won. Except for substituting the Soviet Union in the role of chief adversary the United States pursued a course between 1946 and 1950 that appeared to lend credence to this theory. American military planning in these years was shaped largely by World War II experience and the priority afforded to Europe over the Pacific and Far East. In 1950 the defense of western Europe still held first claim on American military resources, and plans were devoted almost exclusively to general war. Furthermore, reflecting its coalition effort, the United States sought to strengthen nations that might be helpful to it in any crisis with the Soviet Union, its most likely opponent in a time of increasing frictions throughout the world.

The Soviet Union and its allies were apparently superior to the United States and its allies in conventional military strength, for except in nuclear weapons the United States military power dropped sharply in the postwar years. Russia, on the other hand, kept powerful military forces in being and strengthened and modernized those of its satellite nations. Thus, the United States was resolved to contain Russian influence and prevent threats to world peace and the independence and stability of other nations by resorting to collective security arrangements and acting through the United Nations.

Beginning in 1948, the United States gave military assistance to a number of friendly nations in Europe, the Middle East, and Asia, to enable them to resist communist encroachment and, if necessary, to join effectively with the United States in any war with the communist bloc of nations. More significant was United States sponsorship of the North Atlantic Treaty Organization (NATO), which in April 1949 bound the United States, Canada, and ten nations of western Europe together to prevent the communist seizure of western Europe. As the most powerful single nation in NATO, the United States assumed a considerably enlarged obligation in Europe.

The successful explosion by the Soviet Union of a nuclear device in September 1949 nullified to some extent the American atomic advantage and intensified efforts by the United States Government to build stronger collective security arrangements. But this event came too late to affect specific defense plans in 1950.

Strategic planning after World War II was carried on at the joint level and approved by the President. Within the joint plans, each military service prepared its own war and emergency plans. By 1950, broad national military policy called for meeting an all-out Russian attack with a strategic offensive in western Eurasia and a strategic defensive in the Far East.

The Army's Place in the National Defense Structure

The Secretary of the Army, appointed by the President, directed the activities of the Army. The Chief of Staff, the top military man, advised the Secretary and acted for him in carrying out approved Army plans. The Army staff in Washington, D.C., responsible to the Chief of Staff, planned and supported Army operations and activities throughout the world. The Chief of Army Field Forces, stationed at Fort Monroe, Virginia, conducted the training of Army units.[1]

The President, as Commander in Chief of all the military forces, exercised his control through a chain of command extending downward through the Secretary of Defense, the Joint Chiefs of Staff (JCS), and the commanders of certain unified and specified commands. The Secretary of Defense, a member of the President's Cabinet, was responsible for directing the services and for advising the President on military matters. Under his jurisdiction the Army, Navy, and Air Force were organized into separate departments.[2]

The Joint Chiefs of Staff, composed of

[1] (1) For detailed explanation of changes in Army organization just prior to the outbreak of the Korean War, see Analysis and Explanation of Army organization bill, DA, Feb 50. (2) The Secretary of the Army was served by an under secretary, two assistant secretaries, and such Army personnel as required. The Chief of Staff's immediate military assistants in 1950 included the vice chief of staff, two deputy chiefs of staff, a comptroller, four assistant chiefs of staff, and a secretary of the general staff. The relationship between the Secretary of the Army and the Chief of Staff kept the Army under civilian control while leaving as much latitude as possible for military planning and operations by the military experts. The Chief of Staff and his deputies co-ordinated and controlled the operations of the Army at home and abroad as well as planning for future operations. The chain of authority from the Secretary of the Army through the Chief of Staff extended to the Chief, Army Field Forces, to the army commanders in the continental United States, and to the various army commanders overseas. The continental United States was divided into six continental army areas and the Military District of Washington.

[2] The powers and authorities of the Secretaries of the Army, the Navy, and the Air Force were much less than those enjoyed by their World War II predecessors, the Secretary of War and the Secretary of the Navy. The Secretary of Defense was an important member of the National Security Council (NSC), a body which had also been established in postwar years, and which was charged with advising the President on the integration of domestic, foreign, and military policies relating to national security and with seeking the most effective co-ordination among the services and other government agencies in areas involving national security. For details of membership, functions, and responsibilities of the Department of Defense and of the National Security Council, see: National Security Act, 1947, PL 253, 80th Congress, 27 Jul 47; National Security Act Amendments, 1949, PL 216, 81st Congress, 10 Aug 49; Timothy W. Stanley, *American Defense and National Security* (Washington: Public Affairs Press, 1956); Truman, *Memoirs*, II, 58–60; Statement, Gen George C. Marshall, *MacArthur Hearings*, pp. 583–84; Wilber W. Hoare, Jr., "Truman (1945–1953)," in Ernest R. May, ed., *The Ultimate Decision, The President as Commander in Chief* (New York: George Braziller, 1960).

the Chief of Staff, U.S. Army; Chief of Staff, U.S. Air Force; Chief of Naval Operations, and a chairman appointed by the President, comprised the top advisory body in the United States Government composed exclusively of military men. They were designated by law as the principal military advisers to the President, the National Security Council, and the Secretary of Defense. Subject only to the authority of the President and the Secretary of Defense, the JCS was specifically charged with the preparation of strategic plans and strategic direction of the military forces; the preparation of joint logistic plans and the assignment of logistic responsibility; review of the major requirements of military forces in the light of prepared plans; and the establishment of unified commands in strategic areas.[3]

After World War II, American armed forces in major overseas areas were brought under the operational control of the Joint Chiefs of Staff through the formal establishment of unified commands, which included contingents of all the military services. Operating under the strategic direction of the JCS, each of these commands was directly supervised by a particular chief of staff who acted as the executive agent of the JCS.

In 1950 the major overseas unified commands established by the JCS were the Far East Command, the Alaskan Command, the Caribbean Command, the Pacific Command, and the European Command. Within each of these, individual service commanders commanded the forces of their respective services—Army, Navy, or Air Force—but they were under the over-all supervision of a designated commander in chief from one of the services, and he was named by and responsible to the JCS.

Army Strength and Deployment—1950

In June 1950, the strength of the active Army stood at about 591,000 and included ten combat divisions. About 360,000 troops were stationed within the zone of the interior (ZI). The remaining 231,000 were disposed in overseas commands, most of them performing occupation duties. The largest group overseas (about 108,500) was located in the Far East. In Europe, approximately 80,000 U.S. soldiers were stationed in Germany, 9,500 in Austria, and 4,800 in Trieste. Slightly more than 7,000 were assigned to the Pacific area and about 7,500 to Alaska. In the Caribbean were about 12,200 troops. Several thousand more were assigned to military missions throughout the world.[4]

[3] In their capacities as members of the JCS, the individual members represented the entire military establishment and not their respective services. The Secretary of the Army, for example, had no direct control over the Chief of Staff of the Army in the latter's role as a member of the JCS. The chairman of the JCS had no vote, but presided over the meetings and deliberations of the body. He frequently represented the entire membership before the President, the NSC, and the Secretary of Defense. Although not a member of the NSC, the chairman of the JCS usually accompanied the Secretary of Defense to the meetings of the NSC and explained or defended the views of the JCS, sometimes against the opposition of the Secretary of Defense. For details of the composition, functions, and responsibilities of the JCS in 1950, see National Security Act 1947, PL 253, sec. 211B, 80th Congress; National Security Amendments, 1949, PL 216, 81st Congress; Stanley, *American Defense and National Security; MacArthur Hearings*, p. 904; Hoare, "Truman (1945–1953)," pp. 185–94.

[4] (1) STW 1037, Weekly Estimate of Army Command Strength as of 26 June 1950, 2 Jul 50, AGO Stat and Acc Br, copy in G–3 Deployments Br. (2) These figures are at slight variance with those

THE JOINT CHIEFS OF STAFF AT PENTAGON, *November 1949. Left to right: Admiral Forrest P. Sherman, General Omar N. Bradley, General Hoyt S. Vandenberg, and General J. Lawton Collins.*

The force designated to carry out the Army's emergency assignments was called the General Reserve. Except for one regimental combat team (RCT) in Hawaii, this force consisted of five combat divisions and certain smaller units in the continental United States.[5] The major General Reserve units on 25 June 1950 were the 2d Armored Division, 2d Infantry Division, 3d Infantry Division, 82d Airborne Division, 11th Airborne Division (– 1 RCT), 3d Armored Cavalry Regiment, 5th RCT (located in Hawaii), and 14th RCT. In addition, there

contained in STM–30, Strength Report of the Army, 1 July 1950, which gives the following data on Army forces as of 30 June 1950: Total Strength 591,487; Zone of Interior 347,224; Overseas Strength 244,263. (3) Total strength in both compilations excludes the cadet corps at the Military Academy.

[5] For precise definition of General Reserve, see SR 320–5–1, Dictionary of United States Army Terms, Aug 50. See also Directory and Station List, U.S. Army, 30 Jun 50, copy in OCMH.

were smaller combat support and service support units.[6]

Besides the General Reserve in the United States and Hawaii, four tactical divisions and one RCT were located in the Far East Command. In Europe the Army maintained one tactical division, one RCT, three cavalry regiments, and one separate infantry regiment. One infantry battalion was in Alaska, and two separate regiments were in the Caribbean area.[7]

The authorized strength of the Army, as opposed to its actual strength, was 630,201. Budget planning in the spring of 1950 contemplated a reduction of this figure to 610,900. The proposed cut would have eliminated one of the Army's ten tactical divisions; specifically, it would have reduced the number of divisions in the FEC from four to three.[8]

The strength of the United States Army in 1950 was much less than American military leaders wished. But government economies in the aftermath of World War II allowed no increase.

Army Training

Training programs were hampered by lack of funds, and this, together with the absence of a sense of urgency, detracted from the combat readiness of Army forces in being in 1950.[9] Until 1949 basic training lasted only eight weeks, and graduates sent overseas usually had to undergo further basic training before they could be assigned to units. The Army put in a 14-week training cycle in March 1949 and, although this cycle did not provide for branch training (i.e., artillery, engineers), it included a sufficient amount of basic subject material to give an adequate foundation on which to build individual and unit training.[10] This came rather late for the Korean War.

Army Supply Status

The Army had sufficient stocks of most items of matériel and equipment to support its peacetime program. Certain imbalances—resulting from the cessation or curtailment of production, the surplus property disposal program, and the breakdown of distribution systems—existed, but these presented relatively minor problems and were usually localized.

From the standpoint of war-readiness, the Army's supply position was much more serious. Army procurement after World War II was limited mainly to food, clothing, and medical supplies. The shift of American industry away from military production forced the Army to operate almost exclusively with older and obsolescent equipment. Nor was money available for new procurement. The Army computed its requirements carefully, basing them on minimum essentials, only to find that appropriations

[6] Memo for Gen Collins, 9 Jul 50, sub: Status of Major Units of the General Reserve Which Have Not Been Committed to FECOM, unnumbered notebook of Far East Br, G–3, DA, in G–3, DA files.

[7] (1) JSPC 853/6, 4 Jul 50, App C to Incl B, in G–3, DA files. (2) Four training divisions also were stationed in the United States.

[8] Army Tentative Plans, FY 1952, Part I, p. 55.

[9] For information in detail on Army training in the postwar era, see: Annual History, Office, Chief of Army Field Forces (OCAFF), 1 January–31 December 1949 (hereafter cited as Annual History of OCAFF), Part I, ch. I, pp. 5–9, ch. VI, pp. 2–3, 5–6, ch. IX; *ibid.*, 1950, vol. II, ch. XIV; Rpt of Activities AFF, 1945–49, pp. 8, 10, 54–55. All in OCMH.

[10] Annual History of OCAFF, 1949, ch. VI, pp. 5–6.

habitually fell far short of meeting them. For the fiscal year 1948, for instance, the Ordnance Department estimated it would need $750,000,000 to cover procurement of essential ammunition and equipment, storage and distribution of ordnance material, maintenance of stand-by plants and arsenals, training, and research and development. The Bureau of the Budget cut this figure to $275,000,000, and the Congress reduced the appropriation in final form to $245,532,000.[11]

Maintenance of available equipment assumed greater importance as World War II items wore out under constant use or deteriorated in storage depots. Rapid demobilization had hurt the Army's maintenance program by reducing personnel and facilities to levels allowing proper storage and continuing maintenance on no more than a token basis. At the same time, replacement parts and assemblies became critical in many classes of equipment.[12]

Machine guns and towed artillery were in plentiful supply, but heavy construction equipment, newly developed radios, self-propelled artillery, newer tanks, and antiaircraft guns were critically short. Installations in the United States supporting the current 10-division Army required more than 38,000 commercial-type motor vehicles, but in 1950 only 27,000 were on hand, and 23,000 of these were six or more years old. There were fewer than 900 serviceable light M-24 tanks in the United States, 2,557 unserviceable ones; 1,826 serviceable medium M4A3 tanks, 1,376 unserviceable ones. There were only 319 new M-46 General Patton tanks.[13]

Development of new weapons and vehicles continued, but at a decelerated pace. New models being developed in the spring of 1950 would not be available for issue before the end of 1952. Other research projects indicated many desirable improvements in weapons and equipment, but funds were unavailable to complete development and production.[14]

Ammunition stocks in the United States were far out of balance. Training activities, both of the active Army forces and the civilian components, normal deterioration, and transfers to foreign countries under military assistance programs, had eaten away much of the stockpile remaining at the end of World War II, while economy budgets prevented significant new procurement. The result was a woefully inadequate reservoir of several types of ammunition.[15] In sum, the shortages of men and supplies combined with inadequate training to affect adversely the combat readiness of the Far East Command just as they hindered the effectiveness of the U.S. Army elsewhere.

The Far East Command

On 16 December 1946 the Joint Chiefs of Staff designated General MacArthur

[11] Statement, Maj Gen Everett S. Hughes, 14 Mar 47, *Hearings Before House Subcommittee on Appropriations*, 80th Congress, 1st Session, pp. 941, 967.

[12] G-4 Review of the Month, 1 Apr 48, pp. 1, 29, in G-4, DA files.

[13] (1) Army Presentation Before JCS on Review of Service Establishment, Phase II, Part III for FY 1951 Budget, 29 Jul 49, pp. 143-47. (2) DF, Supply Div to Control Office, 11 Jul 51, sub: Supply Sit in REC and U.S. as of 25 Jun 50, with 7 Incls, in G-4, DA files.

[14] *Ibid.*

[15] Summary Sheet, CSCLD/16027, DCofS G-4 (Gen Reeder) to CofS, 3 Apr 50, sub: Ammunition Reserve, in G-4, DA files.

Commander in Chief, Far East Command, effective 1 January 1947. No specific boundaries were established, but forces placed under General MacArthur's command were located in Japan, Korea, the Ryukyu Islands, the Philippines, the Mariana Islands, and the Volcano and Bonin Islands. These determined in a vague manner the geographic limits of the Far East Command.[16]

The area was vast. It extended over 265,000 square miles of island area inhabited by almost 100,000,000 people. Because of the preponderance of sea over land within the Far East Command and because of the terrain and climatic conditions, varying from sub-Arctic to tropical, the military garrison was compartmented into geographical groups. The primary land area and the area containing the largest number of U.S. troops was Japan.[17]

MacArthur's authorities and responsibilities as CINCFE were defined by directives issued by the Joint Chiefs of Staff. Three general missions were assigned him. The first pertained to occupation of former enemy territories in which he discharged U.S. occupation responsibilities in Japan, Korea, and former Japanese islands. The second broad mission was to support U.S. policies within the areas controlled by his forces. Third, CINCFE was to prepare to meet a general emergency at any time. The top headquarters within the Far East Command was General Headquarters (GHQ) located in Tokyo, Japan. This was essentially an Army headquarters, staffed almost entirely by Army personnel, and resembling the structure of General MacArthur's World War II headquarters.[18]

The Navy and Air Force felt that their activities within the Far East were being directed by the Army staff under an Army commander. But General MacArthur considered his authority over naval and air forces too limited. He complained that he could not exercise sufficient control over the internal organization of these services in his area, direct the troop control of their units, or supervise fully their logistical operations.[19]

As Commanding General, United States Army Forces, Far East (USAFFE), General MacArthur controlled all Army units and personnel within his area. Since this function was inherent in the broader designation of CINCFE, he

[16] Study, Requirements, Means Available, and Procedures Evolved to Accomplish CINCFE Missions (hereafter cited as FEC Papers), Paper 1, 26 Oct 49, p. 2, in G-3, DA file P&O 333 Pacific, F/w-6/3.

[17] *Ibid.*, pp. 5-7.

[18] (1) *Ibid.*, Paper 5, pp. 2-6. (2) The directive from JCS which established the command originally had stated, "Each unified commander will have a joint staff with appropriate members from the various components of the services under this command in key positions of responsibility." General MacArthur had not gone all the way in meeting the spirit of unification. But a joint committee of top-ranking Army, Navy, and Air Force officers was an integral part of GHQ and met each week, though only to advise the Chief of Staff, FEC (an Army officer), in "coordination of interservice matters." Additionally, frequent co-ordinating conferences were held by MacArthur with the commanders of major air and naval elements within his command. Another concession to the principle of unification of command within GHQ was the establishment of the Joint Strategic Plans and Operations Group (JSPOG) to "assist and advise the Commander-in-Chief Far East, on matters pertaining to the exercise of unified command over Army, Navy and Air Forces allocated to the Far East Command." The group consisted of three Army officers, three Navy officers, and two Air Force officers, but hardly constituted a joint staff as envisioned by the JCS instructions of December 1946. See JCS 1259/27, 14 Dec 46, and USAF in the Korean Conflict, USAF Hist Study No. 71, p. 9.

[19] FEC Papers, Paper 12, 1 Oct 49.

DAI ICHI BUILDING, TOKYO

neither used the title commanding general, USAFFE, nor established a separate staff. Because there were within his command a major air force and a major naval headquarters, Far East Air Forces (FEAF) and Naval Forces, Far East (NavFE), respectively, some resentment developed because the coequal Army headquarters, AFFE, was absent. That all Army combat forces were assigned to subordinate Army commands had the effect of placing these lesser headquarters on the same level with FEAF and NavFE. General MacArthur defended this peculiarity in the command structure by saying that imposing an Army headquarters between subordinate Army units and GHQ FEC would duplicate the functions of GHQ and detract from the essential and cohesive relationships between CINCFE and the Supreme Commander, Allied Powers (SCAP).[20]

[20] (1) *Ibid.*, Paper 13, p. 4. (2) A succinct and fairly accurate description of the FEC structure was rendered by a representative of the Department of the Army, Army War Plans Branch, who visited the command in October 1950. He said: "Although a lack of balanced representation from the three services keeps GHQ FEC from being classified as a joint headquarters in the commonly accepted sense, certain joint features do exist. . . . Intelligence is

In June 1950 GHQ, FEC, located in Tokyo, Japan, with main offices in the Dai Ichi Building, had Maj. Gen. Edward M. Almond as chief of staff and Maj. Gen. Doyle O. Hickey as deputy chief of staff. The major subordinate Army commands were Eighth Army, commanded by Lt. Gen. Walton H. Walker; Headquarters and Service Group, GHQ, commanded by Maj. Gen. Walter L. Weible; the Ryukyus Command (RYCOM) under Maj. Gen. Josef R. Sheetz; and the Marianas-Bonins Command (MARBO) headed by Maj. Gen. Robert S. Beightler. In the Philippines, the Thirteenth Air Force controlled U.S. installations through PHILCOM (AF), a small and rapidly diminishing headquarters commanded by Maj. Gen. Howard M. Turner, USAF. Naval Forces, Far East, were commanded by Vice Adm. C. Turner Joy. Far East Air Forces (FEAF), came under Lt. Gen. George E. Stratemeyer. FEAF and NavFE headquarters were located in Tokyo in buildings separate from GHQ, FEC.

GENERAL WALKER

FEC Strategic Planning and Korea

General MacArthur's basic plan to meet a general emergency in the Far East was to defend the Japanese islands. Operations were to be offensive-defensive, with air and naval forces assuming the tactical offensive to protect the withdrawal of forces from outlying areas and to deny to the enemy the control of the sea and air approaches to Japan. The main body of Army forces would be concentrated on Okinawa, the Marianas, and the Kanto Plain of Honshu. Those Army forces located in Korea were to be precipitately withdrawn.

Regarding Korea, the JCS had advised the State-Army-Navy-Air Force Coordinating Committee (SANACC), successor to the State-War-Navy Coordinating Committee, in January 1948, that the withdrawal of the U.S. occupation forces from South Korea would most likely lead

correlated in Army, Navy and Air Force Group with Theater Intelligence Section, G-2; planning is coordinated through JSPOG; a joint committee (composed of the Chiefs of Staff of the three Services) coordinates on the higher level. The Far East Command is a unified rather than a joint command with command lines following straight service seniority channels throughout as opposed to command responsibilities on a joint basis by geographical area; e.g. there is no joint commander of the Ryukyus or in Marianas-Bonins Command. CINCFE commands all major Army commands as theater commander and commands all Navy and Air Force commands through the Senior Commanders of those services." See Memo, Lt Col Stevens, AWPB G-3, for ACofS G-3, 17 Oct 50, sub: Rpt of TDY in FEC, in G-3, DA file 333 Pac, Case 7.

to communist domination of the entire nation. And since it was nevertheless intended to evacuate American troops, eventual Russian control of Korea would have to be accepted as a probability, even though establishing a ROK constabulary force might serve as a temporary deterrent.[21]

The definitive write-off of Korea as an important strategic area came when the Joint Chiefs of Staff asserted that no military security guarantee should be extended to the Republic of Korea because such action would risk a major war in an area where Russia would have nearly all the natural advantages. As a result, the President, on 4 April 1948, approved a policy that stated: "The United States should not become so irrevocably involved in the Korean situation that an action taken by any faction in Korea or by any other power in Korea could be considered a 'casus belli' for the United States." From that moment, Korea was of secondary importance to U.S. planners and policy makers.[22] General MacArthur had been relieved of his responsibility for defending Korea when the last American tactical units had been withdrawn from that country in 1949.

In mid-1949 General Omar N. Bradley, then Army Chief of Staff, challenged the national policy toward Korea. On the eve of the withdrawal of the last American combat troops from the peninsula, General Bradley suggested taking the Korean question again to the National Security Council. He feared that U.S. withdrawal might be followed by an invasion from the north. He had had his staff review the courses of action open to the United States in such an eventuality, and as a result he recommended that, if an invasion took place, the U.S. nationals be evacuated and the aggression immediately be presented to the United Nations Security Council as a threat to the peace. A U.N. composite military force might be considered as a last resort.[23]

Bradley's fellow members of the Joint Chiefs of Staff were reluctant to bring this matter again before the National Security Council. They said:

> From the strategic viewpoint the position of the Joint Chiefs of Staff regarding Korea, summarized briefly, is that Korea is of little strategic value to the United States and that any commitment to United States use of military force in Korea would be ill-advised and impracticable in view of the potentialities of the over-all world situation and of our heavy international obligations as compared with our current military strength.[24]

This concept dominated American planning for the Far East. By 1950, the United States decided that, in the event of a Soviet attack in the area, American Forces would conduct a strategic defense. Specific missions charged to the Far East Command were: (1) defense of the Ryukyus and Japan; (2) protection of air and sea lanes in the FEC; (3) denial of Formosa to the enemy; (4) support of the Pacific Command, the Alaskan Command, and the Strategic Air Command; (5) assistance to the Republic of the

[21] JCS 1483/50, Rpt by JSSC, title: U.S. Policy in Korea, 30 Jan 48.

[22] SANACC 176/39, 22 Mar 48, title: U.S. Policy in Korea.

[23] JCS 1776/4, 23 Jun 49, Incl, Memo, CSA to JCS, 20 Jun 49, sub: Implications of a Possible Full-Scale Invasion From North Korea Subsequent to the Withdrawal of U.S. Troops From Korea.

[24] JCS 1776/4, 23 Jun 49.

SECRETARY OF STATE ACHESON

Philippines in defense of the islands; and (6) provision for the safety of U.S. personnel in Korea. American airmen were to destroy or neutralize enemy air power.[25]

That Korea was considered of little strategic worth to the United States had scarcely been a matter of public knowledge until 12 January 1950, when Secretary of State Dean Acheson said so in a speech at the National Press Club in Washington. Outlining the defensive strategy in the Far East, he excluded Korea and Formosa from the American defensive perimeter. Referring obliquely to Korea, Mr. Acheson stated:

> So far as the military security of other areas in the Pacific is concerned, it must be clear that no person can guarantee these areas against military attack. . . . Should such an attack occur—one hesitates to say where such an armed attack could come from—the initial reliance must be on the people attacked to resist it and then upon the commitments of the entire civilized world under the Charter of the United Nations which so far has not proved a weak reed to lean on by any people who are determined to pro-

[25] FEOP 1-50, GHQ FEC, vol. I, 1 Feb 50, in G-3, FEC files.

tect their independence against outside aggression.[26]

In the light of Secretary Acheson's remarks, it appeared that the United States had no intention of fighting for South Korea. In the view of many observers, his statement was an invitation to Communist China, North Korea, and Russia that they could invade the republic with impunity.

MacArthur's Forces

The general decrease in Army strength that took place in 1947 was reflected sharply in the Far East. General MacArthur had commanded over 300,000 troops, including 42,000 in the Army Air Forces, in January 1947.[27] Just one year later he had only 142,000 men. When asked early in 1948 if he could maintain 30,000 men in Korea, MacArthur told Army officials that to do so would cause a breakdown in logistic support to the Far East Air Forces and a breakdown in the general effectiveness in his command. The real cause of this situation, he charged, was Washington's failure to send him even half the troops approved for his command.[28]

MacArthur warned of irreparable damage to United States national interests in the Far East unless his command was strengthened. In response, the Joint Chiefs of Staff informed MacArthur that all services were having trouble keeping up to authorized strength and that calculated risks in the allotment of manpower had to be accepted throughout the world. Allocating 134,000 troops (including 28,800 Philippine Scouts) to his command, they ordered him to keep 30,000 troops in Korea until elections had been held there.[29]

MacArthur protested. On 24 February 1948 he charged that his personnel resources were exhausted. He asserted that there was no substitute for Army troop strength and that it was essential to meet the dangers and difficulties that existed in the Far East.[30]

There was actually a further decline. MacArthur's authorized strength for the year beginning 1 July 1949 was to be only 120,000 men. Insofar as combat strength was concerned, the Far East Command reached its lowest ebb at this point, April 1948. The Eighth Army, upon which the combat effectiveness of the command depended, was authorized 87,215 men, but had an actual strength of only 45,561 and a combat strength of 26,494. This combat strength was spread over five divisions and an antiaircraft artillery group, making attainment of any satisfactory degree of combat readiness very difficult. MacArthur's protests continued, but to no avail. Exemplifying the general conditions within the Eighth Army, two regiments of the 25th Division had less than 250 men each.[31]

On 3 August 1948 MacArthur complained that his carefully analyzed minimum requirements for Army strength were being brushed aside. He was noti-

[26] (1) Speech, Mr. Dean Acheson to National Press Club, 12 Jan 50, quoted in *MacArthur Hearings*, pp. 1811–12. (2) See also Acheson, *Present at the Creation*, pp. 354–58.

[27] Strength Reports of the Army, Central Statistical Office, Office, Chief of Staff, 1 Feb 47, copy in OCMH.

[28] Rad, CX 58131, CINCFE to DA, 23 Jan 48.

[29] Rad, WARX 96357, JCS to CINCFE, 21 Feb 48.

[30] Rad, CX 58837, CINCFE to DA, 24 Feb 48.

[31] (1) Rad, WAR 81295, DA to CINCFE, 6 May 48. (2) Rad, C 61072, CINCFE to DA, 29 May 48. (3) Rad, C 61943, CINCFE to DA, 29 Jun 48. (4) Rad, WARX 86492, DA to CINCFE, 27 Jul 48.

fied on 9 November 1948 that the nation's authorities were contemplating a reduction in the strength of his Far East Air Forces. This news brought a sharp rejoinder and a strategic estimate of his position in the Far East Command. He maintained that he could not understand what devious thinking had prompted a proposal for reducing his military strength. He said that it would endanger the nation's military position in the Far East beyond the acceptable point of calculated risk. MacArthur charged that the nation's planners should be contemplating an increase in his naval, air, and ground forces.[32]

Despite MacArthur's insistent protests, the strength level in the Far East Command continued with little substantive change. During visits to Tokyo by the Department of the Army Staff, by the Secretary of the Army, and by members of the JCS during 1948 and 1949, General MacArthur presented his views and protests in person. He said consistently that the support which the Department of the Army was giving to forces in Europe was out of proportion and that more support should and could be given to his command in the Far East.[33]

The flow of replacements to the Far East picked up somewhat in 1949 although budgetary limitations on the Army as a whole enforced restrictions on replacements available to the Far East Command. By late 1949, the shortage of funds had become so pronounced that the Department of the Army decided to reduce the number of divisions in the Army from ten to nine. MacArthur's command was to take the loss and during a discussion with MacArthur in October 1949 General Collins, Army Chief of Staff, told MacArthur so. MacArthur, of course, objected. The Department of the Army reversed its decision and kept ten divisions on duty.[34] But, as noted above, the strength of the Far East Command had dwindled to about 108,500 Army troops by June 1950.

The budget limitations and the low enlistment rate forced the Department of the Army to devise a troop program and troop list which could not be manned at 100 percent strength. This reduced over-all personnel ceiling reflected manning levels which, in turn, caused unavoidable reductions either by paring the strength of all subordinate units or by eliminating certain units entirely. Since administrative requirements continued or increased, combat units suffered more than headquarters units.[35] As reflected in the FEC, this condition caused the elimination of certain basic elements from combat units in order to maintain the units within the command. Each of MacArthur's infantry divisions had only one tank company instead of a tank battalion, and one antiaircraft battery instead of an antiaircraft battalion. Each infantry regiment was short its Table of Organization (T/O) tank company and lacked one infantry battalion; each of the divisional artillery battalions was short one firing battery. Although CINCFE had managed to retain the 4-division structure of Eighth Army, he

[32] (1) Rad, W 92269, DA to CINCFE, 9 Nov 48. (2) Rad, CX 65569, CINCFE to DA, 23 Nov 48.

[33] Rad, WAR 82319, DA to CINCFE, 6 Jan 49.

[34] (1) JCS 1800/54/56, Sep 49. (2) JCS 2079/3, Oct 49.

[35] Rpt of OCAFF Observer Team to FEC, 16 Aug 50, with comments by Chief, OCAFF, in G–3, DA file 333 Pac, sec. I–A, Book I, Case 8/8 (1950).

had had to eliminate the normal corps headquarters and corps special troops (artillery, engineer, and so forth). Service elements of Eighth Army were so inadequate that over 150,000 Japanese personnel were being employed in roles normally performed by service troops.[36]

The ratio of noncombat to combat personnel in the Far East was excessive. This stemmed from the Army's attempts during the postwar years to make the Army an attractive career by leaving the choice of arm or service largely to the individual. The combat arms, and especially the infantry, failed to attract sufficient men to keep their strength on a par with other arms and branches. Also the fact that a substantial percentage of the already inadequate output of stateside training divisions went to service schools for further training reduced the number of men available for assignment to combat-type units except in specialist capacities.[37]

MacArthur's combat forces in June 1950 comprised 4 understrength infantry divisions and 7 antiaircraft artillery battalions in Japan, 1 infantry regiment and 2 antiaircraft artillery battalions in Okinawa. The major combat units were the 1st Cavalry Division (actually infantry) in central Honshu, Japan; 7th Infantry Division in northern Honshu and Hokkaido, Japan; 24th Infantry Division in Kyushu, Japan; 25th Infantry Division in south central Honshu, Japan; and the 9th Antiaircraft Artillery Group in Okinawa. General MacArthur had registered frequent protests that his missions in the Far East required a minimum force of at least 5 full-strength infantry divisions, 23 antiaircraft artillery battalions, and 1 separate RCT.[38]

Eighth Army, the main combat force of FEC, stood at about 93 percent of its authorized strength on 25 June 1950. Each division had an authorized strength of 12,500 men as compared to its authorized war strength of 18,900 and none of the divisions was even up to its peacetime authorization. Each division was short of its war strength by nearly 7,000 men, 1,500 rifles, and 100 90-mm. antitank guns; 3 rifle battalions, 6 heavy tank companies, 3 105-mm. field artillery batteries, and 3 antiaircraft artillery batteries were missing from each division. In terms of battle potential, the infantry divisions could lay down only 62 percent of their infantry firepower, 69 percent of their antiaircraft artillery firepower, and 14 percent of their tank firepower.[39]

Until 1949 the primary responsibility of military units in the Far East Command was to carry out occupation duties. Engaged in these administrative and housekeeping tasks throughout Japan and the outlying areas, units had little time or inclination for combat training. The situation was aggravated by constant understrength and excessive turnover of personnel. This turnover amounted to 43 percent annually in the FEC. Training in the rudimentary functions of the soldier was carried on as time and facilities permitted during the period from 1945 to 1949 with emphasis upon discipline, courtesy, and conduct.

---

36 FEC Papers, Paper 10, p. 7.

37 Rpt of OCAFF Observer Team to FEC, 16 Aug 50.

38 FEC Papers, Paper 10.

39 Mono, 1st Lt Charles G. Cleaver, Personnel Problems, in History of the Korean War, vol. III, Part 2, MHS, Hq, FEC, 15 Aug 52, p. 1, copy in OCMH.

No serious effort was made in these years to maintain combat efficiency at battalion or higher level.

This situation changed markedly in April 1949 when General MacArthur issued a policy directive announcing that the stern rigidity which had characterized the occupation of Japan until that time was to be superseded by an attitude of "friendly protective guidance." As a result of this change in policy, combat divisions of Eighth Army were progressively relieved of the majority of their purely occupational missions and directed to undertake along with FEAF and NavFE an intensified program which would lead to the establishment of a cohesive and integrated naval, air, and ground fighting team. Although large numbers of officers and men were detached from military government and civil affairs activities and returned to their parent combat units, there still remained many administrative features of the occupation which could not be relinquished and which constituted a considerable barrier to the full development of the planned training program.[40]

Main objectives of the new training program announced by General MacArthur on 10 June 1949 called for the rapid integration of Army, Navy, and Air Force components into an efficient team capable of performing its primary military mission. Divisions were directed to complete RCT field exercises and develop effective air-ground combat procedures prior to 31 July 1950 and to complete amphibious landing exercises for one battalion of each division by 31 October 1950. Minimum proficiency levels to be attained were (1) company (battery) levels by 15 December 1949; (2) battalion (squadron or task force) level by 15 May 1950; (3) regimental (group or task force) level by 31 July 1950; (4) division (air force or task force) level by 31 December 1950; and (5) combined and joint operations training to include amphibious exercises concurrently with RCT and division-level training.[41]

In a country so heavily populated and predominantly agricultural as Japan, no land was wasted and the maintenance of large military training areas would have imposed a burden upon the Japanese economy which was not considered justified. Consequently, troops were generally restricted in their training to small posts of regimental size. Divisions could not be concentrated and trained together. On 8 August 1949 an area in the vicinity of Mount Fuji was acquired which would accommodate limited division exercises over very rugged terrain. Every other field training area was exploited to the utmost. Exploitation of the relatively few training areas during favorable training weather, however, required that some units undertake field firing problems and tests ahead of the actual phasing of such training in the Mobilization Training Programs. For example, the 7th Cavalry Regiment of the 1st Cavalry Division completed its battalion tests before completing basic individual training in order to use that division's lone training area.[42]

The Army's Career Guidance Program also worked to the disadvantage of the

---

40 (1) FEC Papers, Paper 3, pp. 2–4. (2) GHQ, FEC Annual Narrative Historical Rpt, 1 Jan–31 Dec 49.

41 FEC Papers, Paper 23, pp. 7–8.

42 *Ibid.*, pp. 8–9.

training program within the FEC according to General MacArthur's staff. Staff visits indicated that a wide variance existed between the experience of regimental commanders and their subordinate commanders. There was a great need for improved leadership of combat units at the company and battalion levels. Many officers possessing the qualities of leadership and training experience necessary for proper development of FEC combat units had been given directed military occupational specialties (MOS) under the Career Guidance Program and could not be placed in command of troops where they were needed. From the standpoint of the enlisted man the same situation seriously affected the flexibility of organization and training. In their efforts to strengthen combat units by transferring men from inactivated service units, FEC commanders ran head on into the Career Guidance Program which prevented assignment of enlisted men from one field to another.[43]

The readiness of combat units within the FEC was not enhanced by the quality of enlisted personnel assigned from the zone of the interior. Replacements arriving from the United States during 1949, for instance, were said by General MacArthur's headquarters to have had a very high percentage of low intelligence ratings and a much larger than usual number of men of questionable character. This situation was reflected not only in training, but in discipline, administrative problems, and a larger number of individual incidents which caused criticism of American behavior. In April 1949, 43 percent of Army enlisted personnel in the Far East Command rated in Class IV and V on the Army General Classification Test. On an average, enlisted men of the FEC were several years younger than their counterparts of World War II. Another factor which intensified the difficulty of training for combat readiness was the incomplete basic training received by recruits before shipment to the FEC. According to an FEC report, recruits were not sufficiently indoctrinated to withstand the inactive period of pipeline experience and had lost much of the benefit of basic training before arriving in the Far East Command.[44]

General Collins, Chief of Staff, U.S. Army, visited the Far East Command in the autumn of 1949 and looked into the training program then in progress. He was generally satisfied with what he saw and with what he was told in conference with General MacArthur. Reporting on his findings to the Secretary of the Army General Collins said:

> As a result of the reductions in strength of personnel . . . and because our troops were primarily engaged in occupation missions until recently, the troops of Eighth Army are not now in fighting condition. However, they have recently been brought back up to strength, are making excellent progress with realistic field training and are planning exercises with close fighter-bomber support by the early spring of 1950. Given

[43] (1) *Ibid.*, p. 10. (2) This complaint from the FEC was verified at a later date by a team of observers sent to the Korean battlefield in the first month of the war. These observers noted that classification and assignment procedures had placed in battlefield command officers and noncoms lacking experience and proficiency. This kind of assignment had often resulted in poor leadership, especially at the regimental and lower levels. The observers concluded bluntly that the career program had been detrimental to combat efficiency. See Rpt of OCAFF Observer Team to FEC, 16 Aug 50.

[44] FEC Papers, Paper 23, pp. 2–3.

General Collins

another six months the divisions I inspected should be in excellent shape.[45]

All units of Eighth Army had completed the battalion phase of their training by the target date of 15 May 1950. An air transportability school had been established and was functioning, pointing toward battalion airlift exercises. At an amphibious training center near Tokyo, one battalion from each division had received training in landing techniques and a joint landing exercise was scheduled for August 1950. Reports on the Eighth Army's divisions which were sent to the Department of the Army in May 1950 showed estimates ranging from 84 percent to 65 percent of full combat efficiency for the four divisions in Japan.[46]

[45] Memo, Gen Collins for Secy Army, 20 Oct 49, sub: Rpt of Visit to Hawaii and FEC, in G–3, DA files.

[46] Rpt on Disposition, Strength, and Combat Capabilities of Major Army Forces in Overseas Commands, 30 May 1950, Rpts Control Symbol WDGPO–6, CINCFE to ACofS G–3, Opns, General Staff, U.S. Army, Washington, D.C., in G–3, FEC files.

FEC Supply Status

Equipment in the hands of MacArthur's troops was for the most part of World War II vintage. Much of it had been through combat, and a good deal of it, particularly the vehicles, had been serviced and maintained under difficulty during the years of occupation.

Adding to the difficulty of the logistic situation was the unusual dependence upon indigenous personnel which had developed within the U.S. Army in Japan during the years following World War II. Basically, this dependence stemmed from the acute shortages of trained American soldiers to perform specialized functions of the type normally carried out by service units. In the absence of sufficient service units and with emphasis transferred to a great extent from field-type operations, the natural result had been to exploit the enormous pool of manpower available in Japan. Japanese workmen carried out duties in support of U.S. Army units and in installations ranging from menial mess-hall tasks to highly technical functions calling for advanced training and great skill. Base areas, depots, and ports were manned by Japanese personnel under Army supervision, while protection of these installations, as well as other less sensitive areas throughout Japan, was largely delegated to Japanese guards.

After the war's ending in 1945, vast quantities of U.S. matériel had been left throughout the islands of the Pacific. This residue of the Pacific fighting—vehicles, signal equipment, armament, and other types of military equipment—was originally treated as excess. In many cases, it was left where it lay when the fighting ceased, abandoned for all intents and purposes, or at best gathered into assembly areas and maintained half-heartedly. Some was sold to foreign governments or domestic firms at a fraction of its intrinsic value. In the Philippines alone, 933,265 tons of such equipment had been disposed of through surplus property channels by the end of 1947.

The main islands of Okinawa, the Philippines, and the Marianas-Bonins contained the bulk of this equipment. Since these areas were part of the FEC, the condition and disposition of the material were matters of concern to General MacArthur. In 1947 he had ordered intensive surveys and the initiation of measures to reclaim as much of it as possible. Investigation by ordnance officers of the command showed that the greater part of all classes of this military equipment had been left in open storage, without adequate safeguards, with practically no proper segregation as to type, and with no attempt having been made to classify or catalogue it.

In the years from 1947 until the outbreak of war in Korea, personnel of the FEC had, therefore, been putting forth every effort to reclaim for military use as much of this valuable equipment as possible. Under a program informally known as Operation ROLL-UP, vehicles, weapons, ammunition, and other types of supplies from the island areas had been segregated, classified, and transported to facilities in Japan for repair and proper storage. Critical shortages in qualified personnel plus the desire to arrive at the most efficient and economical solution to the situation had forced this project to depend upon the use of

Japanese industry under the direction of a small American staff.[47]

The original objective of Operation ROLL-UP was to support the FEC and to equip Eighth Army's infantry divisions at minimum cost and with maximum use of all matériel which could be reclaimed. It was planned that the project would be completed by 30 June 1950. As an indication of the progress attained, 200,000 measurement tons of ordnance supplies were moved to Japan from Okinawa during 1949. All types of vehicles, artillery pieces, and ammunition as well as other items were affected by this program.

One result of Operation ROLL-UP was to prepare FEC repair and rebuild facilities, including Japanese industry, for the great expansion necessary to support extensive combat operations. In addition, thousands of military vehicles were available in substantially better condition than would have otherwise been the case.[48]

A shortage of supervisory personnel slowed the renovation program and made unattainable the goal of completing Operation ROLL-UP by 30 June 1950. When the North Korean attack came stocks of unusable equipment were still piled up in storage shops. An estimated 80 percent of the Army's 60-day reserve of armament equipment was unserviceable on 25 June. The Far East Command had received no new vehicles, tanks, or other equipment since World War II. Almost 90 percent of the armament equipment and 75 percent of the automotive equipment in the hands of the four combat divisions on that date was derived from the rebuild program.[49]

Levels of supply on hand in the FEC by mid-1950 amounted to a 60-day depot level plus 30-day levels in station stocks. But supply resources were out of balance both in quantity and quality. Some weapons such as medium tanks, 4.2-inch mortars, and recoilless rifles could hardly be found in the command. Only a trickle of supplies was moving through the pipelines. Units deactivated in the command had turned in large quantities of equipment, but most of this was unserviceable. Eighth Army was authorized 226 recoilless rifles, but had only 21. Of 18,000 ¼-ton 4 × 4 vehicles in Eighth Army's stocks 10,000 were unserviceable, and of 13,780 2½-ton 6 × 6 trucks only 4,441 were in running condition.

Total ammunition resources amounted to only 45 days' supply in the depots and a basic load of training ammunition in hands of units. The level of perishable food supplies was also 45 days in depot stocks and operating levels at various stations. Petroleum products on hand included a level of 180 days packaged and 75 days bulk at depots, station levels of 15 days each of packaged and bulk, and 15 days with units.[50]

By mid-1950 American forces in the Far East had begun a gradual swing away from their primary concern with occupation duties and had started to look more closely to their combat skills. This shift came about more because of the growing

[47] Administrative History of the Ordnance Section, GHQ, FEC, 1 January 1947–31 December 1949.

[48] Hist Rpt, Ordnance Section, GHQ, FEC, 1 Jan–31 Dec 49.

[49] Mono, Logistical Problems and Their Solutions, Hq, EUSAK, ch. I, pp. 5, 7, copy in OCMH.

[50] MS, Maj James A. Huston, Time and Space, ch. V, p. 41, and ch. III, pp. 176, 186, copy in OCMH.

stability of occupied Japan than from any real fear that time was growing short. That these forces were under-strength, inadequately armed, and sketchily trained concerned mainly their commanders. These commanders, within the limits of their resources, sought to overcome the inertia imposed by the years of occupation and the prevailing, if uneasy, peace. But on the eve of the storm the command was flabby and soft, still hampered by an infectious lassitude, unready to respond swiftly and decisively to a full-scale military emergency.

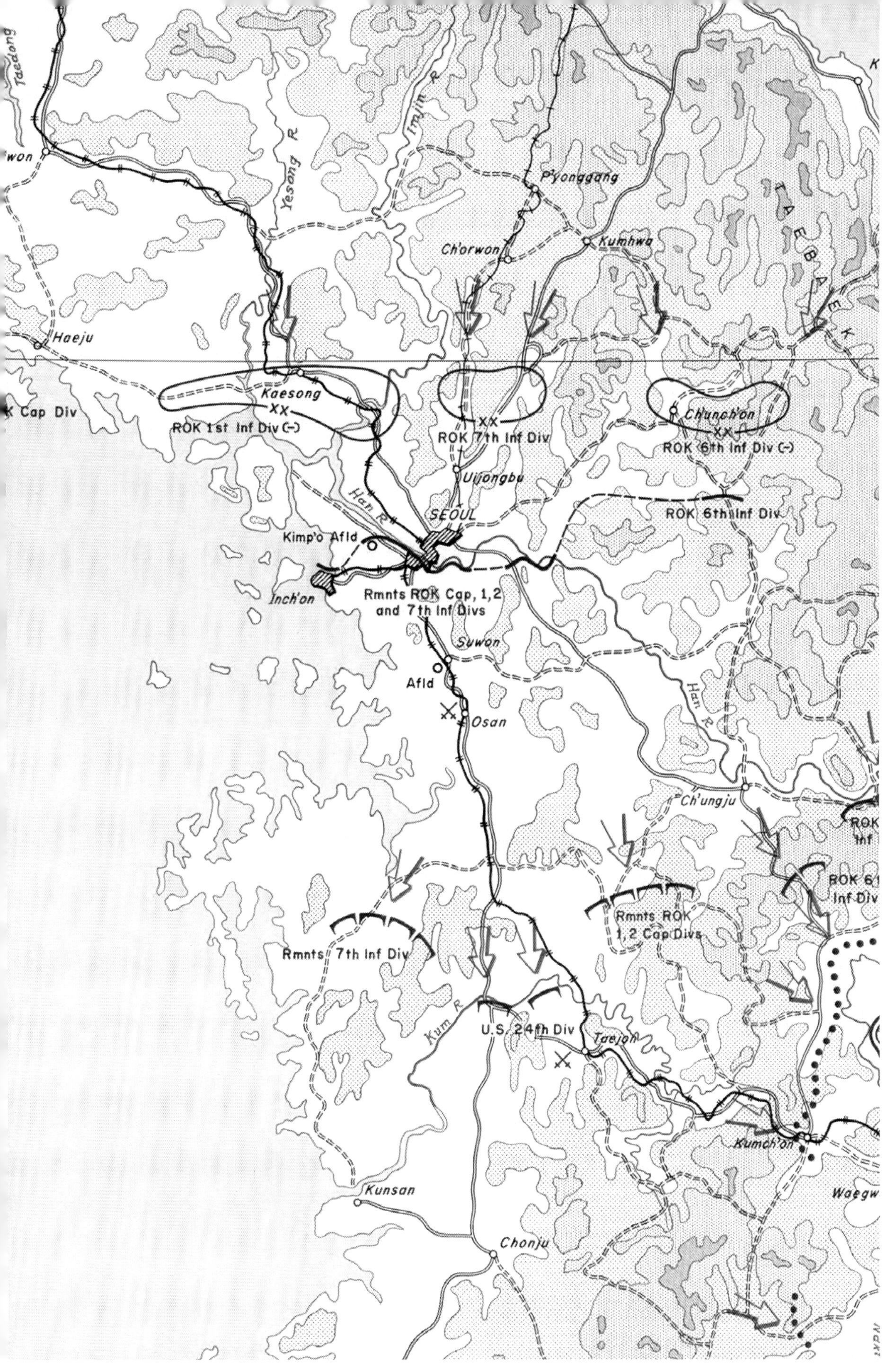

Taedong
won
Haeju
K Cap Div
Yesong R
Imjin R
P'yonggang
Ch'orwon
Kumhwa
TAEBAEK
Kaesong
XX
ROK 1st Inf Div (–)
XX
ROK 7th Inf Div
Chunch'on
XX
ROK 6th Inf Div (–)
Uijongbu
Han R
SEOUL
ROK 6th Inf Div
Kimp'o Afld
Inch'on
Rmnts ROK Cap, 1, 2
and 7th Inf Divs
Suwon
Afld
Osan
Han R
Ch'ungju
ROK 6t
Inf Div
Rmnts ROK
1, 2 Cap Divs
Rmnts 7th Inf Div
Kum R
U.S. 24th Div
Taejon
Kumch'on
Waegw
Kunsan
Chonju

CHAPTER IV

The Communist Challenge

The North Korean Army invaded South Korea at four o'clock in the morning of 25 June 1950—three o'clock in the afternoon of 24 June 1950, in Washington, D.C. (*Map I*) Striking without warning in the predawn dusk, communist units gained complete tactical surprise as they burst across the 38th Parallel swiftly and in strength. Co-ordinated columns of Russian-made tanks and Russian-trained infantry followed massed artillery fires and rolled back the South Korean defenders, engulfing and destroying whole units as they moved toward their objectives in a well-conceived and carefully prepared military operation. North Korean planes, giving tactical support, were virtually unchallenged.[1]

News of the invasion reached Seoul within an hour, before 0500. American officers there were alerted by 0630 and began to arrive half an hour later at their duty posts. Belief that the attack was nothing more than a border raid soon faded. By 0800, it was obvious that many North Korean troops were involved at many separate points. The use of armor and the major orientation on the approaches to Seoul were ominous. ROK defenders at Ch'unch'on in central Korea threw back the first attacks; but on the east coast, near Kangnung, an enemy amphibious landing was unopposed.

The Intelligence Failure

Agencies of the United States Government failed to forecast adequately the North Korean attack. No report sufficiently valid or urgent reached Washington officials before 25 June 1950 indicating that the attack would come when it did. Some information sent to

[1] (1) Unless otherwise cited all material in this chapter dealing with events in Korea comes from the following sources: Daily Opns Rpts, G–3, GHQ, FEC, Jun 50; DIS, G–2, GHQ, FEC, Jun 50; Interv, Dr. Gordon Prange with Lt Col A. J. Storey, Oct 50; Interv, Maj James F. Schnabel with Lt Col Leonard Abbot, Oct 50; Interv, Maj Schnabel with Capt Frederick Schwarze, former ACofS G–2, KMAG, 17 Nov 53. (2) The international communist bloc later charged that the South Korean Army had invaded North Korea, thus triggering a North Korean counterattack. Two documents captured following the fall of North Korea have been authenticated as official attack orders issued by North Korean military authorities to their commanders several days before the assault. Both documents, Reconnaissance Order No. 1, issued in Russian to the Chief of Staff of the North Korean 4th Division and discovered in Seoul on 4 October 1950, and Operations Order No. 4, North Korean 4th Division, were issued on 22 June 1950. See ATIS Res Supp Interrog Rpts, Issue 2 (Documentary Evidence of North Korean Aggression), Part 2.

Washington from the Far East reflected a strong possibility of action toward the end of June, but faulty evaluation and dissemination prevented it from reaching the right people in the proper form. The invasion therefore took all the American political and military leaders by surprise.

The reasons for this intelligence failure are easy to understand. The United States had written Korea out of its national defense plans, and as a result indications from Korea received less attention than those from areas considered more vital to American interests. There was nevertheless an intelligence effort in Korea. KMAG officers worked closely with their ROK Army counterparts in assembling data on North Korean activities. They sent this information to Washington periodically and on occasion made special reports. Other agencies and units in the Far East reported to appropriate officials in Washington.[2] KMAG, not General MacArthur, had the responsibility of securing intelligence data on Korea. When General Collins visited Tokyo in early 1950, he asked whether MacArthur could furnish the JCS information on some areas beyond his sphere of responsibility. MacArthur answered that he had promptly furnished such reports whenever specific items had been developed but that he was reluctant to submit unsupported estimates. If the JCS wanted to give him new intelligence responsibilities, he said he would be glad to have them. He was confident that he had enough personnel to handle them.[3]

Maj. Gen. Charles A. Willoughby, the FEC G-2, had on his own initiative already established a surveillance detachment in Korea called the Korean Liaison Office. In addition, according to General Willoughby, "The Embassy in Seoul maintained military attaché groups—Army, Navy, and Air, as well as their own diplomatic and political specialists whose sole business was to gauge the trend of events." [4]

Significant troop movements and concentrations, forward stockpiling of supplies, border evacuation, and North Korean Army reinforcement in men and matériel were some of the meaningful indications reported to Washington from the Far East before the June attack. But this information was poorly evaluated in the field and at higher echelons. Secretary of State Acheson later testified:

> Intelligence was available to the Department prior to the 25th of June, made available by the Far East Command, the CIA, the Department of the Army, and by the State Department representatives here and overseas, and shows that all these agencies were in agreement that the possibility for an attack on the Korean Republic existed at that time, but they were all in agreement that its launching in the summer of 1950 did not appear imminent.[5]

Since October 1946, when General Hodge had first reported that the North Koreans intended to attack South Korea, dozens of such reports had poured into Tokyo and Washington. Upon the outbreak of border fighting, the reports

---

[2] Sawyer, *Military Advisors in Korea*, pp. 37ff.

[3] Notes on Visit of JCS to FEC, 29 Jan–10 Feb, in G–3, DA file P&O 333 Pacific, sec. I, Case 7/4.

[4] Maj. Gen. Charles A. Willoughby and John Chamberlain, *MacArthur, 1941–1951* (New York: McGraw-Hill, 1954), p. 354.

[5] *MacArthur Hearings*, pp. 123, 350, 436, 1832, 1990–91.

gained credence. By late 1949, talk of a North Korean invasion was almost routine in intelligence circles.[6] By early 1950, there was a pattern of growing urgency. But it went undetected, or at least unheeded, against the more riotous background of threatening communist activities in other parts of the world—in Asia, western Europe, and the Middle East.

On 30 December 1949, General Willoughby sent to Washington several reports that indicated a North Korean invasion in March or April 1950. But his own personal evaluation was that "such an act is unlikely." On 19 February 1950, he passed on two agent reports, which he also discounted, one saying that the North Koreans would attack in March, the other in June. On 10 March, the Korean Liaison Office sent him an agent's report that the North Korean invasion schedule had been set back from March or April to June 1950. Late in March Willoughby said:

> It is believed that there will be no civil war in Korea this spring or summer. . . . South Korea is not expected to seriously consider warfare so long as her precipitating war entails probable discontinuance of United States aid. The most probable course of North Korean action this spring and summer is furtherance of attempts to overthrow South Korean government by creation of chaotic conditions in the Republic of Korea through guerrillas and psychological warfare.[7]

Intelligence in Washington was more concerned with what appeared to be the greater danger in Southeast Asia. Indochina seemed a much more likely target for a communist take-over. In March 1950, Maj. Gen. Alexander R. Bolling, the Department of the Army G–2, stated: "Recent reports of expansion of the North Korean People's Army and of major troop movements could be indicative of preparation for aggressive action." These preparations could be completed by late spring 1950. This forecast was, however, vitiated by the next comment. "Communist military measures in Korea will be held in abeyance pending the outcome of their program in other areas, particularly Southeast Asia. If checked or defeated there, the Soviet might divert effort toward South Korea. In that event, invasion by the People's Army would be probable." [8]

The Office of Special Investigations, USAF, told Headquarters, Far East Air Forces, in mid-April that Russia had definitely ordered an attack on South Korea by the North Korean People's Army. But in early May 1950 the American Embassy in Seoul reported little likelihood of a North Korean invasion in the near future.[9]

In May 1950, the Department of the Army G–2 said, "The movement of North Korean forces steadily southward toward the 38th parallel during the cur-

[6] The author, upon being assigned to G–2, GHQ, FEC, in November 1949, attended a briefing for newly arrived officers in the Dai Ichi Building in Tokyo. Discussing the military situation in the Far East at that time, the briefing officer, a major from the G–2 section, quite frankly stated that the feeling in G–2 was that the North Koreans would attack and conquer South Korea in the coming summer. The point was not emphasized particularly and the fact seemed to be accepted as regrettable but inevitable.

[7] DIS, GHQ, FEC, No. 2669, 30 Dec 49; No. 2720, 19 Feb 49; No. 2754, 25 Mar 50; No. 2900, 18 Aug 50; and KLO No. 518, 25 May 50.

[8] Int Div, GSUSA, DA, Weekly Intelligence Rpt, 17 Mar 50.

[9] (1) OSI Rpt (49) 52–12A–4–1, 17 Apr 50. (2) Rad, Seoul 456, Drumright to State, 4 May 50.

rent period could indicate preparation for offensive action." On 23 May, in another routine summary, he stated, "The outbreak of hostilities may occur at any time in Korea and the fall of Indochina to the Communists is possible this year." [10]

A report forwarded routinely on 19 June 1950, six days before the North Korean assault, provided Washington with strong evidence of an imminent enemy offensive—extensive troop movements along the 38th Parallel; evacuation of all civilians north of the parallel for two kilometers; suspension of civilian freight service from Wonsan to Ch'orwon and the transportation of military supplies only; concentration of armored units in the border area; and the arrival of large shipments of weapons and ammunition. But no conclusions were drawn from these indications.[11] On the same day a report from General Willoughby in Tokyo concluded, "Apparently Soviet advisers believe that now is the opportune time to attempt to subjugate the South Korean Government by political means, especially since the guerrilla campaign in South Korea recently has met with serious reverses." [12]

The Department of the Army G–2 protested charges made later that he had failed to interpret properly the information sent to him from the Far East Command. "An analysis of reports received by G–2, DA," General Bolling told General Collins,

> shows that all reporting agencies were aware of [the North Korean] capability to invade the Republic of Korea. There has been much publicity originating from Tokyo and quoting Willoughby that he had informed the Department of the Army that North Korean troops would invade South Korea in June. The statements made by Willoughby are correct in part, but he failed to indicate [in the publicity] his conclusions that definitely discount the report referred to. In short, there is no intelligence agency that reported a definite date for the opening of hostilities or stated that an invasion was imminent. In fact, the general tenor of reports indicated that the North Korean regime would continue to employ guerrillas and psychological warfare together with political pressure rather than resort to the overt employment of military forces.[13]

American intelligence failed to predict the time, strength, and actual launching of the attack because of reluctance to accept all the reports rendered by Koreans, a distrust of Oriental agents and sources, and a belief that the South Koreans were prone to cry wolf. Situations similar to that in Korea existed in virtually every other land area around the periphery of the USSR. Some appeared to be greater potential danger spots and diverted the focus of interest from Korea. Signs which marked the prelude of the North Korean attack had become accepted as routine communist activity. The increased troop movement and activity in North Korea in the spring of 1950 followed a pattern established

---

[10] Memo, ACofS G–2, DA, for Gen Wade H. Haislip, 24 Aug 50, in G–2, DA file SO 24366.

[11] Sec G–2, FEC, files, M.I.S., Item No. 684595, 19 Jun 50.

[12] (1) DIS, GHQ, FEC, No. 2842, 19 Jun 50. (2) General Willoughby later insisted that "Washington" had been fully informed of what to expect in Korea and should not have been taken by surprise. See Willoughby and Chamberlain, *MacArthur, 1941–1951*, pp. 350–54. See also Douglas MacArthur, *Reminiscences* (New York: McGraw-Hill, 1964), pp. 323–24.

[13] Memo, Gen Bolling for DCofS for Admin, DA, 18 Oct 50, in G–3, DA file CofS 091, Case 28.

by the communists in 1947 when they initiated an annual rotation of completely equipped units from the parallel.

The forwarding of reports in a routine manner detracted from the significance of the data in many cases.[14] In Congressional hearings immediately after the North Korean attack, Maj. Gen. Lyman L. Lemnitzer, director of the Office of Military Assistance, was subjected to sharp questioning about the failure of the Department of Defense to anticipate the attack. Telling the Secretary of Defense of this experience, General Lemnitzer stated:

> I believe that there are lessons to be learned from this situation which can point the way to better governmental operations and thus avoid costly mistakes in the future. . . . I recommend that . . . a clear-cut interagency standing operating procedure be established now to insure that if (in the opinion of any intelligence agency, particularly CIA) an attack, or other noteworthy event, is impending it is made a matter of special handling, to insure that officials vitally concerned . . . are promptly and personally informed thereof in order that appropriate measures may be taken. This will prevent a repetition of the Korean situation and will insure, if there has been vital intelligence data pointing to an imminent attack, that it will not be buried in a series of routine CIA intelligence reports.[15]

In the final analysis, the controversy over the intelligence failure in Korea is academic. The United States had no plans to counter an invasion, even had it been forecast to the very day. The only planned reaction was to evacuate U.S. nationals from the country.

[14] Interv, Maj Schnabel with Capt Schwarze, 17 Nov 53.

[15] Memo, Lemnitzer for Secy Defense, Jul 50.

MacArthur's Reaction

GHQ learned of the attack six and one-half hours after the first North Korean troops crossed into South Korea. The telegram bearing the news from the Office of the Military Attaché in Seoul reported:

> Fighting with great intensity started at 0400, 25 June on the Ongjin Peninsula, moving eastwardly taking six major points; city of Kaesong fell to North Koreans at 0900, ten tanks slightly north of Chunchon, landing twenty boats approximately one regiment strength on east coast reported cutting coastal road south of Kangnung; Comment: No evidence of panic among South Korean troops.

A message ninety minutes later gave confirmation. General MacArthur immediately informed Washington and, within a few hours, sent the first comprehensive situation report on the Korean fighting.[16]

As the news from Korea worsened later that first day, General MacArthur warned Washington officials, "Enemy effort serious in strength and strategic intent and is undisguised act of war subject to United Nations censure." But he hardly realized how strong it was. His situation report showed only three North Korean divisions along the entire border.[17]

American Ambassador to Korea Muccio conferred with President Rhee, who said that the ROK Army would be out of ammunition within ten days. Muccio quickly cabled MacArthur for replenishment. The Ambassador had already

[16] (1) Rad, ARMA 21, USMILAT Seoul to DA, Infor CINCFE, 25 Jun 50. (2) Rad, ARMA 22, USMILAT Seoul to DA, Info CINCFE, 25 Jun 50. (3) Rad, C 56772, CINCFE to DA, 25 Jun 50.

[17] Rad, C 56777, MacArthur (Personal) to Irvin, 25 Jun 50.

directed the acting chief of KMAG, Colonel Wright, to request an immediate shipment of ammunition for 105-mm. howitzers, 60-mm. mortars, and .30-caliber carbines.[18]

Before the day was out, General MacArthur ordered General Walker to load the MSTS *Keathley*, then in Yokohama Harbor, with 105,000 rounds of 105-mm. ammunition and 265,000 rounds of 81-mm. mortar, 89,000 rounds of 60-mm. mortar, and 2,480,000 rounds of .30-caliber carbine ammunition. He wanted the *Keathley* to reach Pusan no later than 1 July. He directed FEAF and COMNAVFE to protect the *Keathley* en route and during cargo discharge. In his information report to the Department of the Army, MacArthur said that he intended "to supply ROK all needed supplies as long as they show ability to use same." [19]

These actions MacArthur took independently. He received no authority from the JCS to supply the ROK until the following day, at 1330, 26 June.

The United States Responds

MacArthur's immediate reactions—to send supplies, these to be protected by air and naval escorts—were as far as he could go on his own authority. Certain basic decisions had to be made in Washington, and the key man was the President of the United States, Harry S. Truman. President Truman was at his home at Independence, Missouri, on the evening of 24 June when Secretary of State Dean Acheson telephoned him the news of the invasion. The President agreed with Acheson that the United Nations Security Council should be asked to convene at once in order to consider this threat to world peace.

Acheson called the President again the next morning, a Sunday, apprising him of the dangerous nature of the developing crisis. The President decided to leave for Washington without delay, and he asked the Secretary of State to meet with the service secretaries and the Joint Chiefs of Staff immediately to work out a plan for his consideration.[20]

At 1400 that afternoon, responding to the call of the United States Government, the United Nations Security Council convened. The USSR representative was absent, for he had begun a boycott of that body in January 1950 because of the United Nations refusal to replace the Chinese Nationalist representative with a Chinese Communist. Ernest A. Gross, Deputy Representative of the United States, briefly outlined salient events in the establishment of the ROK and the continuing opposition of the communists toward unification of Korea, then denounced the unprovoked aggression. He submitted a resolution designed to bring about an immediate cessation of hostilities and a restoration of the 38th Parallel boundary by the withdrawal forthwith of North Korean armed forces to it, and calling upon "all members to render every assistance to the United Nations

[18] (1) Rad, USMILAT to CINCFE, sent about 1800, 25 Jun 50. (2) Rad, USMILAT to CINCFE, sent about one hour later, 25 Jun 50.

[19] (1) Rad, 252130, CINCFE to CG Eighth Army, 25 Jun 50. (2) Rad, C 56775, CINCFE to DA, 25 Jun 50.

[20] (1) Truman, *Memoirs*, II, 331–43, gives a general background of Presidential action and considerations in the first few days of Korean fighting. (2) See also Acheson, *Present at the Creation*, pp. 402–13.

in the execution of this resolution and to refrain from giving assistance to the North Korean authorities." The Security Council adopted the resolution by a vote of nine to zero, with one abstention.

Meanwhile, officials of the Departments of State and Defense had met in impromptu session on Sunday morning. Department of State representatives outlined a plan for supporting the ROK with munitions and equipment and with U.S. naval and air forces.[21]

Early on Sunday evening, shortly before the President arrived in Washington, the Joint Chiefs of Staff held a teletype conference with General MacArthur. They notified MacArthur of the tentative plans made by Defense and State officials to ship supplies and equipment, which MacArthur had already started, and to extend his responsibility to include operational control of all U.S. military activities in Korea. They said he might also be directed to commit certain forces, principally naval and air, to protect the Seoul-Kimp'o-Inch'on area to assure the safe evacuation of American nationals and to gain time for action on the measures then before the United Nations. Most significantly, they alerted him to be ready to send U.S. ground and naval forces to stabilize the combat situation and, if feasible, to restore the 38th Parallel as a boundary. This action, they said, might be necessary if the United Nations asked member nations to employ military force.[22]

No decision on Korea could properly be made without a careful analysis of USSR intentions. The United States believed Russia to be the real aggressor in Korea, in spirit if not in fact, and effective measures to halt the aggression might therefore provoke total war. Hence, a decision to meet force with force implied a willingness to fight a full-scale war with Russia if necessary. The determinant for Korea was, then, as always: "What will Russia do?" [23]

The possible reactions of nations other than Russia were also important. Each alternative open to the United States was accompanied by a strong chance of alienating nations upon whose continuing friendship and support American policy was based. Inaction would be condemned by some nations as a betrayal of the ROK Government. It would gravely impair American efforts to maintain prestige in Asia as well as in other areas, and would cause such nations as Great Britain, Italy, and Japan to re-examine the wisdom of supporting the United States. On the other hand, if the United States took unilateral military measures against the North Korean attackers, Russian charges of imperialistic action and

[21] (1) U.N. Doc S/PV/473, 25 Jun 50, Statement to the Security Council by the Deputy Representative of the U.S. to the U.N. (Gross). (2) U.N. Doc S/1501. (3) Rpt to Senate Committee on Armed Services and Senate Committee on Foreign Relations, *Record of Actions Taken by JCS Relative to the U.N. Operation in Korea From 25 June 1950 to 11 April 1951*, 30 April 1951 (hereafter cited as *JCS Rpt on Korea*), pp. 5-6.

[22] Telecon, TT 3417, CINCFE and JCS, 2330Z, 25 Jun 50.

[23] American determination to resist communist expansion is clearly reflected in President Truman's later thoughts. He feared that if South Korea was allowed to fall no other small nation would dare resist threats and aggression by their stronger Communist neighbors. Not to challenge this aggression would mean a third World War, just as similar failure to challenge aggression had led to World War II. He also saw clearly that the very foundations and principles of the United Nations were at stake. Truman, *Memoirs*, II, 332.

Louis A. Johnson

defiance of the United Nations would appear valid to many nations. The effect would be to anger these nations and to render them more susceptible to Russian points of view.

The most sensible course seemed to be a co-operative effort among members of the United Nations to halt the aggression. But South Korea needed help at once; and the United Nations could hardly act swiftly enough. Furthermore, communist members of the United Nations could be expected to oppose joint action.

President Truman and his key advisers gathered at the Blair House in Washington on the evening of 25 June for an exchange of views. Five State Department members, the Secretaries of the military departments, the Secretary of Defense, and the Joint Chiefs of Staff were present.[24]

At this meeting, the policy-makers discussed the major problems facing the United States in the Far East. Foremost in their minds was a consideration of Soviet intentions and American capabilities. Louis A. Johnson, Secretary of Defense, believed strongly that Formosa was more vital to the security of the United States than Korea, and at his direction General Bradley, now Chairman of the Joint Chiefs, read a memorandum on Formosa prepared by General MacArthur. At the insistence of Secretary of State Acheson, questions of Formosa were postponed temporarily, and the attention of the group was redirected to Korea.[25] Acheson recommended that General MacArthur furnish supplies and ammunition to the ROK at once and that he be directed to evacuate U.S. nationals by any means required. When no one of-

---

[24] (1) This group included all members of the National Security Council except the Vice President and the chairman of the National Security Resources Board. (2) Unless otherwise cited, material for this portion covering the background of governmental decisions was derived from the following sources: *JCS Rpt on Korea*; Albert L. Warner, "How the Korean Decision Was Made," *Harper's*, CCII (June 1951), 100–103; Beverly Smith, "Why We Went to War in Korea," *Saturday Evening Post* (November 11, 1951); *MacArthur Hearings*, pp. 931, 1049, 1475, 2579–81, 2584; and Truman, *Memoirs*, II, 332–36. See also Collins, *War in Peacetime*, pp. 13–14, and Acheson, *Present at the Creation*, pp. 404–07.

[25] The Secretary of Defense later recalled that the only really violent disagreement which ever arose between himself and the Secretary of State took place at this meeting over the issue of the relative importance to American security of Formosa and Korea. Johnson insisted that Formosa take first priority in the evening's considerations, while Acheson insisted that Korea should be the prime topic. President Truman settled the dispute in favor of Acheson. See *MacArthur Hearings*, p. 2580.

fered to comment on Acheson's proposals, Johnson asked each defense representative in turn for an expression of opinion. The responses came forth, and "A major portion of the evening was taken in the individual, unrehearsed, unprepared and uncoordinated statements of the several Chiefs and the Secretaries." [26]

Earlier that day General Collins, the Army Chief of Staff, had received from General MacArthur a comprehensive report on developing events in Korea, and he outlined this to the group. All members of the Joint Chiefs of Staff emphasized the weakness of the American forces in the Far East and the absence of a general plan for defending South Korea.

Collins then suggested and the President approved that General MacArthur be authorized to send a group of officers as observers to Korea. Mr. Truman also approved a proposal that the Seventh Fleet be ordered to the waters off Formosa and Korea at once, and Admiral Forrest P. Sherman, Chief of Naval Operations, left the meeting to start this movement.[27] General Hoyt S. Vandenberg, Air Force Chief of Staff, also left the room to initiate a concentration of jet aircraft on Formosa.

The President ordered that all U.S. intelligence agencies throughout the world be alerted to recheck Soviet plans and intentions. He called also for urgent study to determine what would be needed to destroy Soviet Far East air bases if Soviet planes intervened in Korea.

Finally, President Truman called upon each man for his personal views. Everyone felt that whatever had to be done to meet the aggression in Korea should be done. No one suggested that the United Nations or the United States back away from the challenge. Vandenberg and Sherman had said that American air and naval aid would be sufficient to stop the North Koreans, but Collins believed that, if the ROK Army broke, American ground forces would be required.[28]

General Bradley summed up the prevailing opinion. He said that the United States would have to draw the line on communist aggression somewhere—and that somewhere was Korea. He did not believe that Russia was ready to fight the United States, but was merely testing American determination. President Truman agreed emphatically. He did not expect the North Koreans to pay any attention to the pronouncement of the United Nations, and he felt that the United Nations would have to apply force.[29] Before the meeting adjourned at 2300, President Truman approved the actions proposed by Secretary Acheson and already set in motion by General MacArthur.

The Joint Chiefs of Staff and Secretary of the Army Frank Pace, Jr., called General MacArthur into teleconference immediately after the meeting and informed him of the decisions reached. MacArthur was to send all arms and equipment needed to hold the Seoul-Kimp'o-Inch'on area, with enough air and naval cover to

[26] *MacArthur Hearings*, p. 2580.

[27] President Truman identifies the proposal to move the Seventh Fleet as having originated with Secretary of State Acheson. Johnson, however, testified before a Congressional committee that the move had been recommended by him and that the President had immediately approved his recommendation. See Truman, *Memoirs*, II, 334; *MacArthur Hearings*, pp. 2580–81.

[28] Truman, *Memoirs*, II, 335.

[29] *Ibid.*

insure safe arrival. He was to use air and naval forces to prevent the Seoul-Kimp'o-Inch'on area from being overrun, thereby insuring the safe evacuation of U.S. dependents and noncombatants. He was also told to send selected officers of his staff into Korea as a survey mission.[30]

The commitment of air and naval units to Korea established a precedent for the later commitment of U.S. ground troops. It was done without sanction of or reference to the United Nations and in the full knowledge that U.S. air and naval forces might engage in open conflict with North Korean units. Although generally viewed as less vital than President Truman's later decision of 30 June to support the ROK with U.S. ground forces, the authority to employ the Air Force and the Navy on 25 June rendered the later decision one of degree rather than one of principle. General Ridgway, who was present during the transmission of initial instructions to General MacArthur by teleconference, recalls in his memoirs:

> I was standing by General Bradley at the telecom when the directive went out authorizing the use of air and naval forces to cover the evacuation of American personnel from the Seoul and Inchon area, and I asked him whether this was deliberately intended to exclude the use of ground forces in Korea. He told me, "Yes." [31]

The officers to be sent to Korea as a survey mission were to send back information and also to furnish overt evidence to ROK authorities that they had not been abandoned. The Joint Chiefs of Staff informed General MacArthur that the Secretary of State wished KMAG liaison officers to stay with ROK units so long as these units remained effective fighting forces. Answering a request from KMAG, General MacArthur said that immediate action was being taken and that substantial logistic support was on its way to the ROK forces.[32]

The ROK Army acquitted itself well in some areas, poorly in others. In sectors where they were well led and properly deployed, the ROK Army units fought bravely and well. Elsewhere, they fell back before the better-trained and better-equipped North Koreans without offering determined or effective resistance. All across the front the enemy's superior concentration of force, his well-planned tactics, his armor and artillery supremacy, and his consistently high caliber of leadership forced a general withdrawal.

Four of the eight existing ROK divisions had been deployed widely throughout the interior and southern sections of South Korea, while the four divisions along the 38th Parallel had about one-third of their strength in defense positions and the remainder in reserve ten to thirty miles below the parallel. No ROK division was able to assemble its full combat strength in time to stem the North Korean drive on Seoul. At Kaesong and Munsan-ni, in the Uijongbu corridor, and at Ch'unch'on, the ROK soldiers put up a good fight but were overwhelmed. An abortive ROK counterattack in the vital Uijongbu corridor

---

[30] Telecon, TT 3418, JCS and OSA with CINCFE, 260355Z Jun 50.

[31] General Matthew B. Ridgway, *Soldier* (New York: Harper, 1956), p. 192.

[32] (1) Telecon, TT 3418, 260355Z Jun 50. (2) Rad, CX 56796, CINCFE to KMAG, 26 Jun 50. (3) Rad, CX 46852, CINCFE to KMAG, 27 Jun 50.

failed on 26 June, and North Korean entrance into Seoul seemed assured.[33]

Emergency Evacuation

The unexpectedly rapid and powerful communist onslaught exposed some 1,500 American civilians to immediate peril. The majority were families of AMIK personnel, most of them in the Seoul area. Additionally, more than a hundred women and a sizable number of male employees were working at Department of State, ECA, and KMAG installations.

According to the evacuation plan drawn in July 1949 by GHQ and named CHOW CHOW, the CG Eighth Army, CG FEAF, and COMNAVFE were assigned responsibilities to evacuate U.S. civilians, U.S. military personnel, and designated foreign nationals. The plan estimated that North Korean forces would require at least ninety-six hours to overrun the Seoul-Inch'on area.[34]

In the early morning of 26 June (Korean time) Ambassador Muccio ordered all dependents of U.S. Government and military personnel evacuated. Two commercial freighters at Inch'on, SS *Reinholt* and SS *Norge,* were available, but the *Norge* was too dirty to be used and nearly 700 passengers were evacuated on the 26th aboard the SS *Reinholt,* a vessel normally accommodating only twelve passengers.[35] From the morning of 27 June (Korean time), FEAF transports and commercial aircraft brought out others during two days of flights, and the remaining surface evacuation was from Pusan.

A total of 2,001 people—1,527 of them U.S. nationals—were evacuated, all of them to Japan, 923 by air and the remainder by surface transportation. Most Americans evacuated were members of AMIK, U.S. Government employees, military personnel, and their dependents. Missionaries comprised the next largest group of American evacuees.[36]

Mounting in intensity, the battle for South Korea raged into its third day on 27 June, with Seoul the prime objective of the North Korean attack. The communists apparently judged that with the ROK capital in their hands the rest of South Korea would yield easily. By the evening of 27 June, the main North Korean forces were fourteen miles north of Seoul. Midnight found the northern defenses of the city under small arms fire with armor rumbling toward the outskirts. At 0300, on 28 June, all Americans remaining in the city were ordered to leave. The first artillery fire struck Seoul around 0600, 28 June. By that night the city had fallen to the invaders.

ADCOM Arrives in Korea

General MacArthur's survey group entered Korea at 1900, 27 June, and at that time he assumed his newly authorized control of all U.S. military activities in Korea. Maj. Gen. John H. Church, who headed the group, which was designated GHQ Advance Command and Liaison Group (ADCOM), had instruc-

[33] For a detailed account, see Appleman, *South to the Naktong, North to the Yalu,* Chapters III and V.

[34] There is a striking similarity between the evacuation on 26 June 1950 and the plan for evacuation prepared in GHQ almost a full year before. See Staff Sec Rpt, G–3, GHQ, FEC, 1 Jan–31 Oct 50, p. 14, and supporting Doc 8.

[35] War Diary, EUSAK, sec. I, Prologue, 25 Jun–12 Jul 50, p. 4.

[36] Staff Sec Rpt, G–1, GHQ, FEC, 1 Jan–31 Oct 50, p. 61.

GENERAL CHURCH

tions to make contact with Ambassador Muccio and ROK officials and to send MacArthur reports on the developing situation. A concomitant mission was to instill an enthusiastic will to fight among ROK soldiers and officials.[37]

Ambassador Muccio met the group at the Suwon airport, south of Seoul, and Church established a temporary command post in the town of Suwon. After a frustrating period of communications failures and general confusion, Church made contact with General Chae Byong Duk, Chief of Staff, ROK Army and suggested they establish a joint headquarters. Chae agreed.[38]

Church told Chae that he had to use any organized group in the vicinity to resist the entry of North Koreans into Seoul by street-to-street fighting. He recommended straggler points between Seoul and Suwon to stop the retreating ROK soldiers and to reorganize them into effective units. He insisted that the Han River bordering Seoul on the south be defended at all costs.

On 28 June, Chae gathered about 1,000 ROK officers and 8,000 men and organized them into units near Suwon. Then he dispatched them to defensive positions on the south bank of the Han River.[39]

That evening, Church felt "a reasonable defense of the Han River line from the south bank could be accomplished." But if the 38th Parallel were to be restored, he believed, American ground forces would have to be used. He radioed this opinion to MacArthur together with an admittedly fragmentary report of the situation.[40]

Developments in Washington

Amidst disheartening reports from Korea, President Truman and his advisers met again at the Blair House in Washington at 2100, EDT, 26 June. The group was substantially the same that had gathered previously. The President had received a personal and vehement appeal for help from Syngman Rhee, and General Bradley made known MacArthur's latest dispatches forecasting the early fall of Seoul.[41]

---

[37] (1) Opns Instructions to Gen Church, GHQ, FEC, 27 Jun 50. (2) Rad, CS 56850, CINCFE to KMAG, 27 Jun 50.

[38] Rpt, Gen Church, sub: Activities of ADCOM, 27 Jun–15 Jul 50, copy in OCMH.

[39] *Ibid.*

[40] *Ibid.*

[41] The Joint Chiefs of Staff did not record these meetings. During the hearings on relief of General MacArthur, Senator Harry Cain told General Bradley, ". . . history will not be able to relate the circumstances surrounding the beginning of the war because the Joint Chiefs of Staff have no notes on the subject." See *MacArthur Hearings*, p. 950.

The progressive decline of South Korean resistance and the increasingly obvious evidence of North Korean military strength led Secretary of State Acheson to recommend that American air and naval forces be permitted to engage in combat operations to support the ROK. He proposed also that the U.S. Seventh Fleet be ordered not only to protect Formosa from attack but also to prevent an attack from there on the mainland. The President approved these measures, and after an hour the group adjourned.

Within a few minutes after adjournment, the Joint Chiefs of Staff called General MacArthur into teleconference. They removed restrictions against air and naval operations against North Korean military targets below the 38th Parallel. They informed him about the new missions of the U.S. Seventh Fleet in Formosan waters. They urged him to spread the news that American help was on the way to South Korea in order to maintain South Korean morale.[42]

The air of spontaneity and extemporaneousness which marked the actions of the President and his advisers during the first week of the Korean War is misleading. The key advisers called to informal meetings at the Blair House included all the members of the National Security Council who were available in Washington. Thus, although the sometimes ponderous and always time-consuming normal procedures of the council to develop positions on matters of broad general policy were not followed, the President received its views and advice.[43] He obviously felt no need for Congressional approval, believing that his decisions were within his prerogatives as Commander in Chief. Later objection by Congress that he had usurped its authority was stilled effectively by widespread public approval of Mr. Truman's actions.[44]

Although the President's decisions were decidedly toward complete resistance of aggression, without the slightest tendency to conciliate or appease, the United States, on 27 June, had yet to choose whether to mount a unilateral effort or to promote United Nations action. The advantages of acting under the auspices of the United Nations were apparent to all, but in the absence of specific knowledge on the final attitude of that body, and in a full realization of the need for quick and effective action, American officials pursued an independent course that could later be synchronized with any U.N. plan.

On 27 June, after the ROK Government had appealed to the United Nations for assistance, Warren R. Austin, United States Representative to the United Nations, addressed the United Nations Security Council, denounced the North Korean action, and demanded stronger measures by the body than the proclamation of 25 June, which was having no effect.

The Security Council condemned the North Korean attack as a breach of the peace, called for an immediate cessation

[42] Telecon, TT 3426, CINCFE and JCS, 270217Z Jun 50.

[43] Hoare, "Truman (1945–1953)," p. 191, states, ". . . the President was, for all practical purposes, consulting the NSC, but telescoping its deliberations."

[44] See Collins, *War in Peacetime*, p. 31, and Acheson, *Present at the Creation*, pp. 413–15.

of fighting, and recommended that members of the United Nations ". . . furnish such assistance to the Republic of Korea as may be necessary to repel the armed attack and to restore international peace and security in the area." [45] This resolution confirmed actions already taken by the United States.

MacArthur Visits Korea

Given the grave danger of a complete collapse of morale and fighting spirit among the South Korean people, General MacArthur felt that only a dramatic move would stiffen their resolve to resist. He decided to visit the country as immediate, symbolic proof of American backing. According to General Almond, MacArthur's chief of staff, the visit was also a search for firsthand knowledge of what the Korean Army was doing, what it intended to do next, and what President Rhee and Ambassador Muccio had to say.

Against the advice of his staff officers, who were apprehensive over extremely poor flying conditions and the threat of enemy air attack, General MacArthur flew to Korea. He landed at Suwon Airfield at 1115, 29 June 1950. Five members of his staff and four newsmen were with him.[46]

Although two YAK fighter planes of the North Korean Air Force appeared over Suwon and one dropped a bomb at one end of the runway, MacArthur and his party landed safely. They went to a small schoolhouse where General Church and the American officers of ADCOM awaited them. President Syngman Rhee, Mr. Muccio, and General Chae were also there.

At General MacArthur's request, the meeting opened with a résumé of the current military situation by General Church, who said he had been able to locate only 8,000 of the ROK Army's original 100,000 men. While he was speaking, he received a report that 8,000 more had been gathered and that Korean officers hoped to have another 8,000 by evening.

After a few brief remarks from Muccio, General MacArthur stated, "Well, I have heard a good deal theoretically, and now I want to go and see these troops. . . ." MacArthur and his group, in "three old, broken-down cars," drove thirty miles north to the south bank of the Han below Seoul, where they could see the enemy firing from the city at targets near them. By midafternoon, MacArthur had seen all he needed to and returned to Suwon Airfield, then departed about 1600.

The fall of Seoul and the obvious weakening of the ROK forces demonstrated the need of additional American

[45] Department of State, *Guide to the U.N. in Korea*, Dept of State Publication No. 4299 (Washington, 1951), p. 13.

[46] (1) This account of General MacArthur's visit is based on an interview with Lt. Col. Anthony Storey, General MacArthur's personal pilot, by Dr. Gordon W. Prange, then Chief, Military Hist Sec, GHQ, FEC, FEC, UNC, in 1951, and on an account contained in General Almond's testimony before the Internal Security Subcommittee of the Senate Judiciary Committee on 23 November 1954, contained in *U.S. News and World Report* (December 10, 1954), pp. 86–94; all quotations are as General Almond gave them in his testimony. (2) See also Willoughby and Chamberlain, *MacArthur, 1941–1951*, pp. 356–57, and Maj. Gen. Courtney Whitney, *MacArthur, His Rendezvous With History* (New York: Knopf, 1956), pp. 321–32.

General MacArthur and His Chief of Staff, General Almond, *confer with Ambassador Muccio during their 29 June 1950 visit to the battlefront.*

efforts. Since the United Nations Security Council had called for assistance by member nations to repel the invaders, more, obviously, could be done.

Army officials in Washington who were analyzing the developments in Korea unanimously felt that the USSR had deliberately fostered the outbreak in Korea. General Bolté, then the Assistant Chief of Staff G–3, Department of the Army, reported to Secretary Pace, on 28 June, "There can be no doubt but that the invasion of South Korea is a planned Soviet move to improve their cold war position at our expense." [47] Bolté suggested that the Russians actually were testing United States determination to oppose their expansion. He pointed out that there was no way of knowing whether the Korean aggression was a prelude to a "hot" war, but he reminded Pace of American emergency plans in case a shooting war with the USSR came. These plans relegated the Far East to a position of secondary stra-

[47] Memo, Gen Bolté for Secy Army, 28 Jun 50, sub: Sit in the Far East, in G–3, DA file 091 Korea, Case 25.

GENERAL BOLTÉ

tegic importance but provided for the defense of Japan, Okinawa, and the Philippines. General Bolté was justifiably concerned over the possibility that a massive response to the Korean incident might weaken the Army's ability to defend these islands.

If, the Army G–3 told the Secretary, the American air and naval forces already committed failed to stop the North Korean invasion and if it became necessary to send American ground troops from Japan, the United States garrison there would be reduced to a point where "it would be most doubtful that, in the event of a major war, Japan could be held against Soviet attack." If ground forces sent to Korea from Japan were replaced, "the taking of small reinforcements from the small strategic reserve [General Reserve] in the United States would seriously affect our war readiness in other areas." [48]

President Truman's principal advisers met with him again at 1700, on Thursday, 29 June. Secretary of Defense Johnson presented a draft directive to General MacArthur that implied an American intention to go to war with the Soviet Union. Truman turned it down on the ground that it was too strong. He stated categorically that he did not want to see even the slightest implication of such a plan. He wished to be certain that the United States would not become so deeply involved in Korea that it could not take care of other situations which could well develop.[49]

But when Department of Defense officials requested permission to carry out air operations north of the 38th Parallel, Truman agreed. When Pace cautioned that such operations should be clearly limited, Truman agreed. He pointed out his desire that these aerial attacks in North Korea be restricted to attacks on military targets, since he wished it clearly understood that operations in Korea were only for the purpose of restoring peace and the pre-invasion border.[50]

The Joint Chiefs of Staff then sent General MacArthur additional instructions. He could send his planes into North Korea to bomb "purely military" targets. He had to keep these planes well clear of the frontiers of Manchuria and the Soviet Union. Army ground forces, both combat and service troops,

48 *Ibid.*
49 Truman, *Memoirs,* II, 341
50 *Ibid.*

could, if it became necessary, be sent into the Pusan area to hold the port and the airfield facilities there. Naval vessels could also bombard targets authorized for attack by aircraft.[51] From stocks available in the Far East Command, he was to furnish the Republic of Korea munitions and supplies to keep ROK forces in action. He was to submit estimates of the amounts and types of aid required by the Republic of Korea which he was unable to provide from his own sources. He was to have operational control of the Seventh Fleet but only to neutralize Formosa.[52]

There was a grave note of caution. The Far East commander was reminded that the United States decision to commit naval, air, and limited ground forces in support of the South Koreans constituted no decision to engage in a war with the Soviet Union should Soviet forces intervene in Korea. The Joint Chiefs of Staff concluded their instructions to their field commander by pointing out: "The decision regarding Korea, however, was taken in full realization of the risks involved. If Soviet forces actively oppose our operations in Korea, your forces should defend themselves, should take no action to aggravate the situation and you should report the situation to Washington." [53]

General MacArthur immediately directed his air and naval commanders to carry out intensive operations against the North Korean military machine.[54]

CINCFE's Personal Report

Soon thereafter, General MacArthur dispatched to Washington his frank and, in some respects, gloomy impressions of his visit to Korea. He told Washington officials:

I have today inspected the South Korea battle area from Suwon to the HAN River. My purpose was to reconnoiter at first hand the conditions as they exist and to determine the most effective way to further support our mission.

. . . Organized and equipped as a light force for maintenance of interior order [the

[51] General MacArthur had not waited for this JCS directive to order operations in North Korea. On the flight to Korea, according to Colonel Storey, his pilot, MacArthur had issued orders via his plane radio at 0800 (Korean time), 29 July 1950, saying to FEAF headquarters back in Tokyo, "Partridge from Stratemeyer. Take out North Korean airfields immediately. No publicity. MacArthur approves." This action took place twenty-four hours before the JCS authorized such action in accordance with the Presidential approval. Col. John Chiles, then SGS GHQ, UNC, told the author (September 1955) that he heard MacArthur give this order, dictating it to General Stratemeyer. And one of the newspapermen who was present on the plane, Roy McCartney, recounts the following narrative contained in Norman Bartell, ed., *With the Australians in Korea* (Canberra: Australian War Memorial, 1954), pages 165-79: "On the way to Korea, MacArthur resumed pacing, while weighing out loud how he could 'take out' the airfields from which North Korean Yak fighters were operating. 'Where's the President's directive?' he asked his intelligence chief, Major General Charles A. Willoughby. 'How can I bomb north of the 38th Parallel without Washington hanging me?' Willoughby, it turned out, had left Truman's directive in Tokyo. A half hour later MacArthur emerged from his private cabin and remarked almost casually, 'I've decided to bomb north of the 38th Parallel. The B-29s will be out tomorrow. The order has gone to Okinawa.'" General Whitney describes this incident in his book on General MacArthur and concludes, "Here was no timid delay while authorization was obtained from Washington; here was the capacity for command decision and the readiness to assume responsibility which had always been MacArthur's forte." See Whitney, *MacArthur, His Rendezvous With History*, p. 326.

[52] Rad, JCS 84681, JCS to CINCFE, 29 Jun 50.

[53] *Ibid.*

[54] Rad, CX 56954, CINCFE to COMNAVFE and FEAF, 30 Jun 50.

Korean Army was] unprepared for attack by armor and air. Conversely, they are incapable of gaining the initiative over such a force as that embodied in the North Korean Army.

The Korean Army had made no preparations for a defense in depth, for echelons of supply or for a supply system. No plans had been made, or if made, not executed for the destruction of supplies or materiel in event of a retrograde movement. As a result, they have either lost or abandoned their supplies and heavier equipment and have absolutely no means of intercommunication. In most cases, the individual soldier, in his flight to the south, has retained his rifle or carbine. They are gradually being gathered up in rear areas and given some semblance of organization by an advance group of my officers I have sent over for this purpose. Without artillery, mortars and anti-tank guns, they can only hope to retard the enemy through the fullest utilization of natural obstacles and under the guidance of example of leadership of high quality.

The civilian populace is tranquil, orderly and prosperous according to their scale of living. They have retained a high degree of national spirit and firm belief in the Americans. The roads leading south from Seoul are crowded with refugees refusing to accept the Communist rule.

South Korean military strength is estimated at not more than 25,000 effectives. North Korean military forces are as previously reported, backed by considerable strength in armor and a well-trained, well-directed and aggressive air force equipped with Russian planes. It is now obvious that this force has been built as an element of communist military aggression.

I am doing everything possible to establish and maintain a flow of supplies through the air-head at SUWON and the southern port of PUSAN. The air-head is most vital, but is subject to constant air-attack. Since air-cover must be maintained over all aircraft transporting supplies, equipment and personnel, this requirement operates to contain a large portion of my fighter strength. North Korean air, operating from near-by bases, has been savage in its attacks in Suwon area.

It is essential that the enemy advance be held or its impetus will threaten the overrunning of all Korea. Every effort is being made to establish a Han River line but the result is highly problematical. The defense of this line and the Suwon-Seoul corridor is essential to the retention of the only airhead in central Korea.

The Korean Army is entirely incapable of counter-action and there is grave danger of a further breakthrough. If the enemy advance continues much further it will seriously threaten the fall of the Republic.

The only assurance for the holding of the present line, and the ability to regain later the lost ground, is through the introduction of US Ground Combat Forces into the Korean battle area. To continue to utilize the Forces of our air and navy without an effective ground element cannot be decisive.

If authorized, it is my intention to immediately move a United States Regimental Combat Team to the reinforcement of the vital area discussed and to provide for a possible build-up to a two-division strength from the troops in Japan for an early counter-offensive.

Unless provision is made for the full utilization of the Army-Navy-Air team in this shattered area, our mission will be needlessly costly in life, money and prestige. At worst it might even be doomed to failure.[55]

This message reached Washington an hour before midnight on 29 June. Because of its urgent tone and extremely pessimistic outlook, General Collins consulted with General MacArthur in a teleconference four hours later. He informed the Far East commander that one RCT could be moved to Pusan to guard

[55] (1) Rad, C 56942, CINCFE to JCS, 30 Jun 50. (2) General Whitney states that MacArthur wrote the report during the return flight from Suwon, using a pencil and pad. See Whitney, *MacArthur, His Rendezvous With History*, p. 332.

that port. MacArthur protested that this hardly satisfied the basic requirements. He urged speed in securing permission to place American forces in the battle area.

Lacking the authority to grant this request, Collins told MacArthur he would try to gain Presidential approval. Collins called Secretary of the Army Pace, who called the White House. The President immediately approved dispatching one RCT to the battle area. In less than an hour, word was flashed to Tokyo, "Your recommendation to move one RCT to combat area is approved. You will be advised later as to further build-up." [56]

Throughout this period of intensive search for decisions, culminating finally in the decision to meet the aggressor in ground combat, the President of the United States had been the ultimate arbiter of each step. President Truman had solicited the advice of those best qualified to judge the military effects and requirements of each move taken. General Collins briefed him daily, passing on the views of the Joint Chiefs. But the President made the final choice himself.

Earlier the Joint Chiefs of Staff had not favored the use of American ground forces in Korea,[57] primarily because they knew how unprepared they were for large-scale combat. They were reluctant also to weaken the small General Reserve in the United States, which represented the minimum essential for defense. Deploying any part of the Reserve to the Far East would be a risky, perhaps disastrous, undertaking because of possible Soviet involvement following American action.[58]

General MacArthur quite clearly had tipped the balance in favor of troop commitment. The risks had not changed or lessened, but the nation's leaders became convinced that communist seizure of Korea could not be tolerated. MacArthur's personal appeal, in fact, received even wider recognition on 30 June when he was told, "Restriction on use of Army Forces . . . are hereby removed and authority granted to utilize Army Forces available to you." [59]

---

[56] Telecon, TT 3444, CINCFE and JCS, 300742 Jun 50.

[57] Handwritten Note, to Memo, Dep Secy JCS for JCCS, 28 Jun 50, sub: Preparation of Study.

[58] Louis Johnson, Secretary of Defense when the decision was made, subsequently testified to an almost neutral attitude on the part of himself and his chief assistants. "Neither I nor any member of the Military Establishment in my presence recommended we go into Korea." Johnson recalled, "The recommendation came from the Secretary of State, but I want to repeat that it was not opposed by the Defense Department, all the members of which had severally pointed out the trouble, the trials, tribulations, and the difficulties." See *MacArthur Hearings*, p. 2584.

[59] Rad, JCS 84718, JCS to CINCFE, 30 Jun 50.

CHAPTER V

Emergency Conditions, Emergency Measures

President Truman's decision to send American ground troops against the North Koreans had come in time, but barely. Regardless of American air strikes against their cities, communication lines, and troop columns, and despite naval surface attack against their coastal installations and shipping, the invaders drove the ROK Army down the peninsula. As the vague line of battle receded southward in late June and early July it became clear that the Republic of Korea could not stand by itself.

Armed with Presidential authority, MacArthur sent ground troops into the fight as fast as he could move them. On 30 June, he ordered the 24th Division from Japan to Korea, retaining the unit, for the time being, under his personal control. On the recommendation of his chief of staff, General Almond, he ordered a small task force from the division flown into Korea ahead of the main body to engage the North Korean Army as quickly as possible, sacrificing security for speed. Because it would go by air, he restricted its size to two rifle companies, some antitank teams, and a battery of light artillery. This makeshift unit was to report to General Church at Suwon by 1 July; but, realizing that Suwon might fall at any time, General MacArthur authorized Church to divert the force to Pusan if necessary.[1]

General Church meanwhile struggled to keep the ROK Army in the fight. He had no real authority over the South Koreans, but his status as MacArthur's personal representative gave weight to his advice to the ROK Chief of Staff. In effect, Church took charge of the

[1] (1) Rad, CX 56978, CINCFE to CG Eighth Army, 30 Jun 50. (2) Rad, C 26979, CINCFE to CG Eighth Army, 30 Jun 50. (3) Review Comments, Lt Gen Edward M. Almond, 20 Feb 69. (4) General MacArthur chose this division on the basis of location. The 24th Division was closer to Korea than other combat units in Japan and could be deployed more rapidly. From the standpoint of combat readiness, while there was little to choose from among the four divisions in Japan, the 24th Division had been reported on 30 May 1950 as having the lowest combat effectiveness of the major units. This report gave the following estimates of combat effectiveness for FEC divisions: 1st Cavalry—84 percent combat effective 7th Division—74 percent combat effective; 25th Division—72 percent combat effective; 24th Division—65 percent combat effective. See Memo, U.S. Army Major Units FEC, 3 Jul 50, in G-3, DA files. (5) Interv, author with Brig Gen Edwin K. Wright, ACofS G-3, FEC, UNC, Dec 51.

faltering South Korean Army. Many KMAG officers stayed with ROK combat units, patrolling, feeding information to General Church, and doing whatever they could to stiffen ROK resistance and morale.[2]

American advice could not stop North Korean tanks and artillery. The South Koreans continued to fall back. General Church's command group pulled out of Suwon in the early evening of 30 June to Taejon. Vowing to "run no farther," Church, together with Ambassador Muccio, awaited the small 24th Division task force.

Around midnight, General Almond notified the American Embassy at Taejon that bad flying weather had forced the diversion of the task force to Pusan, where it would land as soon as the weather improved; the first contingents of the main body of the 24th Division would land at Pusan by ship within twelve or fourteen hours. General Almond emphasized that these men were not to be used as "Headquarters Guards" but to fight the North Koreans. He was assured that the railroads from Pusan to Taejon were operating and that there should be no problem in moving these troops to the line of battle. Almond instructed Church to concentrate railroad rolling stock near Pusan to keep it out of enemy hands and to have it ready for the 24th Division.[3]

The small delaying force—part of the 1st Battalion, 21st Infantry—landed at Pusan Airfield on 1 and 2 July, with Lt. Col. Charles B. Smith in command. The artillery battery originally called for had been replaced by two 4.2-inch mortar platoons. A platoon of 77-mm. recoilless rifles and six 2.36-inch bazooka teams had also been added. Because of the poor flying weather many trucks and some soldiers could not be flown in until later.[4]

General MacArthur was concerned that the small force lacked artillery, and on 2 July he ordered General Walker to fly in howitzers from Japan if he had to. It was unnecessary to do so, for elements of the 52d Field Artillery Battalion were already on their way by LST, and they landed in Pusan that evening and moved at once to the battle area.[5]

The commanding general of the 24th Division, Maj. Gen. William F. Dean, flew to Pusan early in the morning of 2 July. After spending 24 hours becoming acquainted with conditions, he telephoned from Taejon to Tokyo and spoke with General Hickey, Deputy Chief of Staff, GHQ. Wanting his initial fight with the North Koreans to be fully coordinated and supported, he told Hickey, "This first show must be good. . . . We must get food and bullets and not go off half-cocked."

A few hours later, MacArthur named Dean commanding general, USAFIK. Dean assumed control of KMAG and all other U.S. Army troops in Korea.

---

[2] Interv, author with Gen Church, 16 Jul 50, copy in OCMH.

[3] (1) Rad, A 041, ADCOM to CINCFE, 30 Jun 50. (2) Rad, JSOB/G G–2 to Capt Hutchinson, 1130, 1 Jul 50. (3) Memo, CofS GHQ, FEC, no signature, 1 Jul 50, sub: Telecon Between CofS GHQ and First Secy of American Embassy, Taejon, 1120. (4) Rad, CX 57009, CINCFE to ADCOM, 1 Jul 50. All in AG, FEC files.

[4] Memo, G–3 GHQ for CofS ROK, GHQ, 021810 Jul 50, in AG, FEC files.

[5] (1) Rad, CX 57073, CINCFE to CG FEAF and CG Eighth Army, 2 Jul 50. (2) Memo, ACofS G–3, GHQ, for CofS ROK, GHQ, 021700 Jul 50. Both in AG, FEC files.

Church's GHQ, ADCOM, served as his temporary staff. At the same time, MacArthur set up the Pusan Base Command, subordinate to USAFIK and under Brig. Gen. Crump Garvin.[6]

The other regiments of the 24th Division—the 34th and 19th Infantry, and the remainder of the 21st Infantry, plus supporting units—moved to Korea rapidly. By 5 July, most of the division was there. To provide more armor General MacArthur ordered Company A of the 1st Cavalry Division's medium tank battalion to bolster the division.[7]

Meanwhile, Colonel Smith's delaying force, after reporting to General Church at Taejon, was sent forward to engage the enemy on sight. Just above Osan, the task force dug hasty positions on the night of 4 July and awaited the approaching North Koreans. Shortly after 0800 on 5 July, the North Koreans appeared. They struck the task force with infantry and about thirty Russian-made T–34 tanks. The Americans stood until they expended their ammunition, then abandoned the field, suffering heavy losses in the process. Their weapons had proved to be almost useless against the enemy armor. Without reserves and with open flanks, the task force remnants withdrew to avoid being surrounded and destroyed.

The pattern of this first engagement was repeated during the following days. All combat elements of the 24th Division closed with the enemy along the main axis of his advance, but the North Korean firepower and greater strength overwhelmed these units at every stand. The men and officers of the 24th Division fought bravely, but their small numbers and inferior weapons left no choice but retreat or annihilation.

General Dean hoped that the 34th Infantry could delay the North Korean advance in the P'yongt'aek-Ch'onan-Kongju corridor. But between 5 and 8 July the regiment, thrown into a fight for which it was unprepared, was cut to pieces. Weak in numbers, completely outgunned, unable to protect its flanks, and short of ammunition the 34th retreated in some disorder, suffering extremely heavy casualties.

The 21st Infantry held at Chonui and Choch'iwon for three days, slowed two enemy divisions, but, after losing heavily in men and equipment, had to give way on 12 July.

East of the main Seoul-Taegu rail and highway lines, the ROK Army tried to stem the North Korean drive through the mountainous central and eastern regions. In bloody hand-to-hand fighting that cost both sides dearly, the North Koreans continued to advance. No defensive line appeared to offer the prospect of a determined stand.[8]

---

[6] (1) Memo, 031140 Jul 50, sub: Telecon Between ADCOM (Gen Dean) and CofS (Gen Hickey). (2) Rad, CX 57153, CINCFE to CG 24th Div, 3 Jul 50. Both in AG, FEC files.

[7] (1) Memo, ACofS G–3, GHQ, for CofS ROK, GHQ, 2 Jul 50. (2) *Ibid.*, 3 Jul 50. (3) Rad, CX 57090, CINCFE to CG Eighth Army, 2 Jul 50. All in AG, FEC files.

[8] (1) Rad, ROB 104, CG USAFIK to CINCFE, 6 Jul 50. (2) For a detailed account of these actions, see Appleman, *South to the Naktong, North to the Yalu*, pp. 59–108. (3) General MacArthur later testified that he had sent the initial task force in the hope of establishing a "loci [locus] of resistance," an "arrogant display of strength" that would fool the enemy into believing that much more American resources were at hand than in actuality. See *MacArthur Hearings*, p. 231.

MacArthur's Estimates

The understrength American division so hastily deployed to Korea was unable to stop the North Korean drive, but this fact did not become evident for several days after the initial encounter at Osan. The situation in Korea could not be accurately evaluated even in Tokyo let alone in Washington, where Army officials could do little but wait impatiently for clarification through General MacArthur's estimates and descriptions. Until these estimates arrived, Washington could neither plan adequately nor gauge the scope of the job to be done. The Army's plans for supporting MacArthur had to be based on requirements established either directly or obliquely by his estimates. Washington authorities had no recourse, in these early days, but to accept his judgment of capabilities and requirements at face value. They knew the limits of the nation's immediate resources. General MacArthur told them what was happening in Korea and what he felt had to be done. In the search for a balance between what they had and what was needed, the nation's military leaders followed advice from the Far East commander which they could not accurately evaluate.[9]

MacArthur's early estimates fell short in appraising the ultimate necessary force, but not in their appreciation of the caliber of the enemy and the seriousness of the threat. The tenor of reports from Church, Dean, and others had already convinced General MacArthur that the situation was indeed serious. The degree of seriousness remained to be determined. He did not immediately arrive at a full appreciation of the strength of the North Korean attack. General MacArthur progressively revised upward his estimate of the strength he would need to defeat the North Koreans.

Late in June, he implied that two American divisions could restore order.[10] But by 7 July his views had changed materially. He told the Joint Chiefs of Staff, "It is now apparent that we are confronted in Korea with an aggressive and well-trained professional army equipped with tanks and perhaps other ground material quite equal to, and in some categories, superior to that available here." The enemy's leadership was "excellent." The North Koreans showed understanding of and skill in tactical and strategic principles—demonstrated by their break across the Han River. To halt and hurl back "this powerful aggression" would, in MacArthur's opinion, require from four to four and one-half full-strength American divisions supported by an airborne RCT and an armored group. To reach this strength level in Korea 30,000 men and officers would have to be sent him from the United States at once. "It is a minimum," he warned the Joint Chiefs, "without which success will be extremely doubtful." [11]

[9] Complementary to the failure of U.S. intelligence agencies to foresee the North Korean assault is the failure to have determined the true quality of the North Korean Army, especially the caliber of its training and the individual worth of the North Korean soldier. General Bradley testified later, "The first few days we did not know just how good these North Koreans were, and it was some time before we could get a good picture. . . ." See *MacArthur Hearings*, p. 893.

[10] Rad, C 56942, CINCFE to JCS, 30 Jun 50.

[11] Rad, C 57379, CINCFE to DA, 7 Jul 50.

Dean echoed this conviction. In a personal letter to MacArthur on 8 July, Dean set forth his views on the enemy strength and on his own most urgent needs. He asked for battle-ready combat teams immediately, troops with full combat loads and extra supplies, ready for co-ordinated action.[12]

North Korean armor had proven extremely effective. In their first engagements, his troops, Dean pointed out emphatically, could not stop enemy tanks. The 2.36-inch rocket launcher, an American antitank weapon of World War II, proved dangerously disappointing against the enemy's heavily armored Russian tanks. The launcher was ineffective against the front and side armor, and American infantrymen quickly lost all confidence in it.[13] Direct fire by artillery was of little help after the pitifully few 105-mm. antitank rounds available at the guns were exhausted. Regular high-explosive projectiles, which composed the bulk of artillery ammunition carried by his batteries, would not penetrate armor deeply enough. Dean stressed the need for getting antitank ammunition to his artillery at once. He described enemy tank tactics as excellent and unusually effective despite terrain which confined tanks mainly to roads. Asserting that "we cannot afford to be out-gunned and out-armored," the hard-pressed American general appealed for American medium tanks and for 90-mm. towed antitank guns.[14]

General Dean warned that the North Korean soldier was a dangerous foe. "I am convinced," he told General MacArthur, "that the North Korean Army, the North Korean soldier and his status of training and the quality of his equipment have been underestimated." [15]

Dean's first-hand account, coupled with graphic evidence of enemy successes on the situation maps in his own war room, brought General MacArthur to the conclusion that he had been much too conservative. On 9 July 1950 he doubled his estimate of the forces needed. "The situation in Korea is critical," he told the Joint Chiefs. "It has developed into a major operation." For the first time he expressed doubt that the Americans could stay in Korea.

> To build up . . . sufficiently to hold the southern tip of Korea is becoming increasingly problematical. I strongly urge that, in addition to those forces already requisitioned, an army of at least four divisions, with all component services, be dispatched to this area without delay, and by every means of transportation available.[16]

To lend validity to this sudden revision, General MacArthur re-emphasized his growing respect for the North Korean Army. He credited the North Korean Army and its employment as being as

[12] Ltr, Gen Dean to Gen MacArthur, 080800 Jul 50, sub: Recommendations Relative to Employment of U.S. Army Troops in Korea, in AG, FEC files.

[13] This weapon, developed during World War II, was much publicized and widely regarded as a "wonder weapon." In reality, the 2.36-inch rocket launcher, or bazooka, did not deserve this reputation. There are relatively few recorded instances in which it was successfully used against German armor. See Hugh M. Cole, *The Lorraine Campaign*, UNITED STATES ARMY IN WORLD WAR II (Washington, 1950), ch. XIV, p. 604. Also, the launcher ammunition used by Dean's men was at least five years old and had deteriorated.

[14] (1) Ltr, Gen Dean to Gen MacArthur, 080800 Jul 50. (2) Rad, ROB 110, CG USAFIK to CINCFE, 6 Jul 50.

[15] Ltr, Gen Dean to Gen MacArthur, 080800 Jul 50.

[16] Rad, CX 57841, CINCFE to JCS, 9 Jul 50.

MEMBERS OF THE 25TH DIVISION *en route to the battle front on 14 July 1950.*

good "as any seen at any time in the last war." Enemy infantry was first class. There were unmistakable signs of Soviet leadership and technical guidance and of Chinese Communist participation. The attack could no longer be viewed as an indigenous North Korean military effort. "To date," he admitted, "our efforts against his armor and mechanized forces have been ineffective." This failure, galling as it was, was not the fault of the fighting men. "Our own troops," he pointed out, "are fulfilling expectations and are fighting with valor against overwhelming odds of more than ten to one." [17] This appeal to Washington for an additional army of four divisions climaxed a series of detailed requests for men and units and marked the upper limit of MacArthur's requests for Korea.

On 5 July General MacArthur had ordered the 25th Infantry Division into combat, and by 9 July its first RCT had cleared Japan for Korea. All regiments of the 25th Division had arrived in or were en route to Korea by 14 July. They went into battle at once. The 1st Cavalry Division was by this time also preparing for an amphibious landing on the east coast of Korea. In order to bring these two divisions and the 24th Division to some semblance of effective fighting strength, MacArthur stripped

[17] *Ibid.*

the remaining FEC combat force, the 7th Division, of trained officers and men. While this cannibalization of the 7th fell far short of building up the other units to a satisfactory war strength, it left the 7th Division a skeleton, temporarily useless for combat.[18]

As the odds grew large that the greater part of Eighth Army would have to fight in Korea, it became apparent that General Walker would have to take personal command there. USAFIK was a provisional headquarters, hastily formed for a specific mission, and could not handle a large operation efficiently. When General Dean proposed on 7 July that his headquarters absorb GHQ ADCOM, General MacArthur had already decided that General Walker would take over.[19]

Five days later, on 12 July, MacArthur named Walker commander of the ground forces in Korea. The USAFIK headquarters was dissolved, and General Church's ADCOM group was ordered to Tokyo.[20]

The extension of Eighth Army's area of responsibility to include Korea introduced the unique situation of an army fighting on one land mass with responsibility for its own logistical support, including port operation and procurement of supply, while administering occupied territory on another land mass several hundred miles away and serving as its own zone of communications. For the sake of convenience, forces in Korea were referred to as Eighth U.S. Army in Korea (EUSAK) and those remaining in Japan were still referred to as Eighth Army or as Eighth Army Rear. General Walker retained command of both.

When Walker assumed command in Korea, he had approximately 18,000 troops spread along a defensive line running along the south bank of the Kum River to a point just above Taejon, there curving northeastward through Ch'ongju and across the Taebaek Range below Ch'ungju and Tanyang, finally bending southward to the east coast of P'yonghae-ri.[21]

Although General MacArthur had hoped to save the 1st Cavalry Division for a later amphibious operation, he yielded to battlefield necessity and sent that unit to Korea in mid-July. The division loaded out of the Yokohama area between 11 and 17 July aboard LST's, other U.S. naval craft, and Japanese-operated cargo ships. The unit was prepared to make an amphibious landing on the east coast of Korea near P'ohang-dong, against enemy opposition if necessary. No enemy appeared, and in the early morning of 18 July the units started coming ashore.[22]

The Build-up

The years of military privation since World War II had left their mark on the ground forces of the United States. Not

---

[18] (1) Rad, CX 57258, CINCFE to CG Eighth Army, 5 Jul 50. (2) Memo, G-4 GHQ for CofS ROK, GHQ, 10 Jul 50, sub: Movement of 25th Inf Div to Korea. (3) Memo, G-4 GHQ for CofS ROK, GHQ, 14 Jul 50. (4) Rad, CX 57692, CINCFE to DA, 12 Jul 50. All memos in AG, FEC files.

[19] (1) Ltr, CG USAFIK to CINCFE, 6 Jul 50, sub: Org of USAFIK. (2) Ltr, CINCFE to CG USAFIK, 1st Ind, 9 Jul 50. (3) G-1 GHQ Log, Item 146, 9 Jul 50.

[20] (1) GO 13, GHQ FEC, 12 Jul 50. (2) Rad, CX 57765, CINCFE to CG Eighth Army, 13 Jul 50.

[21] Appleman, *South to the Naktong, North to the Yalu*, p. 108.

[22] (1) Draft Plan, JSPOG GHQ, FEC, Operation BLUEHEARTS, 2 Jul 50, in AG, FEC files. (2) War Diary, 1st Cav Div, Jul 50.

only were they equipped with outmoded, worn weapons and equipment, but their numbers were scant. Both Army and Marine troops had spread thin in their efforts to perform their interim missions. Aside from scattered elements in the Pacific, the Joint Chiefs of Staff and the Army's leaders had only the understrength General Reserve in the United States from which to draw immediately for fighting men to throw into Korea. Additional ground strength could be developed through Selective Service and through the call-up of Reserve Component forces, but these methods would take time. Thus, when General MacArthur, reacting to North Korean victories, impatiently demanded his due, the nation's military leaders faced a dilemma of considerable complexity and prime importance. The very safety of the nation stood, at times, in the balance.

Demands for combat forces by General MacArthur in July and August 1950 fell into three broad categories: replacements, filler units and individual fillers, and reinforcing units. To meet his demands in any of these categories would affect the balance of United States military strength. Each tied in with problems far broader in scope than General MacArthur's problems in Korea. Within the limits imposed by national policy, as set by the President, the Joint Chiefs of Staff and the Department of the Army made every effort to meet the urgent requirements developing in the Far East.

Replacements

The sources of replacements within the Far East quickly dried up. Men were taken from administrative and noncombatant duties and sent to the combat units. In the United States, every installation was combed for individuals who could be shipped quickly to Korea.

General MacArthur first asked for 5,000 combat and 425 service replacements. On 1 July, he asked that these troops be added to the normal number shipped to his command each month, stipulating that they be qualified and experienced, for they were "going directly into the combat zone in Korea for an indefinite period. . . ." [23] This number could be sent without difficulty, and most would reach Japan within the month, the remainder early in August.

The Department of the Army gave MacArthur special dispensations that would improve the replacement status in the Far East while not enfeebling military strength elsewhere. He could retain enlisted men in his command even though their foreign service tours had been completed. He could keep Reserve officers after their category commitments had expired, if they agreed. He could call to active duty limited numbers of Reserve personnel already in the Far East.[24]

Airlift of replacements from the

[23] (1) Rad, CX 57013, CINCFE to DA, 1 Jul 50. (2) General MacArthur's 2-division estimate was the basis for these figures. This estimate called for deployment of 25,266 combat troops and 9,246 service troops in the combat zone. The formula applied to this battlefield strength to determine replacement needs was taken from FM 101-10, 10 August 1949, and provided a surprisingly accurate figure. United States battle losses in July were 1.3 percent of total strength, whereas the formula forecast had set expected losses at 1.35 percent. See Rad, CX 58760, CINCFE to DA, 26 Jul 50.

[24] (1) Rad, C 57692, CINCFE to DA, 12 Jul 50. (2) Memo, G-1 GHQ for CofS GHQ, 5 Aug 50, sub: Casualties and Replacements. (3) G-1 GHQ Log, Item 41, 5 Aug 50. (4) Rad, C 58232, CINCFE to DA, 19 Jul 50.

United States to Japan began on a modest scale on 18 July. A lift of 80 men a day was gradually expanded to 240 combat soldiers daily. Although sufficient air transport was not immediately available, the Department of the Army and the Joint Chiefs of Staff did everything possible to increase the aerial flow in late July and early August. Replacements were flown to Japan in organized packets of 39 men and 1 officer. Approximately 7,350 replacements reached Japan in July 1950.[25]

Army officials in Washington asked General MacArthur to recheck his figures on 23 July. Perhaps the actual casualties were fewer than the number forecast. Maj Gen. William A. Beiderlinden, the FEC G-1, informed Washington that the actual number of men and officers lost in Korea closely approximated his earlier educated guess. The only discrepancy was an excessive missing-in-action rate, which reflected the ability of the North Koreans to envelop the understrength American units almost at will. Beiderlinden promised to readjust FEC requirements downward whenever this action became possible.[26]

The Department of the Army on 19 July had discarded peacetime strengths and authorized full combat Table of Organization and Equipment (TO&E) strength for all divisions operating in the Far East Command. This increase in authorized men and officers, technically called filler replacements, when added to the number of combat-loss replacements which MacArthur said he needed by 1 September 1950, brought the total replacement requirements of the command to 82,000 men.[27]

Department of Army officials showed General MacArthur the bottom of the replacement barrel on 30 July. All the men and officers eligible for overseas assignment were being shipped to the Far East Command, except for slightly more than a thousand to other joint commands. Despite Presidential approval for the recall of 25,000 enlisted Reservists, a severe shortage of replacements still existed. Individual replacements from the Enlisted Reserve Corps would not be available in quantity for at least two months. All of these men would have to go to General Reserve units. The extensive levies placed upon the General Reserve to furnish FEC replacements had cut the operating capabilities of the emergency force to a dangerous level. For the immediate future, at least, the Army had done about as much

[25] (1) Telecon, TT 3536, CINCFE and DA, 2100, 17 Jul 50. (2) Memo, G-1 GHQ for CofS GHQ, 17 Jul 50, sub: Air Priority, Replacements Versus B-29 Engines. (3) G-1 GHQ Log, Item 40, 18 Jul 50. (4) Rad, FAIRPAC 456, FAIRPAC to CINCFE, 26 Jul 50. (5) G-1 GHQ Log, Item 1, 26 Jul 50. (6) Rad, W 86607, DA to CINCFE, 20 Jul 50. (7) Rad, W 86677, DA to CINCFE, 22 Jul 50. (8) Rad, CX 58760, CINCFE to DA, 27 Jul 50.

[26] (1) Rad, W 87678, DA to CINCFE, 23 Jul 50. (2) G-1 GHQ Log, Item 62, 23 Jul 50. (3) Memo, G-1 GHQ for CofS, 24 Jul 50, sub: Casualty Analysis. (4) G-1 GHQ Log, Item 37, 24 Jul 50.

[27] (1) Rad, W 86450, DA to CINCFE, 19 Jul 50. (2) Memo, G-1 GHQ for CofS GHQ, 25 Jul 50, sub: Replacement Sit in Japan. (3) G-1 GHQ Log, Item 67, 25 Jul 50. (4) General MacArthur had ordered on 11 July the establishment of an Army replacement system by Eighth Army to support both Japan and Korea. A center to receive, process, and allot the anticipated increased numbers of men and officers slated for the Far East opened near Tokyo at Camp Drake on 24 July 1950. (5) Rad, CX 57662, CINCFE to CG Eighth Army, 11 Jul 50. (6) Cleaver, Personnel Problems, p. 82. (7) Memo, Col Grubbs for Gen Beiderlinden, 8 Aug 50, sub: Assignment of Replacements, G-1 GHQ Log, Item 36, 8 Aug 50.

as it could do. It could promise only the most austere replacement support to General MacArthur.[28]

Bringing Divisions to Strength

Another significant effort involved the build-up of MacArthur's divisions from understrength, unbalanced peacetime divisions to fully manned, properly constituted fighting divisions. With only two battalions in each regiment, American forces in Korea could not employ normal tactical maneuvers based on the full firepower and the flexibility of a triangular organization. Nor could they guarantee flank protection. As General Dean said:

> The two battalion regimental organization with which we are operating does not lend itself to effective combat. The same is true, though possibly to a lesser degree of our two battery artillery battalions. Recommend that infantry battalions be sent us to bring all regiments of the 24th Division up to regular triangular organization.[29]

Painfully familiar with the structural weaknesses of his combat divisions General MacArthur appealed to the Department of the Army on 8 July saying, "In order to provide balanced means for tactical maneuver, fire power, and sustaining operations, it is urgently required that infantry divisions operating in this theater be immediately expanded to full war strength in personnel and equipment." The gravity of his concern prompted a second appeal two days later. "I am sure that the Joints Chiefs of Staff realize," he said, "that the division now in action in Korea, and the other two divisions soon to be committed are at neither war strength nor at full author ized peace strength." General MacArthur asked that completely manned and equipped battalion units be sent from the United States wherever possible.[30] He needed 4 medium tank battalions, 12 tank companies, 11 infantry battalions, and 11 field artillery batteries (105-mm. howitzers).[31] If these units could not be sent fully trained and battle-ready as he desired, he wanted trained cadres, followed by filler replacements. Asking that organized units, even if understrength, be sent first, he said he would find filler personnel in his own command.

The Far East Command could provide no trained cadres for new units. Only 60 percent of the first three grades authorized for existing FEC units were available. If noncommissioned officers were taken from divisions already fighting, these divisions would be dangerously weakened. General MacArthur urged all possible speed in sending him units, cadres, and fillers.[32]

The acute shortage of infantry, artillery, and service support units in the General Reserve in the United States turned these relatively modest demands into a problem of major proportions. In marshaling organized combat units to fill out the divisions in Korea and Japan, the Department of the Army

---

[28] Rad, W 87478, DA to CINCFE, 30 Jul 50.

[29] Ltr, Gen Dean to Gen MacArthur, 080800 Jul 50.

[30] (1) Rad, CX 57465, CINCFE to DA, 8 Jul 50. (2) Rad, C 57561, CINCFE to DA, 10 Jul 50.

[31] The four FEC divisions had a total of 12 infantry regiments and 12 light field artillery battalions. The Negro 24th Infantry had 3 battalions and the Negro 159th Field Artillery Battalion had 3 105-mm. howitzer batteries and did not require augmentation.

[32] Rad, CX 57573, CINCFE to DA, 10 Jul 50.

stripped battalions, companies, and batteries from the General Reserve. It pulled trained noncoms from other units and formed provisional cadres for General MacArthur's command. These drastic procedures not only vitiated the combat readiness of the remaining units, but greatly reduced the mobilization base for a later build-up of the Army General Reserve.

The dangers of denuding the General Reserve in the United States came under consideration only as a secondary factor of the larger planning effort: how and where the General Reserve should be tapped to bring FEC units to war strength. The Department of the Army took in stride the decision to accept the great risk of military weakness in the continental United States as it accepted at face value General MacArthur's statement of his needs.[33]

Infantry Strength

The main considerations in selecting infantry battalions for Korea were early arrival and combat effectiveness. Army authorities could have sent eleven cadres for new infantry battalions, but new battalions, even with full cadres and basic-trainee fillers, needed six months to become combat ready. Only in the case of the 7th Division, still in Japan, were three battalion cadres substituted for ready-to-fight units. The General Reserve held only eighteen battalions of infantry at this time. From this small reservoir the Department of the Army finally selected for the Far East Command 2 full battalions and 3 battalion cadres from the 3d Infantry Division; 1 full battalion from the 14th RCT; and 3 battalions from the 5th RCT on Hawaii. The remaining 2 battalions were taken from the 29th RCT on Okinawa. This unit was already part of the Far East Command and its disposition did not affect the General Reserve.

The Department of the Army spared the 82d Airborne Division and the infantry units of the 2d Armored Division. The former unit was not touched because General Collins felt he must keep a completely manned and effective unit for last-resort operations. The armored infantry battalions of the 2d Armored Division were not particularly suited to the type of action taking place in Korea and were passed over for that reason.

The removal of battalions from the General Reserve would reduce the training and mobilization base in the United States by one-sixth. The 3d Division, the 2d Armored Division, because of losses other than in infantry units, and the 14th RCT would be fit only to serve as nuclei around which to build new units. Since it would require from twelve to fourteen months to rebuild these combat units, the Army's ability to carry out emergency missions would be nullified for at least one year.[34]

Division Artillery Units

The same general criteria were used in choosing division field artillery batteries from the General Reserve for shipment to the Far East. Although taking only battery cadres would have placed less strain on Regular Army units, complete

[33] (1) Memo, CofS USA for Gen Bolté, 17 Jul 50, sub: Additional Units to Meet Immediate Requirements of the FEC. (2) MFR, 17 Jul 50, attached to (1). (3) Memo, Study, same sub, 17 Jul 50. All in G–3, DA file 320.2 Pac, Case 17.

[34] Study, Additional Units to Meet Immediate Requirements of FEC, Annex B.

batteries were withdrawn. The 3d Infantry and 2d Armored Divisions each furnished three 105-mm. howitzer batteries. Three batteries were originally scheduled from the 14th RCT and two from the 6th Armored Field Artillery Battalion. With the decision to commit the three batteries of the 5th RCT from Hawaii, the levy on the 14th RCT was reduced to two and that on the artillery battalion was canceled. These eleven artillery batteries were scheduled to reach Korea at about 60 percent strength and at an estimated combat effectiveness of 40 percent.[35] The field artillery mobilization base was cut about 30 percent by these transfers to Korea, and the ability of the Army to support other operations with artillery was cut in half for a full year.

Battalion-sized units could be ready to leave their home stations two weeks after receiving warning orders. But there was no hurry about alerting infantry and artillery units, because all water shipping from the west coast was tied up until about 15 August. The Chief of Transportation, U.S. Army, reporting that 30,000 men and 208,000 measurement tons of equipment were going to the Far East under the most urgent priorities, recommended not shipping the augmentation units until mid-August. General MacArthur was notified that the new infantry and artillery units would reach him before the end of that month.[36]

When the Chief of Staff, GHQ, and the Chief of Staff, Eighth Army, reached agreement in a telephone conversation on 12 July that two battalions of the 29th Infantry on Okinawa should be sent to Korea as soon as possible, General MacArthur ordered the Commanding General, Ryukyus Command, General Beightler, to build these battalions to war strength and send them to Japan without delay.[37] General Walker asked that the two battalions be sent directly to the battle area, bypassing Japan. He said he would give them any training they needed. This request was granted, and on 21 July the two battalions sailed from Okinawa for Pusan, arriving four days later.[38]

General Bolté, the G–3, Department of the Army, had suggested to the Deputy Chief of Staff for Administration, General Ridgway, on 1 July that the 5th RCT stationed in Hawaii, be sent to Korea.[39] Ten days later, when General Collins paused in Hawaii on his way to visit the Far East Command, he looked into the matter. In a teleconference with Ridgway in Washington, Collins asked him to query key staff officers on whether it would be better to send the 5th RCT as a unit or break it down into battalions and battalion cadres to bring other FEC regiments up to war strength. His own

[35] (1) *Ibid.*, Annex C. (2) Rad, WAR 86246, DA to CINCFE, 19 Jul 50. (3) Rad, CX 58506, CINCFE to CG EUSAK, 23 Jul 50. (4) Rad, WAR 87500, DA to CINCFE, 30 Jul 50.

[36] (1) Study, Additional Units to Meet Immediate Requirements of FEC. (2) Rad, CX 58506, CINCFE to CG EUSAK, 23 Jul 50 (passing on data from DA).

[37] CINCFE ordered these battalions sent at full war strength even though his existing troop basis did not allow this.

[38] (1) Memo, CofS GHQ for ACofS G–3, 12 Jul 50. (2) Rad, CX 57798, CINCFE to CG RYCOM, 13 Jul 50. (3) Rad, E 33465, CG Eighth Army to CINCFE, 14 Jul 50. (4) Rad, CX 57894, CINCFE to CG EUSAK, 15 Jul 50. (5) Rad, CX 57799, CINCFE to DA, 13 Jul 50. (6) Rad, WAR 85875, JCS to CINCFE, 13 Jul 50.

[39] Memo, Gen Bolté for Gen Ridgway, 1 Jul 50, sub: Anticipated Requirements of CINCFE.

feeling was that the 5th RCT should be employed as a regiment, not cannibalized. Ridgway and other staff officers agreed, recommending that the regiment be sent to Korea at its existing strength with all possible speed. On 13 July the Joint Chiefs of Staff authorized the commanding general, U.S. Army Pacific, to send the regiment to Pusan at once. The regiment sailed for Korea on 25 July with 178 officers and 3,319 men, entered Korea on 31 July, and went into combat immediately.[40]

By late July, the build-up of FEC divisions to war strength was well under way. Of the 11 infantry battalions required, 8 had been sent or would reach General MacArthur's command within thirty days. The shortage in division artillery of 11 light batteries was also being rectified. Three batteries arrived with the 5th RCT. Three were en route from the 3d Division, 2 from the 14th RCT, and 3 from the 2d Armored Division.[41]

Reinforcement by Major Units

While he had been asking for replacements and filler units, General MacArthur had also been calling for major trained combat units from the United States. Never in this early period did the Department of the Army openly question the validity of any of MacArthur's demands. The continuing success of the North Korean Army was proving vividly that the Far East Command needed fighting units. But as the calls for help mounted they threatened to shrink the General Reserve unduly and had to be considered in terms of national strategy and acted on at a level above the Department of the Army.

The first request by General MacArthur for a major unit from the United States came when he sought a Marine RCT with attached air support elements. Made on 2 July, the request was approved on the next day by the Joint Chiefs, and General MacArthur was told that the Marine unit would be sent to him as soon as possible.[42]

A few days later came his first call for specific major Army units from the General Reserve. He asked, on 5 July, that the 2d Infantry Division, then training at Fort Lewis, Washington, be sent to Korea as soon as possible. He also asked by name for smaller units which, if sent, would further reduce the capabilities of the General Reserve. On 2 July General MacArthur had pointed out that he must have more armored units since his four heavy tank battalions were skeletons with only one company apiece. Two were already in Korea and the remaining two were going. He asked for trained and organized tank companies from the United States to bring these battalions to full strength. He asked also for three additional medium tank battalions.

[40] (1) Telecon, TT 3512, Collins (Hawaii) and Ridgway (Washington), 11 Jul 50. (2) Rad, WAR 85696, DA to CINCFE (for Collins), 12 Jul 50. (3) Rad, WAR 85854, DA to CINCFE, 13 Jul 50. (4) Rad, WAR 85874, DA to COMGENUSARPAC, 13 Jul 50. (5) Rad, RJ 64645, CG USARPAC to CINCFE, 25 Jul 50.

[41] (1) Rad, CX 8506, CINCFE to CG EUSAK, 23 Jul 50. (2) Rad, WAR 86246, DA to CINCFE, 19 Jul 50. (3) Rad, WAR 87500, DA to CINCFE, 30 Jul 50.

[42] (1) Rad, C 57061, CINCFE to DA, 2 Jul 50. (2) Rad, JCS 84876, JCS to CINCFE, 3 Jul 50. For details of movement of Marine and airborne units, see below, Chapter IX.

At the same time he made a bid for an RCT from the 82d Airborne Division and another for an Engineer Special Brigade. The weakness of his antiaircraft artillery defenses impelled him also to seek quick shipment of four additional battalions of antiaircraft artillery. He backed up this request by pointing out that Sasebo, the principal Japanese port of embarkation for Korea, was completely undefended by antiaircraft artillery.[43]

These requests did not surprise Department of the Army officials, but they did pose a serious problem and involve major decisions. General Bolté advised General Collins to take units from the General Reserve and to send them to Korea as reinforcing units. The Chief of Staff accepted this view. General Collins, however, reluctant to tamper with the combat effectiveness of the 82d Airborne Division, recommended that an RCT of the 11th Airborne Division, which was less combat ready, be substituted. He had at first felt that sending four battalions of antiaircraft artillery would be beyond the Army's capability. He told the other members of the Joint Chiefs on 3 July that, as their executive agent for the Far East Command, he had taken action to send two battalions to General MacArthur. This was the maximum deployment of antiaircraft artillery he then believed could be made from the General Reserve without reducing the Army's ability to meet its emergency commitments. He reconsidered this problem in the next few days, decided on 8 July to accept the risks, and released two additional battalions to General MacArthur at once.[44]

While waiting for its recommendations to be considered by the Joint Chiefs of Staff, the Department of the Army suggested to General MacArthur certain priorities for shipping units if their deployment was approved. "It is emphasized," General MacArthur was told, "that final decision by higher authority to furnish major reinforcements requested by you has not yet been taken." The Department of the Army then outlined a proposed shipment schedule for these units. General MacArthur reacted immediately and, citing his most recent appraisal of the deteriorating combat situation, underscored the "impelling urgency" of getting a favorable decision at once. He reversed the proposed order of water shipment and asked that the armored units come first, to be followed by the 2d Division, the antiaircraft artillery battalions, and the Engineer Special Brigade. He asked also that the airborne RCT be flown to Japan at once, together with its supporting airlift.[45]

The Joint Chiefs of Staff decided that the Army should send General Reserve units to General MacArthur. But the issue was so important in terms of worldwide commitments that the JCS on 7 July asked the Secretary of Defense to gain the approval of the President. Mr.

---

43 (1) Rad, O 57218, CINCFE to DA, 5 Jul 50. (2) Rad, C 57093, CINFE to DA, 2 Jul 50. (3) Rad, CX 57152, CINCFE to DA, 3 Jul 50.

44 (1) Memo, Gen Bolté for DCofS for Admin (Gen Ridgway), 7 Jul 50. (2) Memo, Gen Collins for JCS, 3 Jul 50, sub: FEC Requirements for Opns in Korea, in G–3, DA file 320.2 Pac, sec. I–A, Book 1, Case 6. (3) Memo, Gen Ridgway for ACofS G–3, 8 Jul 50. Although the Army Chief of Staff kept the JCS informed of his decisions on the antiaircraft artillery battalions, he did not require their approval to send the units.

45 (1) Rad, WAR 85209, DA to CINCFE, 7 Jul 50. (2) Rad, C 57379, CINCFE to DA, 8 Jul 50.

Truman gave it, and the approved units were immediately ordered to prepare for shipment.[46] By 9 July, the 2d Division, the 2d Engineer Special Brigade, an RCT from the 11th Airborne Division, the 378th Ordnance Heavy Maintenance Company, the 15th and 50th Antiaircraft Artillery Battalions (AW), the 68th and 78th Antiaircraft Artillery Battalions (90-mm.), and the 6th, 70th, and 73d Tank Battalions had been approved for shipment to General MacArthur.[47]

The 2d Division

The deployment of the 2d Division from Fort Lewis, Washington, to the battlefront in Korea began on 8 July when the unit was alerted for shipment.[48] Nine days later, the first elements of the division sailed for Korea. One of its regiments attacked the enemy in the field a single month after the first alert.

The speed with which this division reached Korea as an effective fighting force is remarkable when the scale of the shipment and its many complications are considered. When it began preparing for shipment in early July, the 2d Division was far from combat-ready. General Mark W. Clark, then chief of Army Field Forces, had predicted after inspecting the division in June that it would not be ready to fight for at least four months. The division was approximately 5,000 men short of war strength. Used during the preceding year as an overseas replacement pool, it had undergone a personnel turnover of 138 percent in that period.[49] General MacArthur's first move on being told that the division was coming to his theater had been to ask that it be brought to full war strength before sailing.[50]

In order to comply, the Department of the Army transferred hundreds of men from other units at Fort Lewis to the 2d Division. But putting approximately 1,500 replacements awaiting shipment to the Far East from Fort Lawton into the division evoked an objection from General MacArthur. He remonstrated that all replacements scheduled for his command must come to him directly and not to be used as fillers for the 2d Division. He considered it "imperative that the meager strength authorized units in combat be maintained." [51] The Army had taken this action in order to get the 2d Division to Korea at full war strength as quickly as possible. The 1,340 replacements already assimilated by the 2d Division could not be retrieved. Further diversions were stopped because of General MacArthur's objection, even though Army officials felt that their method would have put the greatest number of

[46] (1) Memo, JCS for Secy Defense, sgd Gen Bradley, 7 Jul 50, in G–3, DA file 320.2 Pac, sec. I–A, Case 6. (2) Rad, W 85359, DA to CINCFE, 10 Jul 50. (3) Note by Secys to Holders of JCS 2147, 11 Jul 50.

[47] (1) Memo, G–3 for Gen Ridgway, 8 Jul 50, sub: Action on Gen MacArthur's Request. (2) Memo, Gen Thomas S. Timberman for Chief, Org and Training Div, G–3, DA, 9 Jul 50. (3) *Ibid.*, 8 Jul 50. All in G–3, DA files.

[48] Rad, WAR 85272, DA to CG Sixth Army, 8 Jul 50.

[49] 2d Div, Comd Rpt, vol. I, 8 Jul–31 Aug 50, prepared by Hist Sec, G–3, Hq, 2d Inf Div, pp. 9–22, copy in AGO Departmental Records, 302.

[50] Rad, CX 57573, CINCFE to DA, 10 Jul 50.

[51] (1) 2d Div, Comd Rpt, 8 Jul–31 Aug 50, pp. 13–14. (2) Rad, AMGA 0720, CG Sixth Army to DA (citing CINCFE radio message), 6 Jul 50, G–1 GHQ Log, Item 6, 15 Jul 50.

men in the Far East Command in the least period of time.[52]

Army officials were anxious not only to meet the requirements set up by the Far East Command but also to do so in the manner designated by General MacArthur. On 19 July, they told him to decide whether he wanted combat replacements or a war-strength combat division. The second increment of the 2d Division, scheduled to sail the next day, would leave at only half strength because men from other stations in the United States could not reach Fort Lewis by sailing time.

The division commander opposed sailing at only half strength, especially when 3,500 men were at west coast ports of debarkation awaiting shipment to the FEC as replacements. Since airlift was very limited, these replacements could not reach the FEC for at least three weeks. Washington asked General MacArthur for an immediate decision as to whether 1,500 of these replacements could be placed with the second increment of the 2d Division when it sailed the next day.[53]

General MacArthur's preoccupation with replacements led him to compromise by agreeing that the maximum number of men from the ports of debarkation could be sent on the same ships as the 2d Division, but not assigned to the division. "Anything," his reply stated, "that will speed up movement of replacements to this theater is desired." Fifteen hundred replacements sailed with the 2d Division on 20 July. General MacArthur had intended to place these men in the 7th Division, but changed his mind. On 28 July he directed that they be assigned to the 2d Division upon reaching Korea.[54]

In the early stages of the division's preparations, General MacArthur had asked that it be shipped to Korea combat-loaded. Each increment would thus land in Korea with its weapons ready to go, with organic vehicles and supporting artillery on the same or accompanying ships, and with each shipload able to operate independently in combat for a reasonable period of time.

While Washington recognized some advantages in combat-loading, there were compelling reasons why it was not practical. The ships being used were not designed for combat-loading. Furthermore, combat-loading would have delayed the division's arrival in Korea by at least two weeks because it was slower than ordinary unit-loading. The procedure also took nearly twice as much shipping space. Since convoys were not being used, unit-loaded shipments would depart as soon as they were loaded. Troops would travel on the same ship as their own equipment insofar as possible. The rest of their equipment and supplies would arrive on cargo shipping loaded for selective discharge to match the unit.[55]

When the assistant division commander of the 2d Division arrived in Tokyo late in July with the advance

[52] Telecon, DA and CINCFE, 16 Jul 50, G–2 GHQ Log, Item 1, 16 Jul 50.

[53] Rad, W 86378, DA to CINCFE, 19 Jul 50.

[54] (1) Rad, C 58193, CINCFE to DA, 19 Jul 50, G–1 GHQ Log, Item 25, 19 Jul 50. (2) Rad, W 86606, DA to CINCFE, 21 Jul 50. (3) Memo, G–3 GHQ for G–1 GHQ, 28 Jul 50, sub: Replacements 2d Inf Div and 7th Inf Div, G–1 GHQ Log, Item 54, 28 Jul 50.

[55] (1) Rad, CX 57546, CINCFE to DA, 10 Jul 50. (2) Rad W 85426, DA to CINCFE, 11 Jul 50.

party, he reported that almost 1,800 enlisted men had been released from the division at Fort Lewis because they were due to be discharged within three months. This information nettled MacArthur and he asked that these men be retrieved and sent to him as replacements. He would see that they rejoined the 2d Division after its arrival.[56] General MacArthur's concern was allayed when he was told that the Department of the Army had already decreed that men having thirty days' service remaining were eligible for shipment to the Far East Command. Port officials had already rounded up most of the men originally released and had shipped them on 20 July. The rest would be shipped out as soon as statutory authority was granted to keep all enlisted men in the service for an additional year.[57]

As fast as ships were loaded they left for Korea. The first regiment of the division unloaded in Korea on 31 July, while another regiment was still being loaded on troop transports in the United States. By 19 August the entire division had reached the Korean peninsula and was on its way into action as a unit.[58]

Supporting Artillery

Lacking nondivisional artillery, MacArthur asked the Joint Chiefs on 13 July to send him light, medium, and heavy artillery battalions. He asked for six 155-mm. howitzer battalions, self-propelled, as the first shipment. He also asked for an artillery group headquarters and a field artillery observation battalion. He pointed out that his division commanders in Korea would be forced, by the extensive frontages, broken terrain, and the limited road nets, to employ their divisions by separate RCT's. With a projected American force in Korea, based upon JCS-approved deployments as of that date, of 4 Army divisions and 1 Marine RCT, there would be 13 American regiments available in Korea. At least ten of these regiments could normally be expected to be in the front lines at any given time. Since only four battalions of 155-mm. howitzers would be present with division artillery units, six more battalions would be required if each of the ten regiments was to have a medium artillery battalion when it was used as an RCT. Two 8-inch howitzer battalions and the 155-mm. guns would be required for general support along the whole front. Light battalions could either reinforce division artillery units, or, if desirable, be committed in support of South Korean units. General MacArthur noted that the profitable extent to which American artillery should be used in support of South Korean forces was under study by his staff. He received no immediate reply and asked again, only four days later, for early arrival of the artillery urgently needed in Korea.[59]

The General Reserve, weak in all its components, was particularly deficient in nondivisional field artillery. Only eleven battalions were in the United

[56] Rad, C 58583, CINCFE to DA, 25 Jul 50.

[57] Rad, W 87191, DA to CINCFE, 27 Jul 50.

[58] 2d Div, Comd Rpt, 8 Jul–31 Aug 50, pp. 23, 27–28.

[59] (1) Rad, CX 57746, CINCFE to DA, 13 Jul 50. (2) Rad, CX 58055, CINCFE to DA, 17 Jul 50. (3) Rad, CX 57796, CINCFE to DA, 13 Jul 50. General MacArthur asked for the 155-mm. gun battalions after a conversation with General Collins on 13 July in Tokyo.

States and all were below war strength. Only four 105-mm. howitzer battalions, five 155-mm. howitzer battalions, one 155-mm. gun battalion, and one 8-inch howitzer battalion could be expected to be partially effective. But Washington Army officials ordered three of the 155-mm. howitzer battalions, the 8-inch howitzer battalion, an observation battalion, and the 5th Field Artillery Group headquarters to Korea.[60]

General MacArthur protested vigorously upon being told that only five artillery battalions of the fifteen he had requested could be furnished him. He pointed out that fifteen battalions were an essential minimum based on ten infantry regiments fighting on the line at any given time. He had now decided that there should be twelve U.S. regiments in action at all times. "Beyond doubt," he predicted, "the destruction of the North Korean forces will require the employment of a force equivalent at least to six United States infantry divisions in addition to ROK ground forces." Fighting in World War II had proven conclusively, according to him, that a field army could sustain a successful offensive against a determined enemy, particularly over difficult terrain, only if it had nondivisional artillery in the ratio of at least one for one as compared to division artillery. While General MacArthur did not spell out these latest requirements, he implied that twenty-four battalions of nondivisional artillery would be needed. He recommended that, since the necessary battalions were not available, they be activated and "an intensive training program of appropriate scale be set in motion at once." [61]

Service Troops

Without an adequate support base behind the battle line in Korea and in the larger service area in Japan, the fighting units could not sustain their desperate defense, much less attack. Although the greatest emphasis was placed on infantry, artillery, armored, and other combat-type units and soldiers during July, the demand for service units and troops increased steadily. Technical service units to supply front-line soldiers, to repair damaged weapons and equipment, to keep communications in operation, and to perform the hundreds of vital support operations required by a modern army, had been at a premium in the FEC when the war broke out. Japanese specialists and workmen performed in large part the peacetime version of service support for the Far East Command. The few available service units had been depleted when specialists and other trained men had been handed rifles and sent to fight as infantry.

Some types of combat and noncombat support were needed more immediately than other types. In view, for instance, of the hundreds of tons of ammunition of all types on its way to the Far East Command for the Korean fighting, ordnance specialists qualified to handle ammunition were needed at once. General MacArthur asked on 11 July that several hundred officers and men qualified for this function be flown to his area

[60] (1) Memo, Gen Bolté for Gen Collins, 9 Jul 50, sub: Strength and Training Status, FA Units, in G-3, DA files, Blue Book, vol. II, Status of Units and Equipment. (2) Rad, WAR 86427, DA to Continental Army Comdrs, Info to CINCFE, 18 Jul 50. (3) Rad, WAR 86558, DA to CINCFE, 20 Jul 50.

[61] Rad, CX 58750, CINCFE to DA, 26 Jul 50.

with all possible haste. The next day he sent a detailed requisition for Army technical service units, showing, in order of priority within each service, the support units needed immediately and those needed later to carry on the essential service support operations in Japan by replacing units scheduled for Korea. Support units coming from the United States did not appear on this requisition of 12 July, but showed up two weeks later on a second requisition.[62]

The stated requirements of the Far East Command for technical service units were viewed in Washington as reasonable and just, but were beyond the capabilities of the Department of the Army to supply. General MacArthur had requested over 200 company-sized units from Chemical, Engineer, Medical, Transportation, and other technical services. This requisition, if filled, would involve shipment of 43,472 men and officers. The Department of the Army had only about 150 company-sized service units in the United States.

Between the extremes of sending only cadres from such units and sending every technical service unit from the United States to the Far East, the Department of the Army charted a middle course. Cadres would have little immediate value in Japan and Korea. But the General Reserve could not be stripped without disastrous effect upon the mobilization base.[63] In order to preserve a minimum mobilization base and still take the edge off the Far East commander's most urgent requirements, Washington officials withdrew cadres for retention in the United States and sent about eighty service support units of company size to the Far East. Although these units were only at about 65 percent strength, their specialized composition and the technical know-how of their men and officers enabled them to function profitably, even at reduced strength.[64]

As the scale of the Korean action became clearer, General MacArthur on 25 July sent a supplemental list of technical service units which would be needed. This list brought the total number of technical service units requested in July to 501, totaling 60,000 men and officers. Officials of the Far East Command knew that they would not receive the bulk of these units for a long time, but they felt that Washington should know their requirements for planning purposes.[65]

The need for combat soldiers remained paramount. Of the service troops sent to Japan as replacements in July, for example, 60 percent were assigned to front-line fighting troops upon arrival in Korea.[66]

The filler units and reinforcing units which the Department of the Army had managed to scrape together for General MacArthur in the first month of the campaign represented the maximum force which the United States was able

[62] (1) Rad, CX 57563, CINCFE to DA, 11 Jul 50. (2) Rad, CX 57693, CINCFE to DA, 12 Jul 50.

[63] For example, General MacArthur requested a corps signal battalion. There was only one such unit in the United States. It would have required nine months to reconstitute such a unit after selected personnel were available.

[64] Study, Additional Units to Meet Immediate Requirements of FEC, Annex D.

[65] Memo, Col Daniel H. Hundley for Gen Beiderlinden, 25 Jul 50, sub: Additional Technical Service Units, G-3 GHQ Log, Item 43, 25 Jul 50.

[66] Memo, G-1 GHQ for CofS GHQ, 5 Aug 50, sub: Casualties and Replacements, G-1 GHQ Log, Item 41, 5 Aug 50.

to furnish. These units might not be enough, but no more were going to be sent until the Joint Chiefs of Staff and other planners had a chance to take a better look at the way things were going. Department of the Army officials told the Far East commander on 21 July that they were in no position even to consider his request for another army of four divisions for the present. Before any decision could be made on that request, American defense officials would have to determine just how far they were going in rebuilding the General Reserve. Then they would have to see if sending additional forces to Korea was as important to national security as having them available for deployment elsewhere in the world.[67]

[67] Rad, CM-OUT 86558, DA to CINCFE, 21 Jul 50.

CHAPTER VI

A New Confidence

The outbreak of war in June 1950 had caught the United States flat-footed. The nation had few forces immediately available and no plans for fighting in Korea. Nevertheless, American leaders had developed in the post-World War II years some policies and principles for meeting communist aggression which they could use as a basis for raising forces and making plans for Korea. These policies and principles provided, broadly, that the United States would work closely with its treaty allies and with other free nations to stop all forms of communist aggression, and that any military action would be taken under the aegis, or at least with the sanction, of the United Nations, if at all possible. The United States earnestly desired to avoid unilateral action, however effective, which might alienate its friends and possibly goad the Soviet Government into extreme action and all-out war. Too, it wished to put to full use the military resources of its allies rather than bear the entire burden single-handedly.

Within hours after word of the North Korean attack reached Washington, the United States had called on the United Nations. The resolutions of 25 and 27 June, drawn up in haste and under pressure, had been steps in the right direction but did not go nearly far enough toward the goal of restoring peace in Korea.

The Security Council resolution of 25 June had called upon members to refrain from helping the North Koreans. The United States Government directed a more specific appeal to the Soviet Union through its embassy in Moscow, asking that it prevail upon the North Korean leaders to halt the fighting. In response, the Soviet Government called South Korea the aggressor and, by implication, refused to mediate.[1]

Faced with Soviet refusal to give even lip service to the United Nations resolution, and with a combat situation that worsened hourly, the United States began carefully to press for a stronger stand and more effective action by the United Nations.

On 3 July the Secretary General of the United Nations, Trygve Lie, circulated a proposed resolution to the delegations of the United States, the United Kingdom, and France. It suggested that the Government of the United States would direct the armed forces of member na-

[1] (1) Leland M. Goodrich, *Korea, A Study of U.S. Policy in the United Nations* (New York: Council on Foreign Relations, 1956), p. 106. (2) *State Dept Bulletin,* XXIII, 575 (July 10, 1950), 46–48.

tions in Korea, but with the help of a "Committee on Coordination of Assistance for Korea." This committee would co-ordinate all offers of assistance, promote continuing participation in Korea by member nations, and receive reports from the field commander. The exact extent of its control was not stated in the proposal.[2]

When, on 4 July, the Department of State sought the views of the Joint Chiefs of Staff on the resolution, the latter opposed forming such a committee. They felt that placing a United Nations committee in the channel between the U.S. Government and the field commander would raise serious operational difficulties. Even though the committee might never try to control military operations, the possibility that it might do so brought the Joint Chiefs together in opposition. They told the Secretary of Defense that, if a committee were needed for political reasons, its powers must be defined and restricted so exactly that it could never take on the nature of a U.N. command headquarters.[3]

The Joint Chiefs of Staff wanted a command arrangement in which the United States, as executive agent for the United Nations, would direct the Korean operation, with no positive contact between the field commander and the United Nations. The major decisions, especially those of political content, must not in any way be made, or influenced, by the officer commanding the U.N. forces in Korea. If the United Nations were to deal directly with the commander on assistance offers, for example, the top levels of the U.S. Government would be bypassed and forces accepted or rejected by a commander, very likely an American, whose outlook would be restricted by his own local situation.[4]

In spite of sympathetic consideration of the proposal by France and the United Kingdom, the United States rejected the projected U.N. committee, and a revised resolution developed. Because the United States occupied a privileged position in the terms of the resolution, it would not have been seemly for the American representative to introduce it. Accordingly, on 7 July, the delegations of France and the United Kingdom brought the draft before the Security Council. Seven votes in favor had been lined up in advance. The resolution therefore passed the Security Council, by a vote of seven to zero, with three nations, Egypt, India, and Yugoslavia, abstaining. The Soviet representative had not yet returned to the council and cast no vote.

This resolution made President Truman executive agent for the council in carrying out the United Nations fight against aggression in Korea. The Security Council recommended that contributing member nations furnish forces to a unified command under the United States. It asked that the American Government select a commander for this unified command and that the United States submit periodic reports on the course of operations in Korea. President Truman designated the Joint Chiefs of Staff his agents for Korea. To General Collins,

---

2 Goodrich, *Korea, A Study of U.S. Policy in the United Nations*, p. 119.

3 Memo, JCS (Bradley) for Secy Defense, 5 Jul 50, sub: Proposed U.S. Position With Regard to Forces in Korea.

4 JCS 1776/19, Rpt by JSSC, 5 Jul 50, sub: Proposed U.S. Position With Regard to Forces in Korea.

Army Chief of Staff, fell the task of serving the Joint Chiefs as their primary representative in Korean operations. At the Army level, General Bolté, the G-3, handled operational details for General Collins. Thus, with authority granted by the United Nations, vested in the President, and running downward through the Joint Chiefs of Staff, the United States Army became responsible for planning and directing the military operations of United Nations forces in Korea.[5]

The Joint Chiefs of Staff recommended that General of the Army Douglas MacArthur be placed in command of United Nations forces.[6] President Truman accepted their recommendation and notified General MacArthur of his appointment on 10 July 1950. On 12 July Department of the Army officials sent detailed instructions to MacArthur. They directed him to avoid any appearance of unilateral American action in Korea. "For world-wide political reasons," they cautioned, "it is important to emphasize repeatedly the fact that our operations are in support of the United Nations Security Council." In furtherance of this, General MacArthur would identify himself whenever practicable as Commander in Chief, United Nations Command (CINCUNC), and whenever justified, would emphasize in his communiqués the activities of forces of other member nations.[7]

Two days later, on 14 July, President Rhee assigned control of his nation's forces to General MacArthur, stating in a letter transmitted through the U.S. Ambassador to Korea:

> In view of the joint military effort of the United Nations on behalf of the Republic of Korea, in which all military forces, land, sea and air, of all the United Nations fighting in or near Korea have been placed under the joint operational command and in which you have been designated Supreme Commander, United Nations Forces, I am happy to assign to you command authority over all land, sea and air forces of the Republic of Korea during the period of continuation of the present state of hostilities, such command to be exercised either by you personally or by such commander or commanders to whom you may delegate the exercise of this authority within Korea or adjacent seas.[8]

Although the Security Council asked the United States to report to the United Nations on activities of the unified command, no procedure was specified. On 13 July the Department of State proposed to the Secretary of Defense that reports be sent to the Security Council each week. These would keep world attention on the fact that the United States was fighting in Korea for the United Nations, not itself. Apprehensive over world reaction to the naval blockade of Korea ordered by President Truman on 30 June, the Department of State was convinced that the Security Council resolutions of 25 and 27 June amply justified the blockade, but wished the actual blockade declaration reported to the Security Council in order to remove any doubt as to its legality. A report from the unified command on the blockade seemed in order.

---

[5] *MacArthur Hearings*, pp. 14, 989, 326, 1259, 1938.

[6] Memo, JCS for Secy Defense, 9 Jul 50, sub: Designation of a United Nations Unified Comdr by the United States.

[7] Rad, WAR 85743, DA to CINCFE, 12 Jul 50.

[8] Rad, State Dept Msg 41, U.S. Ambassador, Taegu, to Secy State, 14 Jul 50, 17 Jul 50, containing text of Ltr, Rhee to MacArthur.

This proposal focused the attention of the Joint Chiefs on the need for a definite arrangement on how and when reports should be made to the United Nations. Late in July they directed General MacArthur to send them a report on the actions of his forces every two weeks. The Joint Chiefs would, in turn, submit the report through the Secretary of Defense to the Department of State for presentation to the Security Council of the United Nations by the American delegation at Lake Success, New York. General MacArthur was assured that he would be consulted in advance if political considerations made it necessary at any time for the Joint Chiefs to alter his reports.[9]

On 24 July 1950 General MacArthur issued orders establishing the United Nations Command (UNC) with general headquarters in Tokyo, Japan. With few exceptions, staff members of the Far East Command were assigned comparable duties on the UNC staff. In effect, the GHQ, United Nations Command, was the GHQ, Far East Command, with an expanded mission.[10]

At the central core of American direction of the operations in Korea on behalf of the United Nations lay the Joint Chiefs of Staff. As advisers to the President, the Joint Chiefs concerned themselves with every aspect of American military power and policy. They had to deal simultaneously with problems at home and abroad, in western Europe and in Korea.

They did not make the national military policy. Yet because they furnished the President, normally through the Secretary of Defense, information and advice to help him set this policy, what they did and what they thought held great importance for the nation and for the Korean War. By the very nature of their work, the Joint Chiefs of Staff had to consider political factors in deliberating national military problems. So closely intertwined were military and political factors in the Korean War that they could not be isolated one from the other.

The mechanical process by which military policy recommendations evolved during the Korean War began with consideration of a particular problem within the military staffs, usually the Army staff, and within the joint staff of the JCS itself. The joint staff consisted of about two hundred officers selected from all the services. These officers developed and furnished recommended positions to the Joint Chiefs of Staff.

Once a final stand on a problem had been discussed and agreed upon by them, the JCS presented their views in a memorandum to the Secretary of Defense. Any political aspects of the matter would be worked out at this level between the staffs of the Defense and State Departments or, on occasion, between the respective secretaries personally. The

[9] (1) JCS 1776/39, Note by Secys, Rpts by U.S. Government to UNSC, 18 Jul 50. (2) *MacArthur Hearings*, Part II, p. 1515. (3) Rad, JCS 84885, JCS to CINCFE, 3 Jul 50. (4) JCS 1775/62, Note by Secys, Rpts by U.S. Government to UNSC, 28 Jul 50.

[10] (1) GO 1, UNC, 24 Jul 50. (2) The United Nations, at no time in the Korean War, sought to interfere in the control of operations which were the responsibility of the United States. General MacArthur later testified to this when he told a Senate investigating committee, ". . . my connection with the United Nations was largely nominal . . . everything I did came from our own Chiefs of Staff. . . . The controls over me were exactly the same as though the forces under me were all Americans. All of my communications were to the American high command here." See *MacArthur Hearings*, p. 10.

Secretary of Defense then presented the views and recommendations thus developed, with a clear statement of any divergencies, to the National Security Council or, if more appropriate, directly to the President. On occasion, the procedure varied but, normally, if there were time things were done in this fashion.

The issues raised by Korea could not be separated from those involved in planning for American defense on a world-wide scale. The withdrawal of men and units from the General Reserve for employment in Korea was incompatible with existing plans. If the Korean outbreak marked the initial stages of an all-out war, it was unsound to tie up large forces in an area of limited strategic significance. But the United States was committed, short of global war, to repelling armed aggression in South Korea. Speculating on 13 July that developments in Korea were part of a general USSR plan which might involve correlated actions in other parts of the world, the JCS planning staff said:

> It is now apparent from Korea that Russia is embarking upon an entirely new phase in her program of world-wide Communist domination. This is a phase in which she is now utilizing for the first time the armed forces of her satellites to impose by military strength a Communist-dominated government upon a weak neighboring state considered incapable of successful military opposition.[11]

A reappraisal of United States objectives and resources thus became necessary. And the Joint Chiefs of Staff constantly faced the major question, "How much of our military strength can we commit to Korea without seriously damaging our ability to meet a global emergency?" A correct solution to this problem would enable them to determine, for instance, if partial mobilization was needed. A second question was, "If we limit our commitments to Korea because of the greater global threat, can we drive the North Koreans behind the 38th Parallel?" [12]

Enemy victories in Korea forced the Joint Chiefs to take action without awaiting answers to the vital questions. Courses of action had to be considered individually as they arose. Decisions on them were greatly influenced by General MacArthur's recommendations, but as each new move weakened the potential means, without lessening the mission, it brought the need for answers to these questions into urgent focus.

By mid-July so much American military strength had been drawn into the Korean War that American military capabilities for action elsewhere had been much reduced. Reserves of trained men and matériel diminished as MacArthur's units were brought up to war strength and given service support and replacement. A further drain upon reserves of critical specialists and equipment would result as operations progressed.[13] A key Army officer commented at this time, "Our ground force potential is so seriously depleted that further significant commitments of even a division or more

[11] JSPC 853/15, 13 Jul 50, in G–3, DA file 091 Korea, sec. I–C, Case 16.

[12] JSPC 853/7/D, 5 Jul 50, in G–3, DA file 091 Korea.

[13] Study, JCS 1924/20, 14 Jul 50, in G–3, DA file 091 Korea.

in size would vitally weaken our national security at home." [14]

The possibility that U.S. troops might be thrown out of Korea was far from academic. The Joint Intelligence Committee (JIC) of the Joint Chiefs of Staff pointed out on 12 July that the understrength U.S. 24th Division was facing 9 North Korean divisions numbering 80,000 men and equipped with a total of from 100 to 150 modern tanks. The enemy not only had a great advantage in numbers of men and in tanks and artillery, but was also well trained, and was fighting determinedly and with great skill. The JIC concluded that the North Korean Army was capable of threatening the security of Pusan within two weeks. Lt. Gen. Matthew B. Ridgway, Deputy Chief of Staff for Administration, had sketched the same depressing picture for the secretaries of the armed services on 10 July. He told these men that, while MacArthur's forces had definitely slowed the enemy, they could not hold unless they were substantially reinforced.[15]

Forced withdrawal of U.S. troops from Korea would be a political as well as a military calamity. It could weaken American alliances and build up communist political influence. It could discredit U.S. foreign policy and undermine confidence in American military capabilities. Voluntary withdrawal could be more damaging than a failure to have sent troops to Korea in the first place. American commitments would be marked as unreliable by other nations and considerable doubt would be cast on American ability to back up commitments in the future. The United Nations actions resulted mainly from U.S. initiative, and withdrawal from intervention on behalf of the United Nations could greatly weaken American leadership within the United Nations.

Failure in Korea could force the United States to revise drastically its policy of general containment of communism by reducing or limiting its commitments and by planning to combat communist expansion only at selected points. The United States would undoubtedly have to start partial military and industrial mobilization to ready its forces for other, almost certain, aggressions; or, in another approach, to begin full mobilization so as to be prepared to threaten full-scale war in case of further Soviet aggression.[16]

First Visit From Washington

President Truman sent two members of the Joint Chiefs of Staff, General Collins and General Vandenberg, to the Far East on 10 July 1950. They were to bring back firsthand information to use in establishing the scope of expansion of the U.S. military program. Immediately upon reaching Tokyo on 13 July 1950, Collins and Vandenberg talked with General MacArthur and key members of his staff. General MacArthur impressed upon them the dangers of underestimating the North Koreans. He described

[14] Quotation from Brig Gen Cortlandt Schuyler, Memo for Gen Lindsay, Adm Ingersoll, and Maj Gen Oliver P. Smith, 14 Jul 50, sub: Estimate of the Korean Sit, JSPC 853/11, in G–3, DA file 091 Korea.

[15] JCS 1924/19, Decision on Estimate by JIC, 12 Jul 50, in G–3, DA file 091 Korea, Case 46. The JIC "Estimate of the Situation" included in JCS 1924/19 was not approved but merely noted by the JCS.

[16] JCS 1924/19, Annex D, 10 Jul 50, in G–3, DA file 091 Korea.

the enemy soldier as a tough, well-led fighter who combined the infiltration tactics of the Japanese with the armored tactics of the Russians in World War II. General MacArthur praised the North Korean Army's ability to march, maneuver, and attack at night. So far, his own forces had not been able to do the equivalent successfully. The North Korean Army exploited its tank firepower to the greatest advantage. Its armored tactics were extremely efficient and approximated, in his words, "the norm of tank effectiveness standard in the Soviet Army." The flexibility of the North Korean commanders had been very apparent in their quick adoption of night operations as a countermeasure against intensified air attacks by American forces.[17]

General MacArthur confessed that the only hope he had seen a week earlier had been "a desperate rearguard action," to slow the North Korean Army by "throwing everything in Japan into the fight." He had done this as fast as he could although his own forces were, as he phrased it, "tailored for occupation duty and not for combat." [18]

By now he had taken a brighter view. He told Generals Collins and Vandenberg that, while he could not predict where the military situation would be stabilized, "that it will be stabilized is indisputable." Originally, he had planned to stand near Suwon and then to envelop the north bank of the Han River. After recapturing Seoul, he would have cut the enemy's line of communications and his withdrawal route. He conceded that his forces were now too far south and too weak to carry out this plan. He had, therefore, postponed its execution until the situation could be stabilized and reinforcements reached him. He placed no blame on General Dean or his men. General Dean had done as well as any man could. The troops had done everything possible, but they were outgunned, outnumbered, and without adequate defense against the enemy's armor.[19]

General MacArthur then outlined his recommendations for winning the fight in Korea. In his opinion, the success of the United States in Korea and the speed of achievement of that success would be in direct proportion to the speed with which the United States sent him reinforcements. All American forces he could spare from Japan would have been sent to Korea by August. If the United States backed this commitment with sufficient reinforcements from the zone of the interior, there would be, in MacArthur's mind, no question as to the result. Without full support, the result would vary in direct proportion to the support received. MacArthur contended that if he were giving advice he would say, "In this matter, time is of the essence." [20]

He expressed extreme impatience with delay or partial measures. The strength of any military stroke depended entirely upon its speed. Accordingly, General MacArthur wanted to "grab every ship in the Pacific and pour the support into the Far East." He would not start modestly and build up, but would make the

[17] Memo, Lt Col D. D. Dickson for Gen Bolté, sub: Rpt of Trip to FEC, 10–15 Jul 50, Tab A: Remarks of Gen MacArthur, in G–3, DA file 333 Pac, sec. I, Case 3. Quotations are taken from the notes kept by Col Dickson.

[18] *Ibid.*

[19] *Ibid.*

[20] *Ibid.*

complete effort at the beginning. In emphasizing these points, the veteran commander said, "Business as usual—to hell with that concept." Admittedly the United States was "playing a poor hand here," but long experience had shown General MacArthur that "it is how you play your poor hands rather than your good ones which counts in the long run." [21]

The question of how much American strength should be saved for areas in other parts of the world obviously interested General MacArthur less than the Joint Chiefs. He believed that winning in Korea would slow down worldwide communism more than any other single factor. He assured his visitors that he fully understood the American obligation to maintain its global military posture. But he made a colorful analogy to point out the error of withholding strength from the Korean battlefront. Assuming the world to be a metropolis of four districts of which District No. 1 was the most important and District No. 4 least so, General MacArthur asked his visitors to consider whether a fire in No. 4 should be allowed to burn uncontrolled because city officials were saving their fire equipment for District No. 1. As he concluded, "You may," he said, "find the fire out of control by the time your equipment is sent to No. 4." A general conflagration should not be handled by attempting to place Korea or the FEC in terms of priority of area. General MacArthur felt that the United States would win in Korea or lose everywhere.[22]

General Collins particularly wanted answers to several specific questions which could help solve the major questions facing the Joint Chiefs. He asked General MacArthur when he would be able to mount a counteroffensive and how many American troops he would need in Korea after the fighting ended. Both questions were keyed to the thorny issue of how much the United States should expand its military program.

General MacArthur insisted that a categorical reply to the first question was impossible. When three divisions had been committed to Korea, he hoped to stabilize the situation. He intended then to infiltrate north and follow any North Korean withdrawal. He was centering his hopes on an amphibious operation. The overland pursuit of North Korean forces was incidental to this operation.

As to the second question, General MacArthur told General Collins that he would not merely drive the invaders across the 38th Parallel. He meant to destroy all their forces and, if necessary, to occupy all of North Korea. "In the aftermath of operations," he said, "the problem is to compose and unite Korea." His troop requirement in the Far East Command under this situation would be eight infantry divisions and an additional Army headquarters.

Not only General MacArthur but also two of his key officers took advantage of General Collins' presence to press for additional forces. General Walker, commander in Korea, and General Almond, chief of staff, FEC GHQ, each emphasized the need for eleven more infantry battalions and 3,600 fillers to be sent by air. The fillers were needed to build up the 7th Division, which General Walker described as "only a crust." General

21 *Ibid.*
22 *Ibid.*

Collins made no on-the-spot commitment since arrangements to meet these requirements were already under way.

From the Tokyo conference, General Collins and General Vandenberg flew to Korea. Collins talked briefly at Taegu with Walker, Dean, and members of the Eighth Army staff. Agreeing with General MacArthur's analysis of the combat scene, Walker told Collins that, barring unforeseen circumstances, he could hold an extensive bridgehead with the troops en route to Korea from Japan. The commander of the battered 24th Division, General Dean, was very worried over his losses. On the day of General Collins' visit, the total of missing soldiers from Dean's 24th Division had risen from 200 to well over 800.[23]

General Collins returned to Tokyo early on 14 July, leaving for Washington the same day. Before leaving, the Army Chief of Staff gave General MacArthur his personal ideas on which major units he could count on having for the offensive which he had in mind. In addition to the four divisions already in the Far East, these units were the 2d Division, the 1st Marine Division, the 4th RCT, the 29th RCT, and an RCT from the 11th Airborne Division.

General MacArthur, after getting Collins' views, told the Chief of Staff that he would make his plans on the basis of the anticipated strength of these units. If Russia or Communist China intervened in force, the plans would have to be changed. He assured Collins that he fully understood the problems faced in Washington and the necessity of maintaining some kind of General Reserve.[24]

[23] Rad, C 57814, Collins to Haislip, 14 Jul 50.
[24] *Ibid.*

Air Operations—July 1950

While possible steps to improve MacArthur's ground strength were being considered, moves to improve air operations in Korea were under way. Since there was no provision in the FEC GHQ staff organization for joint representation of the Navy and Air Force, the central command of air operations over Korea was not possible below the level of General MacArthur himself. Anomalous and inefficient operations sometimes resulted. In early July, as an example, the Navy sent planes from Task Force 77 against targets that FEAF planned to attack the following day. As a consequence, the Air Force medium bombers sat on the ground the next day since it was too late to set up other targets.[25]

Someone obviously had to take over the responsibility, and General Stratemeyer made the first bid for over-all control of air operations in Korea. On 8 July, he told General MacArthur:

It is my understanding that the Navy contemplates bringing into your theater some land-based aircraft; also, as you know, the Seventh Fleet contemplates another strike with air at your direction in North Korea. I request that all land-based naval aviation and carrier-based aviation when operating over North Korea or from Japan, except those units for anti-submarine operations, be placed under my operational control.[26]

[25] For detailed coverage of air and naval operations in Korea, see: Robert Frank Futrell, *The United States Air Force in Korea, 1950–1953* (New York: Duell, Sloan & Pearce, 1961); James A. Field, Jr., *History of United States Naval Operations, Korea* (Washington, 1962); and Commander Malcolm C. Cagle and Commander Frank A. Manson, *The Sea War in Korea* (Annapolis: U.S. Naval Institute, 1957).
[26] Memo, Stratemeyer for MacArthur, 8 Jul 50.

When the Navy objected to Stratemeyer's acquiring control of naval aircraft for operations in Korea, General Almond, the chief of staff, worked out a compromise in a directive issued in MacArthur's name on 8 July whereby Stratemeyer would control all aircraft "operating in the execution of the Far East Air Force mission as assigned by CINCFE." However, when engaged in naval reconnaissance, antisubmarine warfare, and support of naval tasks such as amphibious assault, naval aircraft were to remain under the operational control of COM NAVFE.[27]

U.S. and ROK ground troops needed every bit of close support that could be given them in the first weeks of the Korean fighting. Artillery was at a premium. There were not enough batteries, nor was there enough ammunition. In view of shortages of infantry units and their organic support weapons, the Air Force had to undertake a larger than normal role in ground force support. Unfortunately, the Far East Air Force had an insufficient number of planes of the most desirable types for supporting ground troops in close contact with the enemy. Lacking, too, were men and facilities for air-ground control and coordination.

Drastic measures were taken. Aircraft normally employed in interdiction missions behind enemy lines assumed ground support missions. The use of B–29 bombers as close-support weapons, to the necessary neglect of other functions behind enemy lines, prompted criticism and serious objections by Air Force officials in the Far East. But General MacArthur overrode them on the basis that, if the ground troops were overrun, interdiction of targets deep behind enemy lines would have no significance. He ordered Stratemeyer to send his B–29's "to strafe, if necessary" in order to stop the North Korean drive.

GENERAL STRATEMEYER *(Photograph taken in 1944.)*

Within several weeks after the outbreak of the Korean War, the Air Force established the FEAF Bomber Command as a subordinate element of FEAF. The bomber command consisted of several bombardment groups comprised of medium bombers (B–29's), the aircraft which had been so successful in World War II in the strategic bombing of Japan. In the Air Force concept, this type of bomber should have been employed against strategic targets beyond the area

[27] CINCFE Ltr, 8 Jul 50, sub: Co-ordination of Air Effort of FEAF and U.S. NAVFE.

of ground fighting including such installations as factories, rail yards, warehouses, and other vital points on enemy lines of communication. Nevertheless, because of immediate needs and the lack of other proper aircraft, General MacArthur decided that these medium bombers would operate in support of ground troops wherever necessary. General Stratemeyer had ordered the medium bombers to operate only north of the 38th Parallel. MacArthur overruled him on several occasions in mid-July and ordered the mediums sent against enemy troop concentrations and other tactical targets immediately in front of the Eighth Army lines. MacArthur, on 15 July, also told General Walker that future emergency use of these medium bombers would be ordered by GHQ whenever Walker felt it necessary.[28]

When General Vandenberg and General Collins came to the theater in mid-July, this aspect of the air-ground relationship concerned both of them. Vandenberg did not attempt to interfere since, if Eighth Army troops were driven off the peninsula and the Air Force was meanwhile employing its bombers to bomb remote industrial areas in North Korea, the resultant effect on public opinion would have been most unfavorable. General Collins, on the other hand, expressed great interest in the way the B–29's were being employed and asked to be kept informed.

To tighten his control of the air effort in Korea, General MacArthur on 14 July established a GHQ Target Group, composed of a chairman, a senior Army officer from Willoughby's G–2 section, and Air Force, Navy, and Army members. This group was to advise on the use of Navy and air offensive power "in conformance with the day-to-day situation." The group would recommend targets and priorities which the Air Force and Navy would bomb. The decisions of the target group were passed to the G–3, who passed on the orders to FEAF. Few of the members appointed to the group were experienced pilots and their method of operation consisted of studying maps of Korea, selecting likely targets from these maps, and directing that they be bombed. It was an unwieldy and impracticable method.[29]

According to Air Force officials, this abnormal arrangement was not only unproductive but wasteful. Since the target group performed its function using a standard Army Map Service 1:250,000 map to select targets for medium bombers without checking its information from other sources, an unusual situation developed. Of 220 targets selected by the group between 17 July and 2 August, 20 percent did not exist on the ground.

The FEAF commander called on General MacArthur and the latter's chief of staff, General Almond, on 19 July to complain of this procedure. Stratemeyer followed this visit with a memorandum on 21 July in which he recommended the creation of a target selection committee which would include General Hickey, the FEC GHQ deputy chief of staff, General Willoughby, the G–2, Lt. Gen. Otto P. Weyland, the vice commander for operations of FEAF, and a Navy repre-

[28] (1) Rad, CX 57893, CINCFE to CG EUSAK, 15 Jul 50. (2) Rad, CX57755, CINCFE to CG FEAF, 13 Jul 50.

[29] (1) Check Sheet, Almond to All Staff Secs, GHQ FEC, 14 Jul 50. (2) Interv, Maj Schnabel with Comdr Reilly, JSPOG, GHQ, Nov 51.

Refugees Fleeing From the Combat Area Near Taegu

sentative to be named by Vice Adm. C. Turner Joy. MacArthur approved this recommendation immediately, and FEAF, using the new method, took over the actual selection of targets for interdiction.[30]

The Withdrawal Continues

Meanwhile, the North Korean Army drove hard, aiming to destroy the Republic of Korea and to throw the 24th Division out of Korea before ground reinforcements arrived. At the Kum River line the enemy units again outflanked the 24th Division. The 19th Infantry and its attached artillery lost nearly one-fifth of their men and officers while vainly trying to keep the superior enemy force from crossing the Kum on 16 and 17 July. Having breached American defenses on the last natural barrier before the key railroad center of Taejon, the enemy slashed southward, intent on taking Taejon with a further view, apparently, of capturing the new South Korean capital of Taegu.

General MacArthur's chief of staff, General Almond, contended in a letter

30 USAF Hist Div, Dept of the Air Force, *United States Air Force Operations in the Korean Conflict, 25 June–1 November 1950*, 1 July 1952, p. 13.

to General Collins on 17 July that the North Koreans hoped to capture Taegu mainly for the psychological effect. The enemy commanders, having outflanked the Americans, were attacking as well down the central corridor along the axis Ch'ungju-Taegu, and were pushing back the South Koreans. Almond assumed Collins that General MacArthur was aware of this "vital threat" down the middle. Referring to the plans for the future which General MacArthur had sketched to him three days before, Almond reported:

> Our proposed projects are developing as planned and we are confident that while the enemy stubbornly persists in his efforts to drive us back, we have blunted his principal strikes, and he is bound to be getting more exhausted while we become stronger each day and better organized to stop him. . . . We have no fear of the outcome and thoroughly understand that current conditions are the growing pains precedent to future operations.

General Almond did not believe that Taejon could be held but was not unduly alarmed. "It may not last there," he told Collins, "but the trend is much better."[31]

The 25th Division, although its first elements had reached Korea on 9 July, had not yet met the enemy. Nor had the 1st Cavalry Division, en route to Korea while Almond was addressing Collins. The 24th Division, weakened and disorganized, fell back upon Taejon alone, the enemy hard on its heels.

When President Truman, on 19 July, asked General MacArthur for his estimate of the Korean situation, he received a reply that revealed a new confidence, quite a contrast with the glum prognoses issued earlier in the month. The North Koreans, MacArthur told the President, had lost their great chance for victory. The extraordinary speed with which Eighth Army had been deployed from Japan and the brilliant co-ordinated support by air and naval elements had forced the enemy into "continued deployments, costly frontal attacks and confused logistics. . . . I do not believe that history records a comparable operation." His forces still faced a difficult campaign. They would be hard pressed and could expect losses as well as successes. But the initiative no longer lay entirely with the North Koreans, and United Nations troops held Southern Korea securely. Apparently heartened by the recent promises of reinforcements which would increase his own strength as attrition cut the enemy's strength, General MacArthur assured President Truman, "We are now in Korea in force, and with God's help we are there to stay until the constitutional authority of the Republic is fully restored." [32]

The 24th Division lost Taejon on 20 July in a hard-fought 2-day battle. The division commander, General Dean, was captured after becoming separated from his troops during the withdrawal from Taejon. Division casualties approached 30 percent. On 22 July the 1st Cavalry Division relieved the 24th at Yongdong. In a 17-day losing battle against two superior North Korean divisions, the 24th had fallen back almost 100 miles, and had lost more than 2,400 men missing in ac-

[31] Ltr, Almond to Collins, 17 Jul 50.

[32] (1) Rad, WH 498, Truman (Personal) to MacArthur, 19 Jul 50. (2) Rad, C 58248, MacArthur (Personal) to Truman, 19 Jul 50.

tion and enough matériel to equip a full division.[33]

Two days later General MacArthur reaffirmed his confidence that he could hold the invading communist armies. Called to a teleconference by the Joint Chiefs of Staff on 24 July and questioned on an enemy move around the left end of his line which resembled the start of a double envelopment, General MacArthur admitted that he lacked the strength to prevent it, but saw it as no serious threat. So long as the North Koreans outnumbered the South Koreans and Americans at a particular location they would always be able to mount enveloping attacks. But their main effort continued to be in the center of the line, and the basic question was whether they had sufficient strength to force withdrawals there. If his own forces could hold the center, General MacArthur would have no special worry about the incipient envelopment. "If our center is unable to hold," he said, "our perimeter will have to be contracted." Referring to his recent statements to President Truman which had predicted losses as well as successes, General MacArthur pointed out that the situation was developing in accordance with that estimate.[34]

General MacArthur's piecemeal commitment in early July 1950 of inadequate American forces weak in firepower, mobility, and reserves against a disciplined, determined, and numerically superior

GENERAL DEAN, *soon to be captured, conferring with General Walker, 8 July 1950.*

enemy constituted a basic violation of U.S. military doctrine. The violation could not be avoided and the consequences had to be accepted. Had General MacArthur waited until his ground units were completely combat-ready before sending them against the North Koreans, the entire peninsula would probably have fallen to the communists. But his mission was to assist the Republic of Korea and to prevent it from falling into enemy hands. He parceled out his available means deliberately and in full knowledge of the risk. At the end of July the situation of American forces in Korea remained precarious. By breaking off with the enemy and retreating swiftly, the battered ground units could have evacuated from Pusan with a good deal of their equipment. Once back in Japan, reconstituted and resupplied, these forces could have joined other units

[33] For the full story of the 24th Division's valiant fight on the Kum River line and at Taejon, see Appleman, *South to the Naktong, North to the Yalu*, Chapters X and XI, pages 121–81.

[34] Telecon, TT 3573, Gens Bradley, Collins, Norstad, and Adm Sherman in Washington with Gen MacArthur in Tokyo, 24 Jul 50, in G–3, DA files.

in a later concerted amphibious assault on Korea at a place of the American commander's choosing. But never did General MacArthur seriously consider a course other than a fighting withdrawal to a beachhead perimeter around Pusan, with his men delaying the enemy to the limit of their abilities until reinforcement arrived. Costly though it proved, this course avoided the loss of prestige and political ill effects of voluntary evacuation, at the same time providing a build-up area on the peninsula for later exploitation.[35]

The extraordinary efforts in Washington and Tokyo during July succeeded in strengthening the unified command in Korea and staving off its complete collapse. The full effects of these efforts, because of distances involved, did not become apparent in Korea until July was nearly over. But with the arrival of new men and new equipment, late in the month, backed by the assured arrival of even greater combat strength in the near future, the odds in favor of ultimate North Korean victory dropped sharply.

[35] The North Korean Premier, Kim Il Sung, later remarked on this American tactic as if it were unfair. He said also, in a last appeal to his faltering forces in October 1950, "The first error we committed was, instead of making a complete siege and annihilating the enemy, we gave them enough time to regroup and increase their strength while retreating." See Order from Supreme Commander, NKA, to All Forces, 15 Oct 50, in ATIS Enemy Docs., Korean Opns, Issue 19, 30 Jan 51, Item 1.

CHAPTER VII

Bolstering the Forces

Shaping the Unified Command

Even before the U.N. Security Council passed its resolution on 7 July, some nations had offered military assistance to the United States for use in Korea. The first offer came from the United Kingdom on 28 June 1950, when the British Government announced that it was placing elements of the Fleet at the disposal of U.S. authorities for support of South Korea. The United States accepted the British naval force without hesitation and asked that it report to Vice Admiral Joy, Commander, Naval Forces Far East.[1]

Almost at the same time, but through diplomatic channels, Australia, Canada, and New Zealand offered naval forces and combat aircraft. The Secretary of State passed these offers to the Secretary of Defense, who called on the Joint Chiefs of Staff for their recommendations. The Joint Chiefs of Staff quickly agreed that these forces should be accepted and the Secretary of State took the necessary steps.[2]

These preliminary offers were encouraging proof of allied support, and on 29 June President Truman told the National Security Council that he wanted to see as many members of the United Nations as possible take part in the Korea action. The Secretary of Defense showed greater reserve, feeling that military necessity might weigh more heavily than political considerations in the decisions to accept or turn down forces offered by member nations. Although Secretary Johnson told the Joint Chiefs that they should lean toward accepting forces offered, he qualified this statement by adding, "to the maximum extent practicable from the military point of view." [3]

Since at this early date only vague outlines of the unified command had appeared, forces were being offered to and accepted by the United States, not the United Nations. Meanwhile, the machinery for processing offers of assistance, in the very likely event a unified command was established, came under study by the Joint Chiefs of Staff. They were convinced that military effectiveness, not

[1] (1) Department of State, *United States Policy in the Korean Crisis*, Dept of State Publication No. 3922 (Washington, 1950), pp. 56–57. (2) Memo, JCS for Chairman, British Joint Services Mission, 30 Jun 50.

[2] (1) MFR, Gen J. H. Burns, OSD, 29 Jun 50, sub: Telephone Msg From Mr. Satterthwait of State Dept to Gen Burns. (2) Memo, Secy Defense for JCS, 29 Jun 50. (3) Memo, JCS for Secy Defense, 30 Jun 50, sub: Proffer of Aid by Foreign Govts.

[3] (1) Truman, *Memoirs*, II, 342. (2) Memo, Secy Defense for JCS, 29 Jun 50.

political necessity, should be the main consideration in accepting forces for Korea, and thus sought a controlling voice in passing on military contributions to the unified command. They told the Secretary of State, through the Secretary of Defense, on 30 June, that if, as appeared probable, Generalissimo Chiang Kai-shek offered troops from Formosa for service in Korea, he should be turned down. To make sure that such an offer was not accepted by the field commander unilaterally, they cautioned General MacArthur to refer any Chinese Nationalist offer to the Department of State, saying, ". . . the decision whether to accept or reject the proffer of military aid by foreign governments should properly be made at the highest levels in Washington." This veiled warning reflected the resolve shared by the Joint Chiefs that the field commander should not deal directly with other nations in any way.[4]

The Nationalist Chinese Government, through its Washington ambassador, had, in fact, already offered to furnish to the U.N. unified command 33,000 soldiers. President Truman was, at first, inclined to accept this offer, but was dissuaded in a meeting with his Defense and State advisers. Secretary Acheson warned of the danger of bringing Communist China into the war if Nationalist Chinese troops entered Korea. On the military side, the JCS deplored the low state of training and lack of equipment of Chiang Kai-shek's men, and pointed out that moving them from Formosa would tie up ships and planes which could be better used elsewhere. He remained concerned over the ability of the small available United States forces to stand off the enemy. After further discussion, however, the President accepted the position of the majority that the Chinese offer should be politely declined.[5]

Secretary Johnson, on 1 July, asked the Joint Chiefs of Staff how he should approach the general problem of military assistance from other nations for the Korean fighting. He wanted to know if the United States should actively solicit other nations for troops and, if so, what kind of troops should be sought. The passing of the United Nations Security Council resolution of 7 July made definite standards for accepting or turning down forces mandatory. Johnson received no answer until 14 July, when the Joint Chiefs told him that a number of unknown factors, including combat efficiency and logistics, made a blanket answer impractical. Because of these very factors they urged that, in every case in which a nation volunteered forces, the Joint Chiefs of Staff be consulted.[6]

They saw that some nations which might offer military forces to the unified command might not have the resources to provide effective fighting forces. To accept forces so poorly trained, equipped, and prepared as to be a military liability in Korea would be unwise. Indiscriminate acceptance of troops, without regard to actual combat needs in Korea, could create an unbalanced military team. The Secretary of Defense assured the Joint

[4] (1) Memo, JCS for Secy Defense, 30 Jun 50, sub: Proffer of Aid by Foreign Govts. (2) Rad, JCS 84737, JCS to CINCFE, 30 Jun 50.

[5] Truman, *Memoirs*, II, 342–43 and 348.

[6] (1) Memo, Secy Defense for JCS, 1 Jul 50. (2) Memo, JCS for Secy Defense, 14 Jul 50, sub: U.S. Courses of Action in Korea.

Chiefs of Staff that he would seek their comments on any force offered for Korea.[7]

As they moved to set up military control over the procedure for accepting forces, the Joint Chiefs of Staff questioned MacArthur in mid-July on his standards for foreign units to be integrated into the United Nations Command. By this time, when it appeared that the U.S. reserve of trained ground forces would be strained to its limit, the Joint Chiefs felt that some other nations should be asked to send ground forces to Korea. He recommended, in an immediate reply, that foreign units should be sent at no less than reinforced battalion strength of about 1,000 men, mainly infantry, but having organic artillery support. He would attach these battalions to his American divisions. If service units were furnished, they should be large enough to be usable at once.[8]

The normal channel through which member nations of the United Nations offered military forces and other forms of assistance to the unified command ran from the Department of State to the Department of Defense to the Joint Chiefs of Staff. A nation offering assistance usually approached the Department of State with its proposal, but made no final offer until after preliminary informal talks. During exploratory conversations the Department of State consulted the Secretary of Defense who, in turn, sought the advice of the Joint Chiefs of Staff. The latter officials weighed the offer against needs in the field and the probable effectiveness of the forces offered, keeping in mind General MacArthur's criteria. Their recommendations bore great weight. If they were favorable, the nation then made a firm offer which was accepted.

Offers of ground combat forces came slowly at first, but gradually increased. By 23 August, the United States had accepted forces offered by seven nations, totaling almost 25,000 ground combat troops. Troops of four more nations had been accepted by 5 September.[9] But most of these troops were a long way from Korea and many would not arrive for months.

Rebuilding the U.S. Army

Rushing thousands of men and officers to the Far East left great gaps in the defenses of the continental United States and completely vitiated, for the moment, American plans for emergency operations in western Europe and other areas vital to the free world. Yet nothing substantive had been done to repair the damage. Nor did the Army's top planners have

[7] (1) JCS 1776/23, Rpt by JSSC, 8 Jul 50, title: U.S. Courses of Action in Korea, in G–3, DA files. (2) Memo, Adm Davis, Dir, Joint Staff, for Secy Defense, 14 Jul 50, sub: JCS Views on Proposed State Dept Request for Assistance in Korea From Certain U.N. Nations. (3) Memo, Secy Defense for JCS, 21 Jul 50. (4) JCS History, The Korean Conflict, ch. III, p. 14.

[8] (1) Rad, JCS 85971, JCS to CINCFE, 14 Jul 50. (2) Rad, C 57957, CINCFE to JCS, 15 Jul 50.

[9] (1) Memo, Col Williams, International Br, G–3, DA, for Gen Schuyler, sub: Status of U.N. Aid as of 23 August, in G–3, DA file 091 Korea, Case 77. (2) Rad, number unknown, DA to CINCFE, 5 Sep 50, in G–3, DA file 091 Korea, Case 59/16. (3) For a detailed, comprehensive account of forces contributed to the U.S. command for the Korean fighting, their operations, and problems arising from their employment, see the following monographs: Maj. William J. Fox, Inter-Allied Cooperation During Combat Operations, Military Hist Sec, FEC, 15 Aug 52; and Maj. Sam Gaziano, Problems in Utilization of United Nations Forces, Military Hist Sec, UNC, 10 Dec 53. Both in OCMH.

any basis for planning to reconstitute the reserve forces.

At a meeting on 12 July 1950 with Secretary of the Army Pace, ranking officers of the Army General Staff complained that they were working in the dark. Lt. Gen. Edward H. Brooks, Assistant Chief of Staff, G–1, told the Secretary that he had already scraped the bottom of the barrel to find men for MacArthur. He had stripped the United States of trained specialists. But until someone told him just how much the Army was going to expand in the face of the obvious threat to American security, he had no way of knowing how many new specialists he should train. General Bolté, the Army's Assistant Chief of Staff, G–3, backed Brooks, charging that, without a clear goal, he too was being forced to operate on a "piecemeal basis." The Army's supply chief, Lt. Gen. Thomas B. Larkin, Assistant Chief of Staff, G–4, told the same story. "Hand-to-mouth" described his supply program, he said, until he knew how many troops were going to Korea and how many would be mobilized to replace them.[10]

Siding completely with the Army General Staff, General Clark, Chief, Army Field Forces, told Secretary Pace that definite planning goals must be established for all aspects of the Army's expansion as soon as possible. Pace assured these officers that he would press for definite guidance from above. "It is urgently necessary that a decision be taken as soon as possible as to the forces to be mobilized, because upon this is predicated the vital and related problems of procurement, training capacity, and the degree of required industrial mobilization," he said.[11]

The Joint Chiefs of Staff, assisted by their special planning groups, were of course involved in comprehensive study of these very problems. They were, in certain respects, dependent on the individual services for recommendations and, in this case, required definite proposals from the Army as to the optimum degree of Army expansion.

The General Reserve

The approximate strength of the General Reserve on 25 June 1950 stood at 140,000. One month later only about 90,000 men and officers remained. Of this number, 15,000 were employed in essential operations at posts, camps, and stations in the United States. Not only had the General Reserve lost 50 percent of its units, but also levies for replacements and specialists had reduced most remaining units to cadre strength. Only the 82d Airborne Division, the 3d Cavalry, and certain antiaircraft artillery units retained immediate combat potential. Yet General MacArthur's calls on the General Reserve continued unabated. His requirements exceeded the 50,000 men already sent and he had asked for 32,000 more by 25 July. The strength levels of the Reserve kept dropping steadily. By 6 August the total infantry strength in the Reserve had fallen to 40,000.[12]

---

[10] Min, 20th mtg, Army Policy Council, 12 Jul 50, in G–3, DA file 334 APC, sec. 1.

[11] *Ibid.*

[12] (1) Memo, CofS USA, for ACofS G–3, DA, 26 Jul 50, sub: Depletion of Army's General Reserve by Requirements for Korea, in G–3, DA file 320.2, sec. I, Case 14/3. (2) Memo, Gen Bolté for Brig Gen David A. Ogden, Chief, Org and Trng Div, G–3, DA, 16 Aug 50, sub: Status of Major Combat Units—Continental U.S., 6 Aug 50, in G–3, DA files.

Throughout July, Department of Defense officials were aware of the situation, and national leaders had assumed, before Korea, that mobilization, if required, would be all-out mobilization of national military resources. The action in Korea fell far short of global war, but proved big enough to involve the greater portion of the nation's active ground forces by the end of the first month of fighting. With American Reserve military strength so weakened, some degree of mobilization became mandatory. The nation's military leaders had to decide the degree of mobilization required and also the best method of recruiting additional effective forces swiftly with the least damage to the nation's morale and economy. The solution had to be reached under pressure and in haste.[13]

Authorized Strength

The actual strength of the United States Army had been somewhat less than its authorized strength when the Korean War began. But even had the Army's vacant ranks been filled, it would have been too small to fight the North Koreans and at the same time meet American commitments elsewhere. The first step in expanding the Army to take care of the immediate task in Korea without sacrificing its primary mission was to raise the Army's authorized strength. Those directly concerned saw clearly that the void created in the General Reserve should, in the interest of the nation's safety, be filled as soon as possible. When they selected the 2d Division, the airborne RCT, and the three medium tank battalions from the Reserve in early July, the Joint Chiefs of Staff told their superiors that these units would have to be replaced. Both President Truman and the Secretary of Defense agreed and on 6 July approved an increase of 50,000. From this first increment, which raised the authorized strength of the Army to 680,000, the Joint Chiefs of Staff set aside enough men for two antiaircraft battalions for the General Reserve. They planned to use the rest, when available, as individual replacements for General MacArthur's forces.[14]

When the President raised the Army's authorized strength to 740,500 a few days later, the Joint Chiefs decided to use part of these 60,500 new spaces to bring units going to the FEC to war strength, to furnish more combat and service units for the FEC, and to replace losses in the FEC. But they set aside enough spaces to activate an infantry division to replace the 2d Division in the General Reserve and to form two more antiaircraft artillery battalions.[15]

This memorandum gives a detailed breakdown of the authorized and actual strengths of General Reserve combat units on 6 August. The 82d Airborne Division, authorized 17,490 men, had 15,805, while the 3d Division had only 5,179 of an authorized 18,894.

[13] A comprehensive study of the many and complex problems arising out of the nation's efforts to mobilize its armed strength widely, with analyses of each major personnel action, is contained in a monograph by Maj. Elva M. Stillwaugh, History of the Korea War, "Personnel Problems." Only the most significant measures will be discussed here. The extremely detailed and involved steps taken by Chief, Army Field Forces, in this early period to raise troops and to mobilize units are set forth in OCAFF, Actions in Support of FECOM, 3 July–30 September 1950, OCAFF, Blue Book. Both in OCMH.

[14] (1) Memo, Gen Bradley for Secy Defense, 7 Jul 50, in G–3, DA file 320.2 Pac, Case 6, I–A. (2) Memo, Secy Defense for Secy Army, same file, Case 19. (3) JCS 1800/97, 6 Jul 50, CofS file 230.2, Case 35.

[15] (1) Memo, Secy JCS for ACofS, 10 Jul 50, sub: Personnel Requirements, SM 1477–50, with attached

By 19 July the Joint Chiefs of Staff had asked for and received a third increase in authorized military strength. The Army limit was lifted to 834,000, a jump of 93,500 spaces. Some of this addition, too, was scheduled for the Far East Command as combat and service support units and replacements. The JCS set aside the lion's share for twenty more antiaircraft artillery battalions and other units to augment the depleted General Reserve.[16]

But a paper army wins no battles and deters no aggressor. The Army's authorized strength had to be transmuted into actual strength quickly. Voluntary recruitment, Selective Service, recall of individual Reservists, and ordering National Guard and Organized Reserve Corps units to active service were means used to fill the Army's manpower needs.

When the Korean War began the Department of the Army was relying almost entirely on volunteers to fill its enlisted ranks. Authority existed for procuring new soldiers through the draft under the Selective Service Extension Act of 1950, but the Army had made little use of it. The increased need for manpower caused the Department of the Army to call in late July for 50,000 draftees to be inducted in September.[17]

Recall of Reserves

Congressional action on 30 June 1950 gave the President the authority to order units and individual members of the Organized Reserve Corps (ORC) and units of the National Guard of the United States into active federal service for a period of twenty-one months.[18] On 19 July President Truman delegated this authority to the Secretary of Defense, who further delegated it to the secretaries of the military departments.[19]

In the case of both officers and enlisted men, the Army established and carried out a policy of recalling individuals from the Inactive and Volunteer Reserves. In order to avoid enfeebling Active Reserve units, already understrength in most cases, and to enable these units, if it became necessary to call them into service, to come on duty in some semblance of

handwritten notation, 1130, 13 Jul 50, sgd SGS (Gen Moore), G–3, DA file 320.2 Pac, Case 7. (2) Memo JCS for Secy Defense 13 Jul 50, sub: Personnel Requirements in Support of Current Opns in FEC, 2d Increment, same file, Case 48.

16 (1) JCS 1800/104, Bradley for Johnson, 18 Jul 50, sub: Fiscal Year 1951 Force Requirements. (2) Memo, Johnson for Secys Army, Navy, and Air Force, Asst Secy Defense (Comptroller), and Gen Bradley, 19 Jul 50. (3) The major units for which the 834,000 Army strength would provide were at this time 8 infantry divisions, 1 armored division, 2 airborne divisions, 8 separate infantry regiments, 4 separate armored regiments, 72 antiaircraft artillery battalions, and 90 combat battalions of other types, i.e., armored, field artillery, and engineer. See JCS 1800/101, 18 Jul 50, in G–3, DA file 091 Korea, sec. III, Book I, Case 48/2. (4) The Congress of the United States controls the size of the armed forces. In this emergency period, approval by the President was a temporary measure, the only feasible procedure in view of the need for speedy action. The President immediately asked for and secured Congressional approval in the form of legislation removing all statutory personnel ceilings and expanding budgetary appropriations. See JCS History, The Korean Conflict, ch. v, p. 20.

17 (1) Rpt, sub: Personnel Procurement, pp. 19–20. (2) Memo, Secy Army for OSD, 25 Jul 50, sub: Additional Selective Service Call. Both in Annual Narrative Hist Rpt, ACofS G–1, 25 June 1950–8 September 1951, copy in OCMH.

18 PL 599, 81st Congress.

19 (1) Rpt, sub: Personnel Procurement, Tab A, in Annual Narrative Hist Rpt, ACofS G–1, 25 June 1950–8 September 1951. (2) Memo, 8 Nov 50, sub: Call of Reserves, in CofS, DA file 320.2. (3) Memo, Secy Defense for JCS, 21 Jul 50.

combat readiness, the Army felt that it should not take their officers and men. True, the men and officers in these units had been receiving pay for attending drills and were, or could logically be expected to be, more ready for active service than Inactive or Volunteer Reservists. Nevertheless, when it became necessary to fill Reserve and Regular units it was deemed necessary to draw on the Inactive and Volunteer Reserves. Persons who were members of the Active Reserve, assigned to units, drilling regularly, and receiving current training were not recalled to active duty as individuals.

Membership in the Inactive Reserves meant, in fact, that officers and men had accepted a Reserve status and all its attached obligations but would not, or could not, spend the time required for training in the Active Reserve. The fact that a man was in the inactive portion of the Reserve did not, however, obviate his obligation to serve if his country needed him. Volunteer Reserves were those members of the Active Reserve who were not assigned to mobilization troop basis units.

Another factor bearing on the problem was that an important provision of Public Law 810, 80th Congress, was in the process of being implemented as of 30 June 1950. This provision required those members of the Volunteer Reserve who had not been sufficiently active to earn the specified minimum number of retirement credit points under the above law would be involuntarily transferred to the Inactive Reserve. The screening of the Volunteer Reserve to determine who should thus be transferred had just begun when the Korean War broke out. It was known, however, that a large number of officers in the Volunteer Reserve would be affected.

When the first order went out for the involuntary recall of individual Reserve officers, no real distinction could be made between the Inactive and Volunteer Reserve since there were so many in the Volunteer Reserve who had been as inactive as those assigned to the Inactive Reserve. The first recall program, authorized by the Extension Act of 1950 of the Selective Service Act of 1948, consequently specified that officers be recalled from either the Volunteer Reserve or the Inactive Reserve without establishing a priority or any other distinction between the two categories.

The Army met numerous problems in recalling Reservists. It had no clear picture of the actual number who would be available for duty. It knew, for example, that on 30 June 1950 it had 416,402 in the Inactive and Volunteer Reserves and 184,015 in the organized units of the Reserve. It did not know, however, how many of these were physically qualified for duty. The required periodic physical examinations for Reservists had been suspended in February 1947. Many more Reservists had to be called for physical examination than the number needed because of the large numbers found physically disqualified. Considerable administrative overhead and delay hindered selections. Further, many Reservists had undergone changes in economic status after entering the ORC which made active duty an undue hardship. The result was authorization of large numbers of justifiable delays which caused further difficulty in filling quotas. Records on Reserve officers were inadequate, and virtually did not exist for enlisted men.

Finally, the recall of Inactive and Volunteer Reservists engendered much ill-will from the public, the press, and the Congress.

Since officers, particularly in company-grade and combat arms, were needed badly, the Department of the Army, on 22 July 1950, appealed to Reserve officers to volunteer for active duty. So few responded that, on 10 August 1950, empowered by the Congressional authority, the Department of the Army recalled involuntarily 7,862 male Reserve captains and lieutenants of both the Volunteer and Inactive Reserves. On the same date it announced a program for recalling 1,063 Army Medical Service officers. These first involuntary recalls of Reserve officers were followed several months later by a larger program affecting almost 10,000 company-grade officers of the combat arms.[20]

The shortage of trained enlisted specialists prompted the Department of the Army to recall, also involuntarily, 109,000 enlisted men from the Reserves during August. All of these men were specialists, slated to fill critical positions.[21]

National Guard Divisions

The only source from which the Army could draw complete, relatively ready, divisions other than from the General Reserve was from the National Guard of the United States. General Collins was extremely reluctant to advise the calling up of National Guard divisions until he was sure that no other solution could be found to the grave manpower situation. His reasons for holding back stemmed from his concern over the great impact upon the economy and morale of home areas of selected divisions. The other members of the Joint Chiefs of Staff in mid-July also opposed federalization of any National Guard divisions so long as it could be avoided.[22]

Many National Guard units were not divisional in nature, had specialized functions, and were made up of specialists and other men trained during World War II. These units appeared to be a likely source of strength for MacArthur's forces, and the Joint Chiefs of Staff, although hesitating to call on National Guard divisions, asked for authority to call to active duty some other National Guard units if required. "The Joint Chiefs of Staff," they told the Secretary of Defense on 14 July,

> are of the opinion that the emergence of the Korean situation cannot be fully met or in time by merely strengthening units already in existence or by filling them with untrained men through the Selective Service process or recruitment. Also it has developed that the requirements for units and personnel cannot be met on the basis of voluntary return of Reserves to active duty for which approval presently exists. . . . The Joint Chiefs of Staff request that the Secretary of Defense obtain at once authority for the three Services to call to active duty, within such personnel ceilings as have

[20] (1) Rad, WCL 34125, DA to ZI Comds, 22 Jul 50. (2) Rad, WCL 37558, DA to ZI Comds, 10 Aug 50. (3) Rad, WCL 37577, DA to ZI Comds, 10 Aug 50. (4) Ltr, DA, 15 Sep 50, sub: Recall of Additional Reserve Officers to Active Duty, AGAO-S 210.4 (ORC), 15 Sep 50.

[21] Hist Summary, 7 Nov 51, sub: Distribution of Enlisted Replacements, prepared by Manpower Control Div, ACofS G-1, DA, p. 2, copy in OCMH.

[22] (1) JCS 1924/20, Rpt by JSPC, 14 Jul 50, title: Estimate of the Military Sit in Light of Events in Korea, in G-3, DA file 091 Korea, I-C, Case 16. (2) MFR, Gen Moore, SGS, DA, 15 Jul 50, in G-3, DA files.

been or may be approved, such selected National Guard units and selected units and individuals of the Army, Navy, or Air Force as may be required to meet the demands of the Korean situation.[23]

More significant reasons than the disruption of regional social and economic conditions lay behind the reluctance of American military planners to call up complete divisions. General Collins, in addressing the Army Policy Council on 25 July, admitted that much public sentiment was developing in favor of a rapid Army expansion, including the calling up of the National Guard. He pointed out that, if the Chinese Communist forces intervened in Korea, the United States would have to federalize from three to six National Guard divisions at once. Calling up divisions immediately, perhaps prematurely, might not be wise. Too, there was no point in building up too rapidly, since the ability to meet American commitments was definitely limited by shipping. He contended that federalization of National Guard units would not help the situation in Korea since it would take a long time for these units to become effective.[24]

The Army Chief of Staff was waiting for an agreement by the Joint Chiefs of Staff on the size and make-up of the forces which they wanted to develop. There had been a difference of opinion as to whether a small, balanced, and mobile expeditionary force for emergencies similar to Korea should be created and maintained in addition to forces for Korea and the General Reserve.[25]

General Bolté nevertheless kept urging General Collins to call up National Guard divisions. At a meeting in his office on the morning of 31 July, General Collins decided to accept his G–3's recommendations. Later that day, at a conference of the Joint Chiefs of Staff, he forcefully proposed that four National Guard divisions and two RCT's be called to active duty. Collins said:

> In view of the world-wide international situation and recent developments in Korea, I have now concluded that we can no longer delay in calling into Federal service certain major units of the National Guard. . . . I had hoped that this step might prove unnecessary, but it is my firm conviction that further delay may have grave results on our ability to insure the security of the United States.[26]

The Joint Chiefs of Staff quickly agreed and recommended to the Secretary of Defense that the National Guard units be called to active duty. This action meant lifting the Army's authorized strength from 834,000 to over 1,000,000. On 10 August, the Army received word of approval. President Truman authorized calling into federal service on or about 1 September four National Guard divisions and two National Guard RCT's. These units would be brought to full strength through

---

23 Contained in Memo, Col Keith L. Ware, ASGS, for Asst Secy Army, 14 Jul 50, sub: Proposed Mobilization of Reserve Units and Calling of Selected Reserve Officers to Active Duty, in CofS, DA file 091 Korea, Case 7.

24 Min, 23d mtg, Army Policy Council, 25 Jul 50, in G–3, DA file 334, Case 7.

25 JCS 1924/20, Rpt by JSPC, 14 Jul 50, title: Estimate of the Military Sit in Light of Events in Korea.

26 (1) Memo, Bolté, ACofS G–3, DA, for Ridgway, DCofS for Admin, 7 Jul 50, in G–3, DA files. (2) Memo, Ridgway for Bolté, 31 Jul 50, in G–3, DA file 320.2, sec. I, Case 15. (3) Memo, Collins for JCS, 31 Jul 50, sub: Increased Augmentation of the Army (above 834,000), in G–3, DA file 320.2, sec. I–B, Book I, Case 8/1.

Selective Service by 1 November 1950 and would be ready for operational employment by 14 April 1951.[27]

The question of which National Guard divisions should be called up had been under study for some time. General Collins had, on 21 July, asked the Chief, Army Field Forces, for recommendations. Less than a week later General Bolté asked General Clark for an expanded study of the same problem. In considering the problem, General Clark leaned heavily upon the continental Army commanders, soliciting their recommendations as to which divisions within their areas were best trained, best equipped, and most ready to go. After careful study, General Clark submitted to the Department of the Army his recommendations of six divisions most appropriate to be called on the grounds of training, manning, equipment status, and general fitness. The divisions recommended in order of priority of selection were the 28th Division (Pennsylvania); the 29th Division (Virginia and Maryland); the 31st Division (Mississippi and Alabama); the 37th Division (Ohio); the 45th Division (Oklahoma); and the 50th Armored Division (New Jersey).[28]

On 31 July, General Ridgway notified General Clark that the Secretary of the Army and General Collins were fearful of the political repercussions unless there was a better geographical spread among the divisions selected. Clark said that he and his advisers had considered this point very carefully, but had given more weight to other factors. They had, for example, looked very closely at the leadership in the particular divisions, wishing to avoid the difficulties experienced at the beginning of World War II when some of the National Guard commanders had been relieved after call-up. They had evaluated the comparative state of training of each division and had also taken into consideration the divisional strengths in men and qualified officers. On this latter point, the Chief, Army Field Forces, felt it important to keep to a minimum the number of filler replacements which would have to be transferred into a particular National Guard division to bring it up to full strength. Ridgway then asked Clark to consider the readiness status of divisions on the west coast since it might be desirable to choose one division from that area.

Later the same day, General Clark learned that four divisions would be chosen. He was asked if he had adjusted his recommendations to conform with the necessity for a geographical spread. At that time he recommended that four divisions be chosen from among the 28th (Pennsylvania); the 29th (Virginia and Maryland); the 31st (Mississippi and Alabama); the 37th (Ohio); the 40th (California); and the 45th (Oklahoma).

The National Guard divisions finally called into service as of 1 September 1950 were the 28th, the 40th, the 43d (Rhode Island and Connecticut), and the 45th. Also called were the 196th RCT (South Dakota) and the 278th

---

[27] (1) JCS 2147/3 and Incl, Memo, Secy Defense for Secy Army and JCS, 10 Aug 50. (2) MFR, CofS USA, sub: Request for Four Divs in Korea, in G–3, DA file 320.2 Pac, sec. I–3, Book I, Case 19/7.

[28] OCAFF Rpt, Actions in Support of FECOM, 3 July–30 September 1950, OCAFF, Blue Book, entries of 21 Jul, 27 Jul, and 31 Jul 50, copy in OCMH.

RCT (Tennessee). These units would be brought to full war strength. But General Collins directed Bolté to limit the number of troops called up to support the divisions. He felt that this restriction would not involve great risk, since the Joint Chiefs of Staff had made no commitment to send the new divisions overseas. If it should become necessary to send them to Korea later, they could get by with a far smaller ratio of corps and army support troops than had been needed in World War II. General Collins based this theory on his appraisal of the terrain conditions and limited road nets in the Korean area. If the new divisions reverted to inactive status before deployment, the Army would store their equipment to have it immediately available for another emergency.[29]

The Theater Scene—August 1950

In Korea, meanwhile, ROK and U.S. forces fought off the North Korean Army with stubborn determination. General Walker used his small mobile reserves with great skill and his men, ROK and American, fought bravely. The dearly acquired battle experience and the fresh strength pouring into Korea began to show in greater enemy losses and a slackening of his advance. Nevertheless, the Eighth Army lost ground and fell back toward Pusan.

Walker proved a determined and tenacious commander. He well appreciated the great danger of pulling back upon his base of supply under continuous pressure. He hated to give up any more ground to the North Koreans, but on 26 July, with the enemy pressing in on Taegu where irreplaceable signal equipment was in danger of being lost, Walker called Tokyo and asked permission to move his command post back to Pusan. He did not imply in any way that he wanted to pull his divisions back to the port city.[30]

General Almond, who took Walker's call, told him that he, personally, objected to any such move. To remove the command post to Pusan would damage the army's morale. It might give the impression that the Eighth Army could not stay in Korea and might trigger a debacle.

As soon as Walker hung up, Almond went to MacArthur and recommended that MacArthur fly to Korea and talk to Walker at once. Apparently, Walker's attitude had shaken Almond's faith in the Eighth Army commander's judgment. Almond told MacArthur that he felt the situation in Korea had reached the critical stage and required MacArthur's personal observation. MacArthur pondered briefly, then told Almond that he would make the trip the next day.

On 27 July, MacArthur, with a staff including General Almond, landed in Taegu about 1000. This time, MacArthur did not visit the front line, contenting himself with conferences in Taegu. The most significant conference took place between MacArthur and Walker. Only one other person, General Almond, sat in on this 90-minute meeting.

MacArthur did not mention Walker's request of the day before, nor did he

[29] Min, 26th mtg, Army Policy Council, 2 Aug 50, in G–3, DA file 334 (APC) 1950, Case 5.

[30] Interv, Lt Col Roy E. Appleman with Gen Almond, 13 Dec 50, copy in OCMH.

AMERICAN ARTILLERYMEN *fight off enemy efforts to break through Pusan Perimeter.*

criticize Walker for any of his actions. He merely talked over the tactical situation, emphasizing that Eighth Army must hold its ground. He told Walker that withdrawals would cease. Later, in the presence of several members of the Eighth Army staff, MacArthur said that there would be no evacuation from Korea—there would be no Dunkerque.

On 29 July as a result of MacArthur's visit, Walker issued a widely publicized order, in the form of a public statement during a speech to the staff of the 25th Division. Walker stated that the Eighth Army would retreat no more, that there was no line to which it could retreat, and that, in effect, every man in Eighth Army would "stand or die" along the present line.[31]

The defensive line behind which Walker intended his troops to "stand or die" lay mainly on the Naktong River barrier in the west and fanned out from

[31] (1) *Ibid.* (2) Ltr, Landrum to Appleman, recd 23 Nov 53. (3) War Diary, 25th Div, G-3 Jnl, Jul 50, Div Hist notes.

Pusan. Rectangular in shape, measuring nearly 100 miles from north to south and about fifty miles from east to west, the area quickly became known as the Pusan Perimeter. (*See Map I.*)[32]

Between 1 and 4 August, U.S. and ROK units withdrew behind this line and prepared for a last-ditch stand. Most of the western edge of the perimeter was traced by the Naktong River with the exception of about fifteen miles at the southern end of this line. The northern border ran through the mountains above Waegwan and Uisong to the sea, with the town of Yongdok forming the eastern anchor. ROK troops held this portion of the line.

General MacArthur sent his deputy chief of staff, General Hickey, into the Pusan Perimeter on 6 August to confer with the Eighth Army commander. Walker told Hickey he was worried about the condition of the 24th Division. He appraised that unit's combat worth as negligible after a month of hard fighting. Before it could become effective again, it would have to be completely rehabilitated. His other divisions were in somewhat better condition. The 25th Division, which had seen less action than the 24th and which had been less severely attacked by the enemy, was in fairly good shape. General Walker expressed some doubts as to its offensive capabilities, as he felt it lacked leadership. The Eighth Army commander told General Hickey that, because they were too few, all his army staff members were overworked. That they were not getting enough rest was being reflected in the quality of their work.[33]

The first weeks of August were marked by savage North Korean efforts to break through the Pusan Perimeter. Several enemy penetrations across the Naktong into Eighth Army's lines came perilously close to success, but in each case skillful deployment of reserves along interior lines enabled Walker to contain and beat back the enemy thrusts. Fresh units arriving in the perimeter were quickly thrown into the fight at key points in the perimeter. Elements of the 2d Division arrived from the United States on 31 July, the 5th RCT reached Korea on the same day from Hawaii, and the 1st Provisional Marine Brigade closed at Pusan on 3 August.

The mounting toll of American casualties and the depleted ranks of Walker's divisions underscored the great need for fresh fighting men in Korea And every feasible means of meeting this need was being exploited by the Department of the Army.

Replacement Troubles

By 5 August the Department of the Army had stepped up both air and water transportation to the Far East Command, using military and commercial planes and vessels. Most of the surface shipping space had been taken for units and equipment, but airlift brought 340 replacements each day. Still, the Eighth Army was receiving more casualties than re-

[32] For a complete account of the valiant stand of Walker's forces in the battle of the Naktong during August and September 1950, see Appleman, *South to the Naktong, North to the Yalu*, Chapters XV–XXIV. See also Lynn Montross and Capt. Nicholas A. Canzona, *U.S. Marine Operations in Korea, 1950–1953*, vol. I, *The Pusan Perimeter* (Washington, 1954).

[33] Memo, Gen Hickey, DCofS GHQ, UNC, for Gen Almond, CofS GHQ, UNC, 7 Aug 50, sub: Rpt of Visit to Korea, in CofS GHQ, UNC files.

U.S. GROUND TROOPS *board a transport plane at a Japanese air base for shipment to Korea.*

placements. Losses by 5 August totaled 7,859, but only 7,711 individual replacements had reached the FEC and only part of these had arrived in Korea. General Beiderlinden, MacArthur's personnel chief, took an optimistic view, believing that the near future would bring a marked improvement in the situation. He expected casualties in Korea to decrease as the front stabilized and anticipated a great increase in replacements from the United States by the middle of August. He was counting also on returning to combat many soldiers who had recovered from wounds in FEC hospitals. As an example, the number of men returned to combat from hospitals on 4 August equaled 30 percent of the casualties received on the same day. He told Almond that the Department of the Army appeared to be providing replacements to the limit of its capability. His greatest concern, justified in light of the latest report from Washington, was whether there would be a sufficient reservoir of replacements in the United States to keep supplying the FEC's needs until Selective Service,

National Guard, and Reserve personnel could be called to duty and made available.[34]

The optimism expressed by Beiderlinden on 5 August disappeared with startling speed two days later. General Hickey's talk with General Walker erased the slightly optimistic picture conjured by statistics and promises. General Beiderlinden appealed to General Almond on 7 August, pointing out that every division in Korea was suffering critical shortages of men and officers. Almond approved an urgent call on Washington for 8,000 men to reach the FEC within fifteen days. All infantry regiments in Korea were so weakened that unless these men reached them in two weeks, they would deteriorate so badly that major steps would be necessary to rebuild them. Most urgently needed were infantry and artillery soldiers, and company-grade officers. Almond urged, as a matter of highest priority, that airlift be expanded to get the 8,000 men to the theater by 20 August.[35]

The lack of replacements for Eighth Army's divisions resulted to a degree from the way in which replacements were used after they reached the Far East Command. Less than half of the 16,000 replacements arriving in Japan between 1 July and 15 August went straight to Korea. Some were used to fill the 7th Division, but more were assigned to nondivisional units within Japan. About 25,000 men and officers under control of Eighth Army remained in Japan at this time.[36]

The fighting in Korea prompted staff agencies of GHQ FEC to seek more people. They took experienced replacements, particularly officers, out of the pipeline to Korea. At the same time, GHQ section chiefs kept at desk jobs many of their original men and officers who could have been sent as replacements. At other stations in the replacement stream from Japan to the battlefront, men and officers intended for combat duty were diverted to administrative and rear-echelon service. General Beiderlinden warned fellow members of the GHQ staff about allowing this practice to grow. General Headquarters could hardly justify its strident pleas for replacements if it kept these men from the fighting units. On 15 July he cautioned, "Until a flow of replacements commensurate with current critical needs materializes, it is mandatory that . . . the tendency to augment administrative and rear-echelon service organizations . . . be resisted." He urged the fullest use of Japanese and American civilians in Japan.[37]

This chiding did not deter GHQ section chiefs. General Beiderlinden told the chief of staff, GHQ, in early August that he was still worried by the con-

[34] Memo, G–1 GHQ for CofS GHQ, 5 Aug 50, sub: Casualties and Replacements, G–1 GHQ Log, Item 41, 5 Aug 50.

[35] (1) *Ibid.*, 7 Aug 50, sub: Loss Replacements, G–1 GHQ Log, Item 15. (2) Rad, CX 59519, CINC FE to DA, 7 Aug 50.

[36] (1) Memo, G–1 GHQ for CofS, 16 Aug 50, sub: Replacements for EUSAK, G–1 GHQ Log, Item 45. (2) Memo, Col Grubbs for Gen Beiderlinden, 8 Aug 50, sub: Assignment of Replacements, G–1 GHQ Log, Item 36. (3) Memo, GHQ for Gen Hickey, 12 Aug 50, sub: Replacements for EUSAK, G–1 GHQ Log, Item 54.

[37] Memo, G–1 GHQ for All Staff Secs, GHQ SCAP, and FEC, 15 Jul 50, sub: Utilization of Personnel, G–1 GHQ Log, Item 7.

tinuing trend toward empire-building in the GHQ staff. He felt that, instead of looking for more people, the GHQ staff sections should get more mileage out of those they already had. He hesitated to charge the other staff heads with wasting their resources, but he believed that they could, if they tried, achieve greater efficiency without strength increases. At General Beiderlinden's request, the chief of staff talked with section chiefs, stressing the importance of keeping GHQ manpower requirements at as low a level as possible.[38]

So urgent was the need for front-line soldiers in August that General MacArthur cut out the short, intensive training course which had been set up on 14 July for replacements at Camp Drake. He ordered replacements kept at Drake only long enough to receive their individual equipment. As a result of this ruling, replacements were given no chance to fire their individual weapons. Many men went into the front lines in Korea without having determined the characteristics and proper setting of their rifles or carbines.[39]

General Collins sent General Ridgway to Korea in early August to find out from MacArthur what specific requirements had developed since General Collins' July visit. General Collins gave Ridgway a personal letter to be handed to MacArthur which, he hoped, would serve to explain the Army's situation and to reassure MacArthur that everything possible was being done on his behalf.

> In order to meet your requirements for four divisions with supporting units [Collins wrote] we decided to recommend to the Joint Chiefs of Staff calling for four National Guard divisions to active duty on or about 1 September 1950. . . . On 1 August I recommended the Joint Chiefs call up those units. The Joint Chiefs of Staff approved, but reserved judgment as to definite commitment of all four divisions to your theater at this time. This was based on the fact that no one can definitely foresee the exact developments of the Korea fighting.
>
> I have felt all along that once the weather clears up and we are able to get effective results from our air attacks, the logistic support of the North Korean forces will rapidly dry up. This might result in your being able to pass to the counteroffensive more nearly according to your original time schedule and your original plans.
>
> You will recall that we agreed that this might be possible with troops already definitely allotted to you which, including the full Marine division, and an airborne combat team would aggregate almost seven divisions. On the other hand, if the North Koreans are continually reinforced from the North you may well require the full strength of units requested. . . .
>
> I am confident that the Joint Chiefs of Staff will be willing to accede to a definite request for these troops when the situation has stabilized and you are able to make more definite plans than is possible now. Meanwhile we will proceed with the training of the divisions quickly. They will be permitted to accept volunteers up until the time of actual induction. . . . Here again I think we must wait and see how the North Koreans react during the next couple of months. I think it is wholly possible that once they begin to fold, and I am sure they will under the pressure of your counteroffensive, that they may go very fast. . . .
>
> Let me assure you again of my warmest support. If there is anything we are doing

---

[38] Memo, G-1 GHQ for CofS GHQ, 2 Aug 50, sub: Requests for Increase of Staff Personnel (Instant Case, G-3), G-1 GHQ Log, Item 36.

[39] (1) Rad, 59867, CINCFE to CG Eighth Army, 12 Aug 50. (2) Memo, GHQ for CofS GHQ, 31 Aug 50, sub: Rpt of Staff Visit to Personnel Pipeline, sgd Col T. A. Seely, GSC, G-1 GHQ Log, Item 14.

now that should be changed or anything further that we could do to back you up in this critical struggle please don't hesitate to call on me.[40]

General MacArthur made his needs known to General Ridgway at once. He repeated the call already made by his staff for 8,000 replacements by 20 August. When Ridgway passed this information to General Collins, he expressed the belief that the Department of the Army could meet the full requirement. The enlisted Reserve specialists, particularly those with prior service, could, with a minimum period of three weeks for processing and training, be sent to the FEC by September and would help cut down the shortages significantly. General MacArthur had suggested that the United States triple its transpacific shipping by using commercial shipping lines.[41]

The principal request which the Far East commander placed upon the Department of the Army through General Ridgway was for the 3d Division. In the relatively near future, Japan would be completely stripped of American combat troops. So that the Japanese islands, doubly vital now as a support base for Korean operations, might not be completely defenseless against a possible Soviet attack, General MacArthur felt that the 3d Division should be sent to Japan by mid-September.[42]

When General Ridgway returned to Washington, he met with the Army Policy Council and, at the request of the Secretary of the Army, reported his observations on the combat situation. Ridgway had come away from Korea convinced that Walker would hold the Pusan Perimeter. Enemy pressure was still great enough to force limited tactical withdrawals from the edges of the perimeter and the actual final line had not yet been developed, but the defensive line would be held successfully and the beachhead kept intact. Regardless of his favorable prognosis, General Ridgway was quick to point out that General Walker had a serious problem. His forces still faced a ruthless and savage foe. Any idea that the North Koreans would weaken or fall back was faulty and dangerous. As an example, General Ridgway cited enemy reaction to the strongest offensive thrust yet made by Walker's forces. Eight American battalions had attacked in the southern sector to stop an enemy move at Pusan. Within an hour after the attack jumped off, the enemy counterattacked fiercely and effectively.[43]

United Nations forces were still too few in number to carry on a defense according to the book. One division held a 21,000-yard front with six battalions. The enemy could infiltrate the thinly defended front at night and attack from the rear the next morning. General Walker had not had time to organize the ground effectively. General MacArthur had told Ridgway that he was pleased with the support given him by Washington, but had asked for more. After Ridgway reported to the council,

[40] Ltr, Gen Collins to Gen MacArthur, 4 Aug 50, in CofS, DA file 323.3 FEC.

[41] Memo, Gen Ridgway for Gen Collins, 18 Aug 50, in G–3, DA files.

[42] The National Police Reserve of Japan (NPRJ) had been formed only recently, while American forces left in Japan after September were mainly service and headquarters troops.

[43] Min, Mtg of Army Policy Council, 8 Aug 50, in G–3, DA file 334 APC, Case 7.

General Collins told Secretary Pace that the request for more men and units was already being studied by his staff, but that he was gravely concerned by the demands.[44]

At a special meeting of the Joint Chiefs of Staff held later the same day to hear General Ridgway's formal report and to consider the Far East commander's needs, General MacArthur's request for another division occasioned a debate. Some members of the Joint Chiefs wanted to send the 82d Airborne or a National Guard division instead of the understrength 3d Division. General Ridgway recommended that the 3d Division be sent since he felt that the combat-ready airborne division must stay in the United States for use in a general emergency. After a 15-minute discussion, the tenor of thought among the Joint Chiefs inclined toward the same view—namely, to send the 3d Division and to fill it up from any and every source. No final decision was made at this time, but General Collins and Admiral Sherman were charged with examining the matter urgently and reaching a recommendation by 10 August.[45]

General Bolté, Army G–3, did not believe that the 3d Division could be filled and sent to General MacArthur without seriously delaying the Army's plans for rapid expansion of training activities in the United States. He told General Collins that the 3d Division could reach the Far East by 15 September, untrained and worthless for combat, but that the training and mobilization base in the United States would suffer as a result. If General Collins could see his way clear to delay the division until December, it could be built up with National Guard and Enlisted Reserve Corps (ERC) fillers without ruining the ZI training base and could arrive in the Far East as a reasonably well-trained division. If General Collins considered it absolutely necessary to give General MacArthur another division by 15 September, the 82d Airborne could be sent. According to General Bolté, the 82d, already at about 85 percent strength, would not need many fillers. Furthermore, it would be ready to fight on arrival. Its departure, of course, would leave the continental United States without a combat-ready division.[46]

General Bolté's views did not prevail. The JCS decided to send the 3d Division to FECOM. On 11 August President Truman approved its removal from the General Reserve.[47]

The 3d Division, although it had three

[44] *Ibid.*

[45] Memos (handwritten), Lt Gen Alfred M. Gruenther, DCofS for Plans, for Gen Bolté, ACofS G–3, 1110, 8 Aug 50; 1125, 8 Aug 50; 1150, 8 Aug 50. All in G–3, DA file 320.2 Pac, Case 19/7. This series of penciled notes sent out of the JCS meeting by General Gruenther reports the progress of the meeting to the Army G–3, so that, in Gruenther's own words, "you won't get crash-landed" and "just to keep you off balance."

[46] (1) Memo, Gen Bolté for Gen Collins, 10 Aug 50, sub: Feasibility of Redeployment of 3d Inf Div to the FEC by Mid-September, in G–3, DA file 320.2 Pac, Case 19/7. (2) Transfer of men and officers from the 82d to the 187th RCT of the 11th Airborne, which was being readied for shipment at this time, had reduced the division to an approximate strength of 15,000. See Memo, Gen Bolté for Gen Collins, 8 Aug 50, in G–3, DA file 320.2 Pac, Case 6/20.

[47] (1) Memo, Ridgway for Collins, 18 Aug 50. (2) Memo, Bolté for Collins, 8 Aug 50, in G–3, DA file 320.2 Pac, sec. I–A, Book I, Case 6/20. (3) JCS 2147/4, Note by Secys, 10 Aug 50, title: Reinforcement of the FEC, in G–3, DA file 320.2 Pac. (4) Memo, Johnson for JCS, 11 Aug 50, sub: Reinforcement of the FEC.

regiments, was very much understrength. Already it had furnished many men, officers, and units to the Far East Command. The division was short 2 infantry battalions, 1 tank battalion, and 2 field artillery battalions. Only drastic measures would place the division in a reasonably effective status, even for occupation duty. By reducing one regiment to zero strength and dividing its men and officers between the remaining two regiments, then assigning a separate regiment from Puerto Rico to the division, the Department of the Army succeeded in building up the division to a semblance of operational strength.

On 10 August, General MacArthur learned that the 3d Division, less one regiment, was being ordered to his command. A supplementary message, explaining that the 65th Infantry from Puerto Rico had been ordered to the FEC, where it would join the 3d Division as its third regiment, followed a few minutes later, but not quickly enough apparently. Before receiving the information on the 65th Infantry, MacArthur fired back a radio objecting to the dispatch of a 2-regiment division and pointing out, ". . . experience indicates the ineffectiveness of a two unit organization whether in battalions, regiments, or divisions." No answer to this reclama was necessary, of course.[48]

Fearful, also, that press reports of the planned movement of the 3d Division might tip his hand and warn the North Koreans of his future plans, General MacArthur asked that no press release be made until the division was actually engaged in combat. "Information of this sort," General MacArthur warned Washington, "practically reveals our strategic concepts to an alert enemy." [49]

Unfortunately, General Ridgway had already alerted the Army Chief of Information, Maj. Gen. Floyd L. Parks, to release the information on the 3d Division to the press. But the information had not yet gone out when MacArthur's warning was received. General Ridgway was opposed to withholding any such news from the public. "I saw no possibility short of instituting a strict censorship," he said, "of concealing the fact and if we acted otherwise, press reaction would be violent and prompt." When he went to General Collins and expressed this opinion, Collins considered a few moments, then decided to go along with MacArthur anyhow. Ridgway was obliged to notify Parks to make no official release on the 3d Division even though both men knew that the news would leak out at once.[50]

General Collins was determined that there should be no misunderstanding as to the great significance of removing the 3d Division from the United States or to certain restrictions on its combat employment. He sent a personal reminder to General MacArthur underscoring both the risk taken by the Army in sending out the division and the need for special handling of the unit on arrival. "In withdrawing this division from the General Reserve," General Collins pointed out, "the Joint Chiefs of Staff

[48] (1) Memo, Ridgway for Bolté, 10 Aug 50, sub: Additional Combat Forces for FEC, in G–3, DA files. (2) Rad, WAR 88401, DA to CINCFE, 10 Aug 50. (3) Rad, WAR 88465, DA to CINCFE, 10 Aug 50. (4) Rad, C 59863, CINCFE to DA, 11 Aug 50.

[49] Rad, C 59820, CINCFE to DA, 11 Aug 50.

[50] MFR, Ridgway, 11 Aug 50, in CofS, DA file 370, Case 12.

have accepted for the next few months a further serious reduction in the United States capabilities to meet other possible demands for combat ground forces, as well as a further serious reduction, during the same period, in the Army's capability to train additional forces for your theater." The Joint Chiefs were sending the 3d Division with the understanding that it would serve for the time being in Japan, as a theater reserve. They were assuming also that General MacArthur would, because of the division's very low combat effectiveness level, permit it "sufficient training time to reach a minimum acceptance training level" before committing it to battle.[51]

Late in August, after comprehensive inspections of the 3d Division, its ranks now swelled from a low of about 5,000 to over 11,000, General Clark, Chief, Army Field Forces, reported the division to be about 40 percent combat-ready. There were no major equipment shortages, and since the division was believed to be structurally sound General Clark felt it could be brought to an excellent state of combat readiness in about two and a half months.[52]

Corps Headquarters

By late July, it had become apparent that U.N. forces, comprising American divisions, ROK divisions, and units expected from member nations of the United Nations, would soon be so numerous that tighter tactical control would be necessary. In anticipation of such a development, General MacArthur, on 19 July, called on the Department of the Army for two corps headquarters. He asked that these headquarters be sent as soon as possible with attached medical and military police units and with two signal battalions. If feasible, these two headquarters should be designated I and IX Corps.[53]

A few days later, General MacArthur revealed that his plans called for using one of these corps headquarters for an amphibious enveloping force, and stated that the operation could be deferred to no later than 25 September. Although General MacArthur had not said specifically what use he intended to make of the other corps headquarters for which he had asked, the Department of the Army planners assumed that it would be placed under Eighth Army to serve in the breakout and exploitation phase following the initial amphibious assault.

Officers of the DA G–3 section conferred on the matter with officers from Army Field Forces and determined that the Army could produce only one corps headquarters by the target date. The available corps (U.S. V Corps) was at 75 percent combat effectiveness. Only one signal battalion, the 4th, suitable for employment with a corps headquarters, was in active service in the United States, and it was at 60 percent strength. A lack of critical signal specialists made its estimated combat effectiveness 50 percent. Chances for a second corps looked slim to G–3's planners, particularly in view of the fact that no other corps signal battalion was on duty in the United States

[51] Rad, W 88954, DA to CINCFE, Collins (Personal) for MacArthur, 12 Aug 50.

[52] Rad, OCAFF 810, Chief AFF to CofSUSA, 24 Aug 50.

[53] Rad, CS 58234, CINCFE to DA, 19 Jul 50. I and IX Corps had served under General MacArthur in Japan but had been inactivated in early 1950 as an economy measure.

and at least six months would be required to train one. They concluded that furnishing one corps headquarters with corps troops to the U.N. commander for use in the planned amphibious operation was the maximum capability of the Army. The tasks for which the other corps was slated would have to be given to Eighth Army.[54]

The Army Vice Chief of Staff, General Wade V. Haislip, disagreed vehemently. In his opinion, a second corps headquarters could most certainly be formed insofar as the staff personnel were concerned. Nor did he accept the G–3's position that it would take six months to train a signal battalion. He pointed out that the signal battalion to be used in defensive operations need not be so highly trained as one slated for offensive amphibious operations and directed G–3 to restudy the problem.[55]

As a result of General Haislip's interest, the Department of the Army told General MacArthur that it would be possible to activate and send to him a second corps headquarters, untrained but having all required staff members. An additional signal battalion could be called into service and made available in six months. Or, if he wished, this battalion could be sent, untrained and at little more than cadre strength, in two months. General MacArthur asked at once for the earliest movement of the first corps (I Corps) and for immediate activation and dispatch of the second (IX Corps). He asked that the second signal battalion be called in and sent to him at once regardless of condition.[56]

On 30 July the V Corps was redesignated as the I Corps and began to prepare for movement, less certain cadre personnel, to the Far East Command in early August. The 4th Signal Battalion was to accompany the new corps headquarters. Meanwhile, in response to a request from General MacArthur that the corps commander and his planning staff come by air to Tokyo to plan the details of the forthcoming amphibious operation, General Coulter, the commanding general, and selected members of his staff landed at Tokyo on 10 August.[57]

The IX Corps, activated by Fifth Army, was to be prepared to move by 15 September. No training time was allowed. The 101st Signal Battalion was called into service on 19 August to meet the requirement for an accompanying signal unit.[58]

In mid-August, General MacArthur was notified that I Corps headquarters and headquarters company, medical, and military police units, and the 4th Signal Battalion at reduced strength were ready to sail for his command. The signal battalion could not be brought to full

[54] Memo, G–3 for CofS, 25 Jul 50, sub: Request for More Troops for FEC, in G–3, DA file 320.2 Pac, Case 16/4.

[55] Memo, Col Morse for Col Howell, 28 Jul 50, sub: Corps Headquarters Requested by Gen MacArthur, in G–3, DA file 320.2 Pac, Case 16/4.

[56] (1) Rad, WAR 87493, DA to CINCFE, 29 Jul 50. (2) Rad, CX 59246, CINCFE to DA, 29 Jul 50.

[57] (1) Memo, quoting DA radio for CG Third Army, Info CINCFE, G–1 GHQ Log, Item 7, 31 Jul 50. (2) Rad, WAR 88025, DA to COMGEN V Corps, 4 Aug 50. (3) Rad, CX 58926, CINCFE to DA, 28 Jul 50. (4) Rad, HICPAC 583, GHQ LNO to CINCFE, 10 Aug 50.

[58] (1) Memo, Gen Bolté for Gen Collins, 29 Jul 50, sub: Request for More Troops for FEC, Tab A. (2) *Ibid.*, 8 Aug 50, same sub, in G–3, DA file 320.2 Pac, Case 16/4.

strength before 1 November. The IX Corps, less its signal battalion, could sail in about a month but would be untrained. The IX's battalion could, if trained Enlisted Reserve Corps fillers materialized as expected, sail for the Far East Command about 1 November, but if trained as a unit in the United States would not be ready until the end of 1950.

Artillery elements of both corps, including the additional nondivisional artillery units which General MacArthur had requested earlier and were being activated from Reserve and National Guard sources, would be only partly trained if they sailed with the other corps elements. The Department of the Army suggested that, since MacArthur's requirement for this artillery was not immediate, the units be kept in the United States and trained until ready to fight.[59]

General MacArthur apparently felt that, in this case at least, a bird in the hand was worth two in the bush. He wanted the corps as fast as he could get it regardless of condition. "Walker is now controlling four United States and five ROK divisions," he pointed out. Believing that the green units could get their training faster under him than in the United States, he asked that they be sent to him as soon as they had been filled to authorized strength. His request applied to all organic and attached elements of both I and IX corps.[60]

Late in August, arrangements were sufficiently advanced for a schedule giving anticipated arrival dates of the corps units to be sent to General MacArthur. The I Corps with attached units, including the 4th Signal Battalion at reduced strength, was on the high seas and due to reach Japan on 3 September. The IX Corps headquarters would arrive in Pusan about 10 October and would be followed within three weeks by the artillery units and the 101st Signal Battalion.[61]

Reorganization, Far East Command

Aware that General Walker could ill afford to divide his attention between the battlefield and his responsibilities in Japan, General MacArthur on 24 August established a new and separate command relieving the Eighth Army commander of all duties not directly related to his combat mission. He directed the establishment of Japan Logistical Command (JLC), FEC, with headquarters located in Yokohama in the buildings vacated by Eighth Army. By this order, responsibilities and functions formerly assigned General Walker within the geographical areas of the four main islands of Japan were delegated to the commanding general of JLC, General Weible. Excluded from his jurisdiction, although within these geographical limits, were posts, camps, and stations assigned to the Commanding General, Headquarters and Service Command; General Headquarters, FEC; COMNAVFE; and the Commanding General, FEAF.[62]

On 28 August, with the concurrence of GHQ, FEC, General Weible established a subordinate command, the Northern, at Sapporo, Japan. The Com-

[59] Rad, WAR 88864, DA to CINCFE, 15 Aug 50.
[60] Rad, C 60346, CINCFE to DA, 17 Aug 50.
[61] Rad, WAR 89882, DA to CINCFE, 26 Aug 50.
[62] GHQ, FEC, GO 22, 24 Aug 50.

manding General, Northern Command, Brig. Gen. Edwin W. Piburn, was made responsible for the island of Hokkaido and certain areas on the northern portion of Honshu. Somewhat later, on 19 September 1950, another subcommand of JLC was set up, designated as the Southwestern Command with headquarters at Osaka, Japan. Brig. Gen. Carter W. Clarke was named commanding general of this new command with a zone of responsibility including the islands of Shikoku and Kyushu and all areas of Japan located southwest of Shiznoka and Nagano prefectures, exclusive of those assigned to the British Commonwealth occupation forces and of posts, camps, and stations under control of the Commanding General, FEAF, and COMNAVFE.[63]

In addition to functions in support of the occupation of Japan, the Japan Logistical Command took over the task of getting all supplies from Japan to Korea. The new agency, actually a communications zone command for the Eighth Army, received requisitions for supplies from Walker's headquarters, placed requisitions on the proper agencies in the United States, and processed and transported all supplies to the combat theater, leaving Walker's forces free to fight without worrying about administrative matters in Japan.

The Chief of Staff, United States Army, toured the Pusan Perimeter in late August, visiting all American divisions and conferring with the army commander. He found the morale of the troops at the front to be uniformly high and the major commanders confident and optimistic. But there had been no letup in the enemy's determined pressure. The point of greatest concern to General Walker was still the slow arrival of replacements in the combat zone. He told General Collins, on 22 August at Taegu, that the replacement flow was replacing only about 75 percent of actual Eighth Army losses and his units were fighting at less strength than that authorized them when they came to Korea.[64]

On the brighter side, the North Korean Army had assumed an unbalanced and vulnerable disposition. By the end of August, virtually all enemy combat troops were south of the 37th Parallel and being supported over long, exposed lines of communications. UNC air and naval units, now in complete command of the sky and sea around Korea, kept these exposed routes under constant attack so that North Korean logistical problems worsened daily.

General MacArthur, foreseeing the enemy's vulnerable disposition, had decided early in the war that the old precept, "Hit 'em where they ain't," fitted such a situation perfectly. The golden chance to strike deep behind the enemy's mass, cut his lines of supply, then smash his front-line divisions by attacking from two directions was enticing to the general who, in World War II, had proved so well the value of amphibious envelopment against the Japanese.

Indeed, a seaborne strike against the North Korean rear had long seemed the logical solution to MacArthur. Of

---

63 (1) JLC GO 10, 28 Aug 50. (2) JLC GO 58, 18 Sep 50.

64 Memo, Col Everett for ACofS G–3, 8 Sep 50, sub: Rpt of Visit to FEC and USARPAC, 19–30 August 1940, in G–3, DA file 333 Pac, Case 5.

course, before such a blow could be struck, General Walker had to halt the North Korean Army short of Pusan and General MacArthur had to build an amphibious force almost from the ground up. By the opening of September, both generals had progressed considerably in meeting these essentials.

CHAPTER VIII

Operation CHROMITE: The Concept and the Plan

MacArthur had decided on an amphibious operation against the enemy even before the first clash between American and North Korean soldiers at Osan. On 2 July he asked Washington for a Marine RCT. On the next day he ordered 1,200 specially trained operators for amphibious landing craft. He asked on 5 July for an engineer special brigade trained in amphibious operations and on the same day called for an airborne RCT "to participate in planned operations from 20 July to 10 August." [1]

MacArthur had conceived these "planned operations" a few days after the North Koreans struck. MacArthur then believed that he could land an assault force from the 1st Cavalry Division and the Marine RCT against the enemy's rear at Inch'on as early as 22 July. This force would envelop Seoul and seize the high ground to the north. At the same time, all forces available to General Dean would attack to drive the North Koreans back against the Han. Maj. Gen. Edwin K. Wright's planning group, JSPOG, worked out the details of this early plan. They assigned to it the code name Operation BLUEHEARTS.[2]

General MacArthur on 6 July called Maj. Gen. Hobart R. Gay, commander of the 1st Cavalry Division, to Tokyo and told him of the plan. Some of MacArthur's staff held high hopes for the operation. General Willoughby, MacArthur's G–2, admonished Gay to step lively or be left behind. "You must expedite preparations to the utmost," Willoughby warned, "because if your

1 (1) Information on these requests is contained in previous chapters. (2) Rad, CM–IN 9573, CINC FE to DA, 3 Jul 50. (3) Rad, C 57248, CINCFE to DA, 5 Jul 50. (4) The Chief of Naval Operations, Admiral Forrest P. Sherman, had cabled COM NAVFE, Admiral C. Turner Joy, that a Marine RCT could be made available for service in Korea, if General MacArthur desired. Joy called upon MacArthur in Tokyo on 2 July. MacArthur, who had just returned from a depressing inspection of the situation in Korea, accepted with alacrity and, according to Joy, with unusual enthusiasm. For an account of this transaction, see Montross and Canzona, *U.S. Marine Operations in Korea, 1950–1953*, vol. I, *The Pusan Perimeter*, pp. 48–49.

2 (1) Draft Plan, Opn BLUEHEARTS, JSPOG, GHQ, FEC, Jul 50, copy in JSPOG, GHQ files. (2) For other coverage of the plans and preparations for the Inch'on landing, see Appleman, *South to the Naktong, North to the Yalu*, pp. 488–500; Field, *Naval Operations, Korea*, pp. 171–83; Lynn Montross and Capt. Nicholas A. Canzona, *U.S. Marine Operations in Korea, 1950–1953*, vol. II, *The Inchon-Seoul Operation*, chs. I through IV; and Col. Robert Deles Heinl, Jr., *Victory at High Tide* (New York: J. B. Lippincott Co., 1968), ch. 2.

GENERAL WRIGHT

GENERAL GAY

landing is delayed, all that the 1st Cavalry Division will hit when it lands will be the tail-end of the 24th Division as it passes north through Seoul." [3]

Operation BLUEHEARTS died a-borning. The failure of the weak American and weaker ROK forces to halt the enemy and the forced commitment of the 1st Cavalry Division before 22 July made the operation, in July or even in August, quite infeasible. It was canceled on 10 July.[4]

[3] Ltr, Gen Gay to Col Appleman, 24 Aug 53, copy in OCMH. Gay recalls that his division did hit the tail of the 24th Division on 20 July, but under quite different circumstances.

[4] During his briefing of General Collins on 13 July, General MacArthur explained why Operation BLUEHEARTS could not be carried out. There is a marked similarity between BLUEHEARTS and the strategic concepts developed later.

The increasingly grave turn of events on the ground strengthened MacArthur's determination to strike amphibiously. He told Generals Collins and Vandenberg of his intentions on 13 July and outlined a tentative strategy. He had not yet chosen a target date nor a definite landing site, but informed Collins and Vandenberg that as soon as the North Koreans had been stopped, he would attack their rear on the west coast. He believed that Inch'on would be the best place to strike. But he was also considering landing beaches at Haeju and Chinnamp'o, both north of Inch'on.

A day later, General Collins talked with some of MacArthur's key staff officers about the proposed landing. The Army Chief of Staff, aware of the tremendous tidal changes at Inch'on, questioned the wisdom of a landing there. Rear Adm. James H. Doyle, assistant to Admiral Joy and a man of much experience in amphibious techniques, agreed that a landing at Inch'on could be extremely difficult and would require considerable preliminary naval bombardment. But he told Collins that it could be done.[5]

Turning to General Almond, Collins asked how the assault troops would cross the formidable barrier of the Han River after landing at Inch'on. Almond pointed out that amphibious trucks, available in the theater, could be used to ferry troops. The crossing would probably be unopposed since General MacArthur would use the airborne RCT to seize and secure the north shore of the Han. General Collins returned to Washington without committing himself, either for or against the planned operation. But he described to his fellow members of the Joint Chiefs of Staff and to his Army staff assistants the broad outlines of the maneuver MacArthur had in mind.[6]

The commitment of the 25th Division and the 1st Cavalry Division against the North Koreans had slowed, but not stopped, the enemy's drive, and did not come in time to prevent the fall of

REAR ADM. JAMES H. DOYLE

Taejon to the enemy on 20 July. The loss of all Korea loomed as a very real possibility. Nevertheless, by that date General MacArthur had discussed his idea with General Almond and General Wright and had ordered detailed plans drawn up for an amphibious envelopment. Primary emphasis, he directed, was to be on Inch'on as the assault site, but he also specified that alternate plans be prepared.

Wright's planning officers at once began to ready the basic framework of a plan for an amphibious assault landing at Inch'on during September and to draw up several alternate plans as well. On 23 July all these plans went to GHQ staff officers most directly concerned with the proposed operations.[7]

[5] General Wright calls Doyle "a real expert on amphibious operations, a real commander in every sense of the word, a thorough planner and an able and enthusiastic executive of those plans. . . ." See Ltr, Gen Wright to Maj Gen E. W. Snedeker, USMC, 16 Feb 56, Marine Corps files.

[6] (1) Memo, Col Dickson for Gen Bolté, sub: Rpt of Trip to FEC, 10–15 Jul 50, in G–3, DA file 333 Pac, Case 3, Tabs A and C. (2) Collins, *War in Peacetime*, p. 116. (3) President Truman, in volume II, page 348, of his *Memoirs*, recalls that on his return from Tokyo, General Collins had serious misgivings about MacArthur's plans for the counterattack.

[7] (1) Draft Plan 100–B, JSPOG, 23 Jul 50, copy in JSPOG, GHQ, FEC files. (2) Plans circulated at the same time were Plan 100–C, calling for a landing at

General MacArthur confirmed the message which General Collins had carried back to Washington on 23 July, when he told the Joint Chiefs of Staff that he meant to use the 5th Marine RCT and the 2d Division for "major amphibious operations" in mid-September. An airborne RCT would drop into the objective area soon after D-day to seize key communications centers immediately ahead of the advancing assault forces. MacArthur did not pinpoint his objective area, but he described in broad terms how the assault would go. After the beachhead had been seized, Eighth Army, by that time augmented by the additional infantry, artillery, and tank battalions, would attack from the south and destroy the North Koreans.

"Although the exact date of D-day is partially dependent upon enemy reaction during the month of August," MacArthur reported to Washington:

> I am firmly convinced that an early and strong effort behind his front will sever his main line of communication and enable us to deliver a decisive and crushing blow. Any material delay in such an operation may lose this opportunity. The alternative is a frontal attack which can only result in a protracted and expensive campaign to slowly drive the enemy north of the 38th Parallel.[8]

General MacArthur's proposals for a September landing reached Washington at a bad time. They came on the heels of the grim news that Taejon had fallen and while the North Koreans were obviously preparing a double envelopment of Walker's defenses. MacArthur's term, "enemy reaction during . . . August," probably struck the Joint Chiefs of Staff as euphemistic. At any rate, they called General MacArthur to a teleconference on 24 July and asked pointedly whether, in the face of increasing enemy pressure and the stepped-up tempo of the fighting all along the front, he still believed it wise to schedule an amphibious landing for mid-September.

Confidently, General MacArthur assured them that, "barring unforeseen circumstances, and with complete provision of requested replacements, if the full Marine division is provided, the chances to launch the movement in September would be excellent." Complete tactical surprise was essential to the success of the amphibious operation, he declared, and warned Washington not to give away his intentions, saying "I cannot emphasize too strongly the necessity for complete secrecy with reference to this matter. The spokesman for the Department of the Army should not reveal our grand strategy in the slightest degree." The Joint Chiefs of Staff derived little assurance from their exchange

Kunsan, and Plan 100–D, calling for a landing on the east coast near Chumunjin. General Wright recalls that alternate landings featuring Wonsan and Chinnamp'o were also under consideration. General Walker, Wright says, wanted a flexible plan with landings scheduled for either coast so that the main effort could be mounted with little advance notice. But from the standpoint of a communications complex which could be used to support the breakout from the beachhead and the pursuit phase, Seoul-Inch'on "stood out like a sore thumb," according to General Wright. See Interv, author with Wright, Dec 51.

[8] Rad, C 58473, CINCFE to DA (for JCS), 23 Jul 50.

[9] (1) Telecon, TT 3573, JCS and CINCFE, 24 Jul 50. (2) Details of MacArthur's request for the "full Marine division" mentioned here are contained in Chapter IX, below.

A Russian-Made T-34 Tank *knocked out in Taejon.*

with MacArthur. They could only watch and wait for new developments.[9]

The predicament of Walker's divisions in Korea concerned General MacArthur far more than was apparent in his reassuring words to the Joint Chiefs of Staff. Walker's slowing but continued withdrawal before the North Koreans threatened to render plans for an amphibious operation in September purely academic. Walker himself was worried and disappointed because his divisions were not stopping the North Koreans. Troops often came close to panic and commanders sometimes nearly lost control. Walker was particularly disappointed over the failure to check the enemy advance down the Taejon-Taegu axis in late July and early August.

Because of the Eighth Army's precarious position, MacArthur took a drastic step which, seemingly, negated his plans for a mid-September landing. He ordered the 2d Division and the 5th Marine RCT, both on the high seas and both scheduled for his amphibious assault, to sail directly to Korea where they entered combat almost at once.

This move by MacArthur caused his own planning staff to urge a reconsideration of the timing of the proposed operation. To launch an attack by mid-

September, with his entire assault force now committed in the Pusan Perimeter, seemed to them almost impossible. If the attack was to be made in September, both the 2d Division and the Marines would have to be taken away from Walker, or only the Marines withdrawn and teamed with the 7th Division for the amphibious landing. Officers of JSPOG pointed out to General Almond that if General Walker needed the 2d Division in August, he would most certainly need it in September. Also, pulling a division out through the cluttered port at Pusan would tie up supplies and seriously hamper support of Walker's forces remaining on the line. These officers believed that any plan based on use of the 7th Division would be "visionary and impracticable." That division, still in Japan, was at less than half strength, and was not expected to reach full strength before 1 October or to be ready for amphibious operations before 1951. They recommended that General MacArthur postpone the target date for the amphibious operation until 15 October.[10]

One of General MacArthur's outstanding attributes, demonstrated quite often in World War II, was a keen sense of timing. He had not hesitated in the past to override the recommendations of his staff whenever he felt his judgment was more correct than its counsel. Nor did he hesitate in this case. Apparently, he not only believed that forces for the operation would materialize in time for the landing in September, but also, that he could not afford to wait beyond that date.

General MacArthur's refusal to abandon his mid-September date was influenced by his knowledge of the Inch'on area as well as by his desire to relieve the pressure on the Pusan Perimeter as quickly as he could. October might well be too late. Low seas were common in the Inch'on area from May through August, with September a month of transition to the high seas which prevailed from October through March. This left September as the only autumn month when conditions were suitable for landing troops and equipment under fire. During only three days, even in September, would the tidal conditions favor a landing. From 15 to 18 September the tidal surges would be high enough to cover the extensive mud flats that fronted Inch'on Harbor and landing craft could be brought in. The next opportunity would not come until mid-October. By that time seas might be too heavy, and there would be little good weather left for the pursuit and breakout phase of the operation.[11]

He confided to the Joint Chiefs of Staff on 29 July that, while the enemy's successes were upsetting his plans nearly as fast as they were made, he was still holding to the September date. "In Korea," he said, "the hopes that I had entertained to hold out the 1st Marine Division [*sic*: Brigade] and the 2d Infantry Division for the enveloping counterblow have not been fulfilled and it will be necessary to commit these units to Korea on the south line rather than . . . along a separate axis in mid-September." He had not given up hope of mounting the waterborne attack even

[10] Memo, JSPOG, for CofS GHQ FEC, UNC, 29 Jul 50, in JSPOG, GHQ, UNC files.

[11] Lynn Montross, "The Inchon Landing—Victory Over Time and Tide," *Marine Corps Gazette* (July 1951), p. 28.

though he now admitted it might have to be staged out of the Pusan Perimeter rather than Japan. And he informed the Joint Chiefs that as soon as the 7th Division could be brought to approximate strength he was going to throw it into the fight.[12]

General MacArthur realized that without full support from Washington the landing could not be made. And sensing, perhaps, a certain coolness among the Joint Chiefs of Staff, or at least an absence of enthusiasm approaching his own, he included an evaluation of amphibious landings with particular emphasis on Korea. "It is essential, in my opinion," General MacArthur told his superiors, "to utilize our own strength in naval and air forces in the form of amphibious envelopment. When and if this can be accomplished, the ground initiative which the enemy now possesses will be wrenched from him and a decisive result made possible."

On 1 August General Walker had ordered his entire force to break contact with the enemy and to pull back behind the Naktong River, there to make a final stand. On 6 August, General Hickey, Deputy Chief of Staff, GHQ, flew into this perimeter, carrying with him a brief of the plans for the amphibious landing. The hard-pressed Walker agreed with the concept and with the detailed provisions of the plan. But members of General Walker's staff, particularly those of his G–3 section, were skeptical of Eighth Army's ability to carry out the co-ordinated frontal assault provided by the plan. They frankly and openly doubted that the divisions then in the Pusan Perimeter could drive through the mountains to the Kum River. Bridges were out all across the Eighth Army front. Walker was seriously short of trucks. But the biggest obstacle, according to the Eighth Army staff, would be the North Korean Army, which would be intact and capable of fierce and sustained resistance even though the amphibious assault in its rear was successfully carried out. Some of Walker's officers felt that the North Koreans would, if driven from the roads, take to the surrounding hills and prevent the American divisions from breaking out to the north. One key officer suggested that Eighth Army take the much longer coastal route up the west coast where roads were good and flank protection would be afforded by the Yellow Sea. Eighth Army officers generally agreed that after the landing in the north Walker would need at least two more divisions before he could break out.[13]

President Truman sent his special assistant, Averell Harriman, to Tokyo on 6 August, primarily to discuss Far Eastern political matters with General MacArthur. General Ridgway and Lt. Gen. Lauris Norstad of the Air Force accompanied Mr. Harriman. While these officials were in Tokyo, General MacArthur took the opportunity to express his views on the situation facing him in Korea, MacArthur believed that speed was the keystone of victory over the North Koreans. He told Harriman and the military officers that the United States could not afford to wait for a slow

[12] Rad, C 58993, CINCFE to JCS, 29 Jul 50.

[13] Memo, Lt J. B. Warren for Gen Wright, 7 Aug 50, sub: Trip to EUSAK, in JSPOG, GHQ, UNC files.

build-up of forces in Korea. The United States must destroy the North Korean Army as early as possible. If not, the Russians and Chinese Communists, MacArthur feared, would be able to strengthen their protégé by shipping in more arms and supplies. MacArthur also saw in a failure to settle the matter speedily, political dangers. United Nations members would grow discouraged and Oriental peoples would be disappointed with, and lose confidence in, the United States.[14]

On 12 August, shortly after these visitors departed, another and more fully developed draft of the landing plan was issued, setting a target date of 15 September. The strategic concept of this plan would be put into effect one month later without substantive change. Without naming major Army units, the plan proposed committing the GHQ Reserve and the 1st Marine Division in an amphibious operation to seize the Inch'on-Seoul area and to cut the main lines of enemy communications and supply to North Korean units in the south. In conjunction with the seaborne assault, the Eighth Army was to break out of its perimeter and drive northwest along the Taegu-Taejon-Suwon axis to link up with the amphibious force. The Navy and the Air Force would carry out vital missions of transportation, security, naval gunfire support, carrier aircraft support, and strategic bombing. The 1st Marine Air Wing would furnish tactical air cover for the landing.[15]

These plans for landing at Inch'on on 15 September met opposition both within MacArthur's own staff and in other quarters. Navy and Marine officers raised objection to the plans. These officers did not oppose an amphibious assault even though they felt that Army planners were minimizing the problems which the Navy and Marine Corps must overcome in carrying and landing the assault forces on D-day. They did not want to land at Inch'on.[16]

Their concern over Inch'on arose from its natural obstacles to military and naval operations. From the standpoint of navigation, sea approaches, and landing beaches, Inch'on ranked among the worst harbor areas in Korea. The Yellow Sea in its periodic surges into the harbor (changes in the sluggish, heavy tide exceeded thirty feet) had created broad mudbanks and tidal flats which fronted the entire harbor. These flats were so soft and the muck so deep they would not support men on foot. Twice a day the tides rolled in to cover these flats. The naval officers believed it would require a 23-foot minimum tide before small landing craft could safely operate over these flats and a 29-foot tide before Navy LST's could come into Inch'on's beaches. This meant that they could land men and supplies only from the time an incoming tide reached

[14] Truman, *Memoirs*, II, 349–51. (2) See also MacArthur, *Reminiscences*, pp. 340–41.

[15] (1) Opn Plan 100–B, 12 Aug 50, in JSPOG, GHQ, UNC files. (2) Special Rpt, U.S. X Corps, Opn CHROMITE, copy in OCMH.

[16] This portion is based on the following: Chronicles by General Oliver P. Smith, USMC, 22–23 August 1950 (hereafter cited as General Smith's Chronicles), copy available in Hist Sec, G–3 USMC, Hq, Washington, D.C.; Special Action Rpt (SAR), 1st Marine Div, 15 Aug–30 Sep 50, copy in same files; Malcolm C. Cagle, "Inchon, Analysis of a Gamble," *U.S. Naval Institute Proceedings*, vol. 80, No. 1 (January, 1954), 47–51. See also, Field, *History of United States Naval Operations*, Korea, pp. 171–83.

twenty-three feet until the outgoing tide dropped again to that level, a period of only about three hours. Troops ashore would then be stranded until the next high tide about twelve hours later. Morning high tide for 15 September was forecast at 0650 and evening tide at 1920. As already noted, the tide on that date would be deep enough for landing craft.

Numerous islands bracketed Inch'on to seaward, forming a natural pocket and restricting naval maneuver to narrow channels. Navigation through these channels, particularly the main Flying Fish Channel, was treacherous even in daylight. The channel was narrow, twisting, and dead-end. If the enemy mined this channel, approach would be virtually impossible.

In order to land, the Marines would have to scale seawalls ranging from twelve to fourteen feet high which fronted the harbor across almost its entire width. The Inch'on area was heavily built-up. The enemy could mount a very effective resistance, taking advantage of buildings for protection. The Marines did not want to land in the middle of a built-up area if they could help it. To complicate matters, Wolmi-do, a 350-foot-high pyramidal island, heavily fortified, dominated Inch'on Harbor. All in all, Navy and Marine planners found Inch'on a poor place to land.

These officers had objected and argued with General MacArthur's staff from time to time in general terms, but when the commanding general of the 1st Marine Division, Maj. Gen. Oliver P. Smith, reported to Admiral Doyle, Commander, Amphibious Group One, on 22 August in Tokyo, these objections suddenly became concrete and specific. General Smith had flown to Tokyo ahead of his division to take command of the landing force under Admiral Doyle who would command the attack force. These two officers and their staffs worked very closely in arranging the details of the amphibious assault on Inch'on.[17]

GENERAL SMITH

On 22 August, General Smith heard

[17] The Special Action Report of the Marine division says of the command relationships and the planning phase, "Although relationships between the division as Landing Force and COMPHIB Group One were clear from the outset and in accordance with . . . doctrine, the command status and command responsibilities for the assault landing phase of CG X Corps, CJTF 7 and COMNAVFE were vague and confusing. None of the latter commands ever appeared under well defined titles and none of the accepted titles which would have been appropriate to these echelons was used."

for the first time that the assault was scheduled for 15 September. He had been told before leaving the United States that the target date was 23 September. He found Admiral Doyle very, very skeptical about landing at Inch'on, across mud flats, over docks and seawalls, and in the face of a city of sizable population. Doyle told Smith that he had sent his reconnaissance parties in at various sites along the Korean west coast to find a better landing site than Inch'on. He had found what he regarded as a better location for an amphibious assault. This area, Posung-Myon, was about twenty miles south of Inch'on and almost due west of Osan. Navy underwater demolition teams had made several trial landings there and had found that beach conditions were much better than at Inch'on and would not restrict the landing to a particular day or hour. The area was not built up and, according to Doyle, was in striking distance of the enemy's lines of communications south of Seoul.

That evening, General Smith reported to the Dai Ichi Building for an interview with General MacArthur. He first met General Almond to whom he briefly raised his objections to Inch'on, without, however, mentioning Posung-Myon. Almond dismissed Smith's protests by telling him that the enemy had no organized forces at Inch'on, that the difficulties to be met there were only mechanical, and that the date and place of the landing had already been fixed. He then ushered Smith into General MacArthur's office where the Marine general received not only a warm greeting, but assurance that the Inch'on landing would be decisive and that the war could be over in one month after the assault. General MacArthur insisted that the North Koreans had committed all of their troops against the Pusan Perimeter, and he shared Almond's view that the Marines would meet no heavy opposition at Inch'on. When Smith objected that 15 September would be too early to assemble his forces, General MacArthur admitted that the landings would have to be somewhat helter-skelter. But he would not consider any date other than 15 September.

These doubts within MacArthur's own headquarters were matched at a higher level by mounting suspicions within the Joint Chiefs of Staff, suspicions arising from ignorance of exactly what General MacArthur was up to. Under the directives given him by the Joint Chiefs of Staff, as well as through precedent in the considerable latitude granted comparable American commanders in the past, General MacArthur had authority to dispose and employ his forces as he saw fit. This authority reflected the fact that planning for major operations of the Korean War and decisions of tactical and local strategic significance originated with General MacArthur. The Joint Chiefs of Staff set for him broad objectives and sometimes voiced their concern over his handling of matters of political significance. They entered into the planning picture most influentially in matters involving allotment of forces and supply. But in the case of the proposed Inch'on landing, the Joint Chiefs of Staff grew increasingly worried during August because MacArthur did not keep them informed of the development of his plans. He submitted no campaign plan to them and, aside from his requisitions for forces, passed along only the bare outline of his plans.

Knowing full well the weakened condition of American military resources at the time, observing the continued successes of the North Korean Army, but ignorant of the exact nature of MacArthur's preparations and plans for an amphibious counterblow, the Joint Chiefs of Staff began to wonder if MacArthur was not getting ready to bite off more than the United States could chew.

In order to determine more precisely what was taking place in Tokyo, the Joint Chiefs of Staff sent two of their members to the Far East. General Collins and Admiral Sherman, accompanied by a staff of Air Force and Army officers, flew to Tokyo on 19 August to talk with MacArthur.[18]

Meeting privately with General Collins and Admiral Sherman upon their arrival in Tokyo, MacArthur covered general aspects of the whole Korean operation, and then staged a full-scale briefing on the proposed amphibious movement for top military and naval officials. This briefing, which took place in General MacArthur's conference room on the 6th floor of the Dai Ichi Building in Tokyo in the late afternoon of 23 August 1950, was attended by Generals MacArthur, Collins, Almond, and Wright of the Army and Admirals Sherman, Joy, Struble, and Doyle of the Navy. Various other officers of lesser rank participated in the briefing.[19]

Just before this briefing, General Smith had approached General Almond on the possibility of landing in the Posung-Myon area instead of at Inch'on. General Almond stated very definitely that he was not interested in a landing there except perhaps as a subsidiary landing in connection with Inch'on. Almond told Smith that the real objective of this operation was to capture Seoul at the earliest possible date. Too, GHQ planning officers had looked into Posung-Myon and did not believe that the area had the necessary road net to support heavy vehicles in any breakout of the area.[20]

Admiral Doyle's planning officers presented the first portion of the briefing. For nearly an hour they covered the problems faced by the Navy in the landing operation, emphasizing the great difficulties and the risks involved. Their remarks were decidedly pessimistic. Ad-

[18] (1) Rad, WAR 89118, DA to CINCFE, 18 Aug 40. (2) General Collins described the purpose of the visit as ". . . to find out just exactly what these plans were. Frankly, we were somewhat in the dark, and as it was a matter of great concern, we went out to discuss it with General MacArthur. We suggested certain alternative possibilities and places and everything of that sort. . . ." Louis Johnson, who as Secretary of Defense at this time claimed to have supported MacArthur wholeheartedly in his proposals for landing at Inch'on, describes the purpose of this visit differently. He stated, "General Collins . . . did not favor Inchon and went over to try to argue General MacArthur out of it." See *MacArthur Hearings*, pp. 1295, 2618.

[19] (1) Ltr, Adm Joy to Col Appleman, 12 Dec 52. (2) Ltr, Gen Almond to Col Appleman, 2 Dec 52. (1) and (2) in OCMH. (3) Walter M. Karig, *Battle Report, The War in Korea* (New York: Rinehart, 1952), pp. 166–67. Karig's work, which both Joy and Almond describe as substantially correct and factual, is used as the basis for this account of the 23 August briefing. Modifications from Joy's and Almond's letters have been applied to Karig's version where appropriate.

[20] (1) General Smith's Chronicles, 22–23 Aug 50. (2) Col John Chiles, SGS GHQ, and later G–3, X Corps, told the author during a conversation at the Army War College in February 1955 that he had examined charts of the Posung-Myon area, and found the routes of egress entirely insufficient for an operation of the scale planned.

miral Doyle concluded this presentation by conceding that the operation was not impossible, but he stated that he did not recommend it.

General MacArthur, already familiar with the views of his naval staff, seems not to have been taken aback by this adverse comment. Taking the floor, he came to the defense of his plans calmly and with great assurance. He omitted any mention of the hazards, dwelling instead upon the reasons why the landing should be made at Inch'on and upon the tactical conditions which favored its success. He pointed out the disposition of the North Korean Army and its vulnerability to an amphibious encirclement.

If there were one vital spot in the enemy's line of communications, the Seoul-Inch'on area was that spot. Almost all of the major rail and highway lines leading from North Korea channeled through that area. Only by seizing Seoul and Inch'on, MacArthur insisted, could he achieve a quick and decisive victory over the enemy. He also pointed out the tremendous political and psychological advantages to be gained by retaking the Korean capital from the invaders.

General Collins and Admiral Sherman had suggested to him that a landing at Kunsan, nearly one hundred miles south of Inch'on, might be just as effective and involve less risk. But MacArthur deprecated Kunsan as a main objective area, maintaining that such a shallow envelopment would not cut the enemy's line of communications nor surround his divisions. It would not lead to quick victory and a bitter Korean winter campaign would have to be fought. Only Inch'on, in General MacArthur's opinion, would do.

General MacArthur did not ask Collins or Sherman to approve his plans, nor did they offer to do so. The briefing was a briefing and nothing more, but the purposes of the Joint Chiefs of Staff had been served. They now knew what MacArthur intended to do and how he intended to do it. They were no longer in the dark.

General MacArthur's able presentation did not completely convince the naval and Marine officers. On the morning of 24 August, these officers, in a meeting which included Admiral Sherman, Admiral Joy, Lt. Gen. Lemuel C. Shepherd and the lesser naval and Marine commanders, assembled in a private airing of their grievances. All present felt strongly that MacArthur should give greater consideration to the Posung-Myon area. They selected General Shepherd, Commanding General, Fleet Marine Force, Pacific who was reputed to enjoy particular influence with General MacArthur, to make a personal appeal for the Posung-Myon area. General Shepherd called upon General MacArthur and presented the Navy-Marine case but to no avail. From that hour, the naval and Marine officers abandoned Posung-Myon and concentrated on Inch'on.[21]

Upon their return to Washington, General Collins and Admiral Sherman explained to their fellow members of the Joint Chiefs of Staff the concept and the state of preparation for the attack on Inch'on. Now that the veil had been lifted, the Joint Chiefs examined the

---

[21] General Smith's Chronicles, 24 Aug 50.

plans carefully. They found no real disagreement with what MacArthur intended to do and, on 28 August, notified him that they approved his plans for an amphibious operation on the west coast of Korea. They suggested, though, that he also prepare plans for an amphibious envelopment in the vicinity of Kunsan.[22]

The Joint Chiefs of Staff very pointedly told MacArthur that, from here on in, they wanted to know what went on in his theater. "We desire such information as becomes available with respect to conditions in the possible objective areas and timely information as to your intentions and plans for offensive operations." [23]

Why had the Joint Chiefs of Staff found it necessary to send MacArthur approval of his plans? General Collins may have felt that the controversy evident at the Tokyo briefing had now been resolved and took this way of clearing any doubt from MacArthur's mind. The Inch'on landing would tie up a major share of the nation's ready combat forces and, while by strict interpretation, the landing would be a purely tactical maneuver at the discretion of the theater commander, failure would have repercussions far beyond Korea. This may have led the Joint Chiefs to identify themselves with the operation by granting approval, at the same time placing them in a better position to call off the maneuver if the risks suddenly appeared too great. Their admonition requiring "timely information" is in line with this latter possibility. Certainly the Joint Chiefs of Staff did not tell MacArthur that they were taking the reins from his hands.[24]

General Shepherd

Orders for the attack followed almost immediately. General MacArthur, on 30 August, issued his operations order for the Inch'on landing, setting forth the objectives and assigning specific missions to his commanders.

He directed the U.S. X Corps, the headquarters of which he established within the theater (see ch. IX), to land on D-day at H-hour on the west coast of Korea to seize Inch'on, Kimp'o Airfield, and Seoul, and to sever all North Korean lines of communication in the area. He ordered co-ordinated attacks

[22] Rad, JCS 89960, JCS to CINCFE, 28 Aug 50.
[23] *Ibid.*

[24] General Collins and Admiral Sherman talked with President Truman on their return, telling him of MacArthur's plans and informing him that they had approved these plans. "It was a daring strategic conception," Truman commented. "I had the greatest confidence that it would succeed." See Truman, *Memoirs*, II, 358.

from the southern perimeter by Eighth Army and all available ground, naval, and air forces, to destroy the North Korean Army south of the line Inch'on-Seoul-Utchin. Admiral Joy, COMNAVFE, would command while afloat. He would furnish Navy and Marine assault forces and would transport follow-up landing forces. Once the lodgment ashore had been seized, Joy would land the follow-up troops on the beachhead. After the beachhead was secured, commanding general, U.S. X Corps, would land, inform the naval commander of his readiness to assume responsibility for further operations, and take command of all forces ashore. The U.S. X Corps would operate directly under General MacArthur until otherwise ordered. MacArthur charged General Stratemeyer, Commanding General, FEAF, with general air support to isolate the objective area and with giving required close support. The principal air effort would support the Eighth Army breakout. If so ordered, General Stratemeyer was to ferry, protect, and drop an airborne RCT. General Walker on D plus 1 would launch a general offensive from his perimeter, making his main effort along the Taegu-Taejon-Suwon axis. Annexes to the operations order gave detailed instructions to all commanders on all phases of the operation, including intelligence, logistical support, and command relationships.[25]

A representative of the Department of the Army G–3, who had been making an inspection tour of the Far East Command and who returned to Washington in early September, reported to General Bolté that "Plans for the contemplated envelopment operation in Korea are well advanced. Nearly everyone in FECOM concerned with these plans is confident that they can be carried out successfully despite serious shortages in combat and service troops and logistic support." The officer pointed up Washington's lack of participation in the planning for Operation CHROMITE: "In order that DA may further integrate its planning with that of FECOM," he said, "working level officers in FECOM charged with preparation of the campaign plan will attempt to obtain General MacArthur's permission to forward a copy of this plan to DA. . . ." [26]

When, by 5 September, the Joint Chiefs of Staff, despite their request of 28 August, had heard nothing more from General MacArthur on his plans, they again called upon him, saying, "Pursuant to the request . . . desire to be informed of any modification which may have been made in your plans for the mid-September amphibious operation." [27]

This terse reminder triggered only a casual reaction from MacArthur. He replied that "the general outline of the plan remains as described to you." He promised that by 11 September, using

[25] (1) Opns Order No. 1, GHQ, UNC, 30 Aug 50, copy with Annual Narrative Hist Rpt, GHQ, FEC, 1 Jan–31 Oct 50, Annex IV. (2) For a more detailed study of this order and of the organization of landing and attack forces, see the following: USAF Hist Study, United States Air Force Operations in the Korean Conflict, 25 June–1 November 1950, ch. 5, copy in OCMH; and Malcolm C. Cagle, "Inchon, Analysis of a Gamble," *United States Naval Institute Proceedings* (January 1954), pp. 47–51.

[26] Memo, Col Everett for Gen Bolté, 8 Sep 50, sub: Visit to USARPAC and FEC, 19–30 Aug 50, in G–3, DA file 333 Pac, Case 5.

[27] Rad, JCS 90639, JCS to MacArthur, 5 Sep 50.

VICE ADMIRAL C. TURNER JOY

officer courier, he would send them a detailed description of his planned operations.[28]

Meanwhile, the Joint Chiefs of Staff had been weighing the possible fruits of success at Inch'on against the certain price of failure. They lacked General MacArthur's complete faith in ultimate victory at Inch'on. They feared a debacle at Inchon from which the U.N. forces might not recover. North Korean gains along the Pusan Perimeter had continued into September and, from Washington, chances of a mid-September victory on the west coast appeared to be diminishing rapidly.

On 7 September the Joint Chiefs of Staff called General MacArthur's attention to the fact that he had committed almost all of Eighth Army's reserves. He could expect no more reinforcements immediately. All available General Reserve units except the 82d Airborne Division had been sent to him already. If the Inch'on landing failed, the U.N. forces would be in grave danger. It would take at least four months before any of the newly called National Guard divisions could reach Korea. The Joint Chiefs called on MacArthur for a new estimate and a reconsideration of Inch'on.[29]

This shadow of doubt cast over his plans only a week before the target date evoked from General MacArthur a forceful protest, couched in the strongest, most expressive terms. He discounted the seriousness of the situation confronting General Walker, who was, at this time, having some of his darkest days.

General MacArthur showed extreme optimism in describing the probable effects upon the enemy of a landing against his west coast rear areas. "There is no question in my mind," he told the Joint Chiefs of Staff, "as to the feasibility of the operation. . . ." He saw the planned operation as the only hope of seizing the initiative from the enemy. If the landing were not made, General MacArthur warned, the United States would be committed to a war of attrition which might drag on interminably and which the enemy, with his greater potential for reinforcement, might win. While conceding that General Walker might have to contract his perimeter, General MacArthur held that the situation around Pusan was not critical. "There is no slightest possibility," he maintained, "of our forces being ejected from the Pusan beachhead." If, as he

[28] Rad, C 62213, CINCFE to JCS, 6 Sep 50.

[29] Rad, JCS 90908, JCS to CINCFE, 7 Sep 50.

believed the Joint Chiefs of Staff were implying, small increments of reserves were fed into the Pusan area merely to strengthen the perimeter instead of being used for the encircling attack, the cost in time, casualties, and matériel would be immeasurably increased. He suspected, too, the Washington military officials were looking at the map too closely and finding bugaboos. They seemed to fear the result if Eighth Army failed to break out and join the landing force at Inch'on on schedule. In General MacArthur's opinion, the success of the operation did not depend on a rapid joining of the two forces. The seizure of the heart of the North Korean distributing system in the Seoul area would "dislocate the logistical supply of his forces operating in South Korea" and ultimately result in the disintegration of North Korean resistance. Both American forces, Eighth Army and the U.S. X Corps, would be self-sustaining because of the complete American control of sea and air. While the prompt junction of forces would be "dramatically symbolic of the complete collapse of the enemy," General MacArthur certainly did not consider it a vital part of the operation. Troops were already embarking for the amphibious sweep, and preliminary naval and air preparations were going ahead on schedule. "I and all of my commanders and staff officers, without exception, are enthusiastic and confident of the success of the enveloping operation," General MacArthur concluded.[30]

Faced with these most vigorous views from a man who was in a position to judge the theater situation more accurately than anyone else, the Joint Chiefs of Staff acquiesced. They went further and obtained President Truman's approval for the landing. On 8 September, they gave General MacArthur the final green light for the landing at Inch'on one week later.[31]

[30] Rad, C 62423, CINCFE to JCS, 8 Sep 50.

[31] Rad, JCS 90958, JCS to CINCFE, 8 Sep 50.

CHAPTER IX

Operation CHROMITE: The Forces

MacArthur planned his bold amphibious venture at Inch'on sustained only by hope, credit, and promises. At no time during his planning did he have the men and guns he would need. The Joint Chiefs of Staff, moreover, frequently told MacArthur that, with the military resources of the United States at rock bottom and because of the short-fused target date on which MacArthur adamantly insisted, the needed men and guns might not arrive on time. The disagreements over time, place, and method of landing stemmed in part from this fact and were certainly of less significance. MacArthur well knew that even with the fullest support by Washington he might not have by his chosen D-day enough trained men and equipment to breach enemy defenses and to exploit a penetration. Trained men, especially those with amphibious training, were at a premium in the United States as well as in the Far East. To assemble, equip, and move these men secretly and swiftly to the battle area by 15 September would require an enormous, finely co-ordinated effort by all involved. The difficulties were appalling, and to surmount them called for extraordinary energy and ingenuity.

The nature and location of the planned landing dictated that it be directed by a tactical headquarters separate from the Eighth Army. General Walker had his hands full in the Pusan Perimeter and could not easily divide his attention, effort, or staff. The size of the landing force, initially set at about two divisions, indicated a need for a corps command. It was for this reason that MacArthur, concurrently with his efforts to bring the two corps headquarters to his theater in late July, had asked that the commander and planning staff of the I Corps be flown to Tokyo.[1] But by the time General Coulter and his skeleton staff reached Japan, a need for the I Corps in the Pusan Perimeter forced MacArthur to send Coulter on to Korea.

Since the amphibious operation could not be made without a corps headquarters, members of JSPOG recommended that their chief, General Wright, ask MacArthur either to organize a provisional corps headquarters locally or to bring from Pearl Harbor to Tokyo the Fleet Marine Force, Pacific (FMFPAC) headquarters, commanded by General Shepherd. General Wright chose the latter course and suggested to General

[1] Rad, CX 58296, CINCFE to DA, 28 Jul 50.

ADMIRAL RADFORD *conferring with General MacArthur during Inch'on planning.*

Almond that Admiral Arthur W. Radford, Commander in Chief, Pacific, be asked if the Marine headquarters could be moved. "There is urgent need" General Wright argued, "to get a headquarters in being for the GHQ Reserve operation. This headquarters must be one that can operate in the field as a going concern with such things as situation reports, operations reports, communications, etc., happening automatically." Forming a provisional headquarters from theater officers did not appeal to Wright. "A provisional command group selected from GHQ officers will not be a going concern unless it has time to get together and train in the field," he pointed out. "This is true no matter how efficient the individual officers are." Too little time remained to form and train such a group since, Wright warned, "With the target date of 15 September, only thirty days remain in which to complete the landing plan, embarkation plan and the embarkation of the assault element." Wright cited amphibious doctrine which set from 90 to 150 days for planning. For this reason alone he felt that the trained headquarters from Hawaii should be used if available.

General Hickey agreed with Wright. Hickey told General Almond:

> Utilization of this headquarters and staff which is already organized and functioning offers many advantages over the hasty throwing together of a provisional Corps headquarters and staff from available personnel. The latter would be at best only a half-baked affair and would contribute to reducing the efficient functioning of GHQ because of the key personnel withdrawn.[2]

General MacArthur did not accept Wright's suggestion. First of all, after the amphibious landing at Inch'on itself, CHROMITE would be an overland campaign. Secondly, and perhaps most importantly, MacArthur wanted the detailed CHROMITE planning accomplished under his own close and constant supervision, and not by a group less subject to his direct view than his own GHQ staff. Wright therefore made no further attempt to bring in the outside headquarters.[3]

General Wright's second attempt to arrange a headquarters proved more successful. "As your advisor on tactical organization and operations for forces," he told Almond on 10 August, "I strongly recommend that we immediately activate a command for the GHQ Reserve." This command, in Wright's concept, was to be very similar to a corps headquarters. Because of its specialized mission the command would not need an artillery headquarters, observation battalion, engineer brigade, or engineer topographical company. He recommended that this headquarters be moved to the field immediately since the target date of 15 September was fast approaching and the group would have to be ready to load aboard ship by 10 September. Only twenty-five days remained in which to complete corps-level plans, to condition units for the field, to develop standing operating procedures, and to give combat training to headquarters personnel.[4]

General MacArthur accepted Wright's recommendation and ordered the formation of a provisional planning staff, forerunner of the actual corps staff, from officers of his own GHQ staff. To conceal its true purpose, he designated this new group as the Special Planning Staff, GHQ. General Almond chose the officers for this staff and on 15 August directed them to begin part-time planning, and to continue to work on their regular jobs only as necessary.

Almond named Maj. Gen. Clark L. Ruffner, who had arrived from the United States on 6 August, as chief of staff of the Special Planning Staff. Ruffner assembled his staff in a bunker-type concrete structure near the Dai Ichi Building on 15 August. As a first step, these officers drew up a troop list and a standing operating procedure for the landing. When General Ruffner asked what forces would be used for the landing and breakout, MacArthur replied, "The 7th Division which is half-understrength, the Marine Brigade in Korea, other marines from the United States,

---

[2] Memo, JSPOG for Gen Wright, 7 Aug 50, with Ind, in JSPOG, GHQ, UNC files.

[3] (1) *Ibid.* (2) Ltr, Lt Gen Edward M. Almond to Brig Gen Hal C. Pattison, Chief of Military History, Hq, DA, 10 Feb 67, OCMH files.

[4] Memo, Gen Wright for CofS GHQ, UNC, 10 Aug 50, in JSPOG, GHQ, UNC files.

and a battalion of Marines from the Mediterranean." [5]

MacArthur had not yet named a commander for the invasion forces. Near the end of the third week in August, General Almond suggested to him that the time had come to appoint such a commander. MacArthur turned to his chief of staff and said, "It is you." MacArthur told Almond that he would continue as chief of staff, Far East Command, "in absentia." He was so confident of ending the war by a quick victory at Inch'on, that he believed Almond could return to Tokyo within only a few weeks after the initial landing. In effect, MacArthur put General Almond, as well as other officers on the new corps staff, on loan to the corps from GHQ for the landing operation.[6]

On 21 August, General MacArthur asked to be allowed to activate, from sources already available in his theater, Headquarters, X Corps. Department of the Army readily granted this authority.[7]

The Special Planning Staff had already prepared its version of the best organization for the new corps headquarters. General Almond approved it. The major deviation from standard corps Tables of Organization and Equipment was the addition of a small transportation section and an area command, headquarters and headquarters detachment, of about ninety officers and men. General Ruffner told General Almond that, since X Corps would be operating separately "until such time as link-up is effected," it would have to carry out some functions normally carried out by an Army headquarters.[8]

The corps was activated without a TO&E, Table of Allowances, or Table of Distribution being prescribed. The staff used published equipment and personnel tables as guides, but modified the structure to enable the corps headquarters to operate as a separate corps along the lines of a field army headquarters. As a result, all equipment drawn had to be requisitioned and such requests had to be approved as items over and beyond authorized allowances. Each requisition, in fact, had to be reviewed personally by the corps G–4, Col. Aubrey D. Smith, and approved by the chief, Supply Division, G–4, GHQ. Limited time, inexperienced people, and the urgent press of planning the impending operation greatly complicated this problem.[9]

General MacArthur formally established the X Corps on 26 August. The Special Planning Staff, GHQ, became Headquarters, X Corps, and General Almond was officially designated commanding general in addition to his duties as chief of staff and deputy commander, Far East Command, United Nations Command. All units or detachments in or en route to Japan and previously designated GHQ Reserve were assigned to X Corps. Next, on 1 September, Mac-

[5] (1) Diary, CG X Corps, Opn CHROMITE, 15 Aug–30 Sep 50, copy in OCMH. (2) Interv, Col Appleman with Gen Ruffner, 27 Aug 51.

[6] Interv, Col Appleman with Gen Almond, 13 Dec 51.

[7] (1) Rad, C 60770, CINCFE to DA, 21 Aug 50. (2) Rad, W 89390, DA to CINCFE, 22 Aug 50.

[8] Memo, CofS Special Planning Staff, GHQ (Gen Ruffner), for CofS GHQ (Gen Beiderlinden), 23 Aug 50, sub: Org, G–1, GHQ Daily Log, Item 26, 23 Aug 50.

[9] Comments, Col Smith, former ACofS G–4, X Corps, contained in Hq, EUSAK, Mono, Martin Blumenson, Special Problems in the Korean Conflict, pp. 51–53, copy in OCMH.

Arthur assigned the code name, Operation CHROMITE, to the planned landing at Inch'on; and, on 6 September, he confirmed in writing what he had already told his major commanders orally, that D-day for Operation CHROMITE was 15 September 1950.[10]

With time running short and an ominous amount of detailed planning and co-ordination remaining, officers of the new corps headquarters worked around the clock. General Almond crammed as much field training and testing into the few busy days before embarkation as he could. On 1 September, his entire corps staff together with coordinators and umpires moved to a wooded area near Camp Drake in suburban Tokyo and set up a field command post. A tactical exercise prepared by General Willoughby was used to test the readiness of the green headquarters. On the second day of the exercise, General Almond, to measure the mobility and flexibility of his staff, ordered the entire group to displace to Atsugi, twenty miles away, with no break in the continuity of the maneuver. Realism in the maneuver was achieved by confronting the staff with situations closely paralleling those expected at the actual landing. Four main situations were presented, covering the breakout from the beachhead, a counterattack by enemy reserves, an opposed river crossing, and the exploitation of the breakout. Results of this maneuver, which ended on 3 September, made it apparent that General Almond's choice of staff officers had been excellent—the staff demonstrated a state of readiness far beyond expectations.[11]

Marine Forces

The vital factor of the landing operation remained the availability of a strong, well-balanced, and specially trained and equipped amphibious striking force, and enough follow-up units to consolidate and exploit the initial landing. The former could come only from Marine and Navy sources, while a full Army division could provide the latter. MacArthur obtained these forces only after two months of making insistent demands on Washington and by taking unusual steps within his own command.

Like its sister services, the U.S. Marine Corps had shrunk in size during the postwar years. On 30 June, the Marine Corps had only 74,279 officers and men scattered widely among security, training, and administrative posts throughout the world. The operating segment of the Marine Corps, 40,000 officers and men, included the Fleet Marine Force, security forces, and Marines afloat. The Fleet Marine Force was, in turn, divided into Pacific and Atlantic sections. Each of these had a reinforced but reduced strength division and an understrength air wing. The Fleet Marine Force, Pacific, held the 1st Marine Division and 1st Marine Air Wing while the Fleet Marine Force, Atlantic, had the 2d Marine Division and 2d Marine Air Wing. The combined strength of the 1st and 2d

[10] (1) GHQ, UNC GO 324, 26 Aug 50. (2) Rad, C 61660, CINCFE to Major Comds and DA, 6 Sep 50. (3) Ltr, CINCFE to All Major Comdrs, 6 Sep 50, sub: Designation of D-day.

[11] Rpt, JSPOG for CofS GHQ, sub: Map Maneuver X, copy in JSPOG, GHQ, UNC files. This exercise revealed that the corps had no proper equipment for bridging the Han River.

Marine Divisions did not equal that of a single war-strength Marine division.

Early in July, the Joint Chiefs of Staff had approved sending a Marine RCT with supporting tactical air to the Far East Command.[12] The 1st Provisional Marine Brigade was activated at Camp Pendleton, California, on 5 July around the 5th Marines, 1st Marine Division, and Marine Air Group 33 of the 1st Marine Air Wing. The provisional brigade began loading from the west coast almost immediately and sailed on 14 July with about 4,500 ground troops. This number included engineers, a tank company, a light artillery battalion, a 4.2-inch mortar company, amphibious elements, and three infantry battalions, and about 1,350 men in the air group. As of 9 July, Admiral Radford judged this Marine force capable of specialized missions, including amphibious landings, "under conditions where appropriate higher echelon agencies are present." [13] The information on the amphibious capabilities of the new force was well received by General MacArthur since it blended admirably with plans then being developed by his staff. He radioed Washington at once, asking that the Marine brigade, "in view of the extensive opportunity for amphibious employment," be expanded to a full Marine division with appropriate air support.[14]

A few days after this request, General Collins arrived in Tokyo where, in a discussion of the need for forces on 13 July, General Almond upped MacArthur's previous request, asking Collins for a 2-division corps of Marines. The Army Chief of Staff replied that the Marines were in the same position as the Army, very short of men, and that even if another Marine division could be built, the Joint Chiefs of Staff had other plans for it. But, before leaving Japan, General Collins told General MacArthur privately that he believed one full Marine division could be sent him.

In Washington, meanwhile, the Joint Chiefs of Staff had in Collins' absence agreed to bring the 1st Marine Division to war strength. This decision received strong backing from Admiral Radford who personally urged the Chief of Naval Operations to give General MacArthur a full Marine division as soon as possible. Admiral Sherman supported Radford, but with reservations. Radford's support nonetheless proved instrumental in bringing the 1st Marine Division to war strength.[15]

On 19 July, General MacArthur called again for the 1st Marine Division, this time stipulating that all units of the division and the air wing should arrive by 10 September. He also asked that equipment and personnel be sent at once to bring the 5th Marine RCT, already on the way, to full war strength.[16]

To fill the 1st Marine Division, the Marine Corps drew men and equipment

[12] Rad, JCS 84876, JCS to CINCFE, 3 Jul 50.

[13] Rad, 0922322, CINCPACFLT to CINCFE, 9 Jul 50.

[14] (1) Rad, CX 57553, CINCFE to JCS, 10 Jul 50. (2) Rpt, Mobilization of the Marine Corps Reserve in the Korean Conflict, Hist Sec G-3, Hq, U.S. Marine Corps, ch. II, p. 1, copy in OCMH.

[15] (1) Memo, Col Dickson for Gen Bolté, 15 Jul 50, sub: Visit to FEC, Tab B. (2) Rad, C 57814, CINCFE (Gen Collins) to DA (Gen Haislip), 14 Jul 50. (3) Note by Secy for JCS, in G-3, DA file 320.2 Pac, Case 28. (4) Rad, 080941Z, CINCPACFLT (Radford) to CNO (Sherman), 8 Jul 50. (5) JCS 1776/25, Memo, CNO for JCS, 9 Jul 50, sub: Recommendations of CINCPACFLT Concerning Support of CINCFE.

[16] Rad, CX 58239, CINCFE to DA, 19 Jul 50.

from all over the United States. So empowered by Presidential authority, the corps called 138 units with a strength of 1,800 officers and 31,648 enlisted Marines, its entire Organized Ground Reserve, to active service. It also brought 6,800 Regulars of the 2d Marine Division from Camp Lejeune, North Carolina, to Camp Pendleton. An effort was made, however, to avoid stripping the Atlantic area completely of Marines. Admiral Sherman felt that denuding the Atlantic area would be too dangerous; and at Sherman's insistence, the Joint Chiefs of Staff informed General MacArthur that they could not send him the full Marine division before November or December. Nor could they determine the extent to which the 1st Provisional Marine Brigade could be strengthened until Admiral Sherman conferred with Admiral Radford in Hawaii.[17]

This threat to his plans drew fire from MacArthur, and he urgently requested the Joint Chiefs to reconsider. Provision of the full division by 10 September he saw as an absolutely vital element of his entire plan. "There can be," he charged, "no demand for its use elsewhere which can equal the urgency of the immediate battle mission contemplated for it." [18]

Unknown to MacArthur, an influential ally had already come to his support. Admiral Radford, before meeting with the Chief of Naval Operations, had sought the advice of General Shepherd. The Marine general spoke out strongly for General MacArthur and recommended that his request for Marine forces be met in the manner desired. General Shepherd believed that the Fleet Marine Force "as a whole" could provide the amphibious striking force and that it could do so without a serious or lasting impact on the Marine force's readiness to meet other commitments. "I feel," he told Admiral Radford, "that there is a serious war in progress in Korea and employment of amphibious forces will prove the key of achievement of a timely and economical decision for our arms." He held that the Fleet Marine Force was ready "at this moment" to send to Korea a force strong enough to lead the counteroffensive amphibious movement, "the task for which Marines are trained and constituted." [19]

Back in Washington, General Bolté added his support to General MacArthur's plea for early arrival of the Marines. He recommended to General Collins that the latter use his influence with the Joint Chiefs to support MacArthur in his call for a full Marine division in the theater by 10 September.[20]

The intervention of Generals Shep-

[17] (1) Rad, JCS 86511, JCS to CINCFE, 20 Jul 50. (2) Wilbur W. Hoare, Jr., The Joint Chiefs of Staff and National Policy, draft MS, vol. IV, ch. V, p. 13.

[18] (1) Rad, CX 58327, CINCFE to JCS, 21 Jul 50. (2) This statement reflects General MacArthur's conviction that "Washington" followed a policy of slighting his command in favor of the western European area. General Whitney's account of this transaction is interesting, if abbreviated. ". . . on July 10," Whitney says, "MacArthur asked the Joint Chiefs of Staff for the 1st Marine Division. Profiting by his experience with Washington's penchant for skeletonizing his forces, he carefully stipulated a division at full strength. He was turned down flat. He patiently tried again five days later, saying: 'I cannot emphasize too strongly my belief in the complete urgency of my request.' He was turned down again." See Whitney, *MacArthur, His Rendezvous With History*, p. 343.

[19] Memo, Gen Shepherd, CG FMPAC, for Adm Radford, CINCPAC, in JSPOG, GHQ, UNC files.

[20] Memo, Gen Bolté for Gen Collins, 21 Jul 50, sub: Augmentation of Provisional Marine Brigade, in G-3, DA file 320.2 Pac, Case 24.

herd and Bolté prompted the Joint Chiefs of Staff to reconsider. On 22 July, they notified General MacArthur that they would review their previous decision. They asked him to help by telling them what he meant to do with the Marine brigade between its arrival date in late July and 10 September. At the same time, they ordered the brigade brought to full war strength and the Marine Air Group enlarged to full squadrons.[21]

Replying immediately, General MacArthur said that the 1st Provisional Marine Brigade, when it arrived on about 1 August, would be kept in Japan as GHQ Reserve, "To be used in Korea only in event of a critical situation." Meanwhile, he would train, outfit, and prepare the brigade for major amphibious operations in September.[22]

The Joint Chiefs of Staff had, meanwhile, been weighing General MacArthur's need for a full Marine division by 10 September against the dangers in cutting Marine strength in other parts of the world. Admiral Sherman proposed and the other Joint Chiefs approved a compromise by which the Marine strength in the Far East Command would be built up to two war-strength RCT's by mid-September. Even this solution, which would put only two-thirds of a Marine division in Korea by 15 September, would greatly reduce Marine security forces in the United States and cause an extensive call-up of Reserves. The Joint Chiefs, in a teleconference on 24 July, told MacArthur that, "We have now determined it is practicable to further augment the Marine Brigade after its arrival in Japan and bring it to division war strength less one RCT by mid-September. We have directed that this be done. The third RCT cannot be furnished until winter." General MacArthur did not care for this compromise and remonstrated at once. "Subtraction of an RCT from the Marine division," he contended, "tends to jeopardize the entire conception and would involve risks that cannot be determined finally at this time. I regard the third RCT as essential." But Washington officials stood firm. They explained, with forbearance, that the only trained Marine battalions left after sending two regiments to the Far East Command would be one battalion in the 2d Marine Division, one afloat in the Mediterranean, and a battalion of school troops at Quantico, Virginia. These they considered the minimum for absolutely essential needs in the Atlantic.[23]

Still unhappy with the new arrangements, MacArthur shelved the matter for the time being. Other developments were pressing. Whereas the 1st Provisional Marine Brigade had been headed for Kobe, Japan, mounting pressure by the enemy against Walker's perimeter and signs of a strong enemy force sweeping down the west coast to outflank Eighth Army forced MacArthur to abandon plans to keep the Marines as GHQ Reserve in Japan. On 25 July, he ordered the ground elements of the brigade diverted to Pusan, and to be prepared to execute a rapid nontactical debarkation. Units and equipment peculiar to am-

[21] Rad, JCS 86778, JCS to CINCFE, 22 Jul 50.
[22] Rad, C 58473, CINCFE to DA, 23 Jul 50.

[23] (1) Memo, CNO for JCS, 24 Jul 50, sub: Deployment of Fleet Marine Forces to the FEC, cited in Hoare, The Joint Chiefs of Staff and National Policy, vol. V, ch. IV, p. 14. (2) Telecon, TT 3573, JCS and CINCFE, 24 Jul 50.

phibious operations were kept on board ships and taken to Kobe. Upon landing at Pusan on 3 August, the ground troops of the 1st Provisional Marine Brigade were attached to General Walker's Eighth Army and went into a Reserve assembly near Masan.[24]

While hastily assembling another RCT in the United States for shipment to the Far East Command for use by 10 September, the Joint Chiefs on 10 August decided they need not wait until winter to send General MacArthur the third regiment of the Marine division. On that date, they authorized the formation of the final regiment, the unit to arrive in the Far East Command during September. In order that the Joint Chiefs of Staff appreciate the impact of their decision, Admiral Sherman sketched for them the drastic measures that the Marine Corps had to take to give MacArthur a full division. ". . . it will involve," he told them, "moving to the FEC the Marine battalion now in the Mediterranean, one battalion now at Camp LeJeune, and an RCT, less two battalions, to be formed at Camp Pendleton. So doing will eliminate the capabilities of the Fleet Marine Force in the Atlantic for several months." The battalion from the Mediterranean would have to come directly from Suda Bay through the Suez Canal and be hastily augmented with men sent directly to the Far East Command.[25]

The 1st Marines' additional rifle companies and platoons to bring the 1st Provisional Marine Brigade (5th Marines) up to war strength, and support and service units for the division had been building up at Camp Pendleton. These loaded at San Diego between 14 and 24 August and reached Japan between 28 August and 2 September. The third regiment was activated as the 7th Marines on 17 August at Camp Pendleton. Two understrength battalions of the 6th Marines from Camp Lejeune and individual Regulars and Reserves were assigned to the new regiment. Its other battalion, the peace-strength battalion from the Mediterranean, sailed directly to Japan from its post with the fleet. A third rifle company and third platoons for the battalion's other two companies formed with the main body of the 7th Marines.[26]

Admiral Sherman, during his visit to the Far East Command in late August, queried his Washington headquarters on the arrival date of this final component of the division. He was touring the battlefront in Korea when the discouraging reply reached him. "The limiting factor," Sherman learned, "is the readiness of Marine Corps troops, which cannot be advanced ahead of an already tight schedule." Owing to the need for training, the two Marine battalions from the United States could not reach the Far East Command until 19 September, while the battalion coming from the Mediterranean would arrive in Korea on 12 September. "It is impossible," Admiral Sherman was told, "for the entire

[24] (1) Rad, CX 586628, CINCFE to COMNAVFE, 28 Jul 50. (2) Rad, CX 58763, CINCFE to COM NAVFE and CG Eighth Army, 26 Jul 50. (3) Rpt, Mobilization of the Marine Corps Reserve, Hq U.S. Marine Corps, ch. II, p. 1, copy in OCMH.

[25] (1) Hoare, The Joint Chiefs of Staff and National Policy, vol. IV, ch. V, p. 14. (2) Memo, CNO for JCS, 11 Aug 50, in G-3, DA file 320.2 Pac, Case 30/2.

[26] Lynn Montross, "The Inchon Landing—Victory Over Time and Tide," p. 28.

Marine Division to arrive in Japan by 10 September."[27]

While aware of the problems facing the Marine Corps in readying units for shipment, Admiral Sherman was equally aware of MacArthur's problem. He ordered the expediting of the departure from the United States of the 7th Marines' RCT elements. Granting that a division commander could best judge his division's training requirements, Sherman nevertheless told naval officers in Washington that they must take account of the requirements of the Korean campaign and the great need for bringing the division up to strength as early as possible after the Inch'on landing. "It must be assumed," Admiral Sherman radioed his staff, "that the operation will not be delayed and if two battalions are late, the division will fight without them."[28] But for all of Sherman's urging, the 7th Marines with accompanying troops did not embark until 3 September, and reached Korea on the 21st, too late for the landing.

A minor controversy centered around General Walker's very natural unwillingness to release the 1st Provisional Marine Brigade. The brigade had been in almost constant action since its arrival, attacking and counterattacking in the southern sector of the Pusan Perimeter, and had proved to be a mainstay of General Walker's defense. When General Smith, commander of the 1st Marine Division, reached Tokyo on 22 August, he had assumed the brigade would be released to him. He had already ordered liaison officers exchanged between his division headquarters and the brigade staff, and key officers of the brigade had come to Tokyo for briefing on the landing operation. On 30 August, Smith asked Almond for the brigade at once. According to Smith, General Almond appeared very reluctant to commit himself. He apparently did not want to decide, in his capacity as chief of staff, GHQ, on a definite date at which the brigade would be released to the 1st Marine Division to operate under himself as commanding general, X Corps. General Smith, after his talk, made his request more official, sending a radio to commanding general, X Corps, asking for the brigade by 1 September. General MacArthur's headquarters on 1 September ordered the brigade made available to the 1st Marine Division on 4 September, but apparently because of objections raised by General Walker, rescinded the order the same day.

At a showdown meeting on 3 September, General Smith, backed by Admiral Joy, Vice Adm. Arthur D. Struble, and Admiral Doyle, again made his demand for the brigade to General Almond. General Ruffner and General Wright were also present. Almond proposed that the Marine brigade be left with General Walker. He offered to give the 1st Marine Division the 32d Infantry Regiment, 7th Division, as a replacement unit. General Smith refused to accept at the last minute an untrained and untried Army unit for a specially trained and tested regiment of Marines. He felt that it would be unfair to the 32d Infantry and to his own division. He doubted also if it would be physically possible to make the substitution. Shipping had

---

[27] (1) Rad, C 60782, CINCFE (Sherman) to JCS (CNO), 21 Aug 50. (2) Rad, C 60823, CINCFE to CO EUSAK for Adm Sherman, 21 Aug 50.

[28] Rad, 221009, COM 7th Fleet to CNO, 22 Aug 50.

already left for Korea to pick up the Marine brigade and would have to return if it were to pick up the 32d Infantry. Naval officers unanimously opposed Almond's solution. Admiral Struble then hit upon a compromise. He suggested that one of the 7th Division's regiments be sent to Pusan, remaining aboard ship as a floating reserve. This Army regiment would be available to General Walker in extreme emergency and the Marine brigade would be released to the 1st Marine Division. Almond agreed to this plan.

General Wright flew to Eighth Army headquarters in Taegu on the next day, telling Walker of the new arrangements. He relayed instructions from General MacArthur to pull the Marine brigade out of the line not later than the night of 5–6 September and to send it straight to Pusan. To compensate in some measure for the loss of this valuable force, the 17th Infantry Regiment would arrive in Pusan Harbor before 7 September.

Wright tendered further compensation when he told Walker that as soon as the first RCT of the 3d Division, the 65th Infantry, arrived in the theater it would be sent directly to Pusan for assignment to Eighth Army. This RCT would arrive in Korea between 18 and 20 September. Then, unless the 17th Infantry had already been committed to meet an emergency, it would be sent to rejoin its parent 7th Division in the Seoul-Inch'on objective area. General Walker complied with his orders and withdrew the 1st Provisional Marine Brigade from the perimeter on the night of 5–6 September. On 12 September it sailed from Pusan as the 5th Marines' RCT, to rendezvous with the 1st Marine Division at Inch'on.[29]

The 7th Division

Even before he realized that the 7th Division would have to make up his major Army component for Inch'on, General MacArthur had begun to rebuild this depleted unit as much as he could. In mid-July, when the 2d Division was still slated for Inch'on, General MacArthur had ordered 20 percent of all combat replacements from the United States diverted to the 7th Division in Japan. He had also halted all further levies against the division for men and equipment. By stabilizing the division, by feeding in such resources as could be spared from Eighth Army, and by intensive training, he hoped to make the 7th Division strong enough to fight effectively in Korea by October. On 26 July, MacArthur ordered General Walker to prepare the 7th Division "by intensified

[29] (1) Ltr, Gen Smith to Col Appleman, copy in OCMH. (2) Rad, CX 61738, CINCFE to CG EUSAK, 1 Sep 50, with penciled notations on copy in 8th Army file AG 322, 24 Aug–13 Dec 50. (3) MFR, 4 Sep 50, sub: Visit to EUSAK, by Maj Gen Edwin K. Wright, in AG, DA files (CofS), FEC, UNC. (4) Col. John C. Chiles, formerly SGS GHQ, FEC, told the author on 17 February 1955, that he had been present in the Dai Ichi Building during the conference. According to Colonel Chiles, when General Almond telephoned General Walker that he would have to release the 1st Provisional Marine Brigade, General Walker became extremely excited and stated that he could not take the responsibility for the safety of the Pusan Perimeter if the brigade was taken from him. Admiral Doyle, on the other hand, said that he could not accept the responsibility for the Inch'on landing unless he was given the brigade. According to Colonel Chiles, General MacArthur personally made the decision.

training and re-equipping for movement to Korea at the earliest practicable date." This instruction illustrates the dual function then charged to General Walker. While directing his divisions in combat against the North Korean Army, Walker, at the same time, remained responsible for the training and rebuilding of the 7th Division nearly a thousand miles away. The division then stood at less than half strength, with only 574 officers and 8,200 enlisted men. Moreover, many of the division's enlisted men had had little training, and few of the specialists and experienced noncoms taken from the division to patch up units going into combat in early July had been replaced.[30]

Desperately short of men himself, General Walker urgently appealed to General MacArthur on 29 July for the 7th Division's 32d Infantry to be flown into his perimeter. This appeal came shortly before the 5th RCT, the 1st Provisional Marine Brigade, and the 9th RCT of the 2d Division landed at Pusan. Knowing that these three regiments were to arrive and aware of the low combat potential of the 32d Infantry, General MacArthur denied this request, explaining that granting it "would completely emasculate present plans for the entire 7th Division, which is being reconstituted and will move to Korea, probably in late September." [31]

By 4 August, MacArthur saw clearly that if the amphibious force for the Inch'on landing included an Army division, his own command would have to provide it. He therefore called upon Walker to rebuild the 7th Division by 15 September. Walker was to let MacArthur know at once of any difficulties in getting the necessary material and people. MacArthur himself assisted the rebuilding process by moving to the division from Okinawa 1,600 men originally intended for a third battalion of the 29th Infantry Regiment. He also diverted to the division an antiaircraft artillery automatic weapons battalion newly arrived from the United States, as well as two companies of combat Engineers, and sent a rush call to the ZI port of embarkation asking that the three infantry battalion cadres destined for the division be sent without delay.[32]

MacArthur held little hope that the key men transferred from the division to Korea could be replaced in kind, either from the United States or from Japan. Efforts to recover these specialists reached a new high on 7 August, when General Hickey visited Korea and sought the return of 7th Division specialists. Walker made a careful survey to determine if he could give up any of these men, but

[30] (1) Rad, CINCFE to CG Eighth Army, 19 Jul 50, G–1, GHQ Daily Log, 19 Jul 50, Item 62. (2) Rad, CINCFE to CG Eighth Army, 22 Jul 50, G–1, GHQ Daily Log, 22 Jul 50, Item 38. (3) Memo, G–1 GHQ for CofS GHQ, 24 Jul 50, sub: Replacements for 7th Div, G–1, GHQ Daily Log, 24 Jul 50, Item 36. (4) Rad, CINCFE to CG Eighth Army, 26 Jul 50, G–1, GHQ Daily Log, 26 Jul 50, Item 52. (5) Rpt, CG 7th Div to CINCFE, 27 Jul 50, G–1, GHQ Daily Log, 27 Jul 50, Item 52.

[31] (1) Rad, CX 20657 KCO, CG, EUSAK to CINCFE, 29 Jul 50. (2) Rad, CINCFE to CG EUSAK, 30 Jul 50, G–1, GHQ Daily Log, 30 Jul 50, Item 26.

[32] (1) Memo, G–1 GHQ for CofS GHQ, 9 Aug 50, sub: Assignment of 29th Inf (less two battalions), G–1 GHQ Daily Log, 9 Aug 50, Item 53. (2) Rad, CINCFE to SFPE (Stoneman), 12 Aug 50, G–1 GHQ Daily Log, 12 Aug 50, Item 60. (3) Memo, G–1 GHQ for CofS GHQ, 10 Aug 50, sub: Replacements for 7th Inf Div, G–1 GHQ Daily Log, 10 Aug 50, Item 19.

because of the low ebb in Eighth Army's fortunes and strength at the time, found their release impossible.[33]

The lack of specialists and trained men for the 7th Division was on General MacArthur's mind when he talked on 7 August with Harriman, General Ridgway, and General Norstad. MacArthur furnished a complete list of the specialists he needed but who could not be found in his command and asked why the Department of the Army did not quickly recruit experienced noncommissioned officers from among the many who had served in World War II. These men could be sent to him by fast ship and by air.[34]

Three days later, MacArthur informed the Department of the Army of the unusual steps he had taken to refurbish the 7th Division. He estimated that 30 percent of all replacements arriving in the theater before 10 September would be diverted to the 7th Division so that it would be only 1,800 men understrength by the CHROMITE target date. He had already exhausted all other sources of replacements.[35]

The high priority given the 7th Division worked hardships on the American divisions in Korea. All artillery replacements and all infantry replacements having certain qualifications were channeled to the division. These actions, while weakening other units, proved effective in bringing the 7th Division to a reasonable level. By 7 September, shortly before loading for the invasion at Inch'on, the division lacked only 1,349 officers and men of its full war strength.[36]

Compensating, numerically at least, for this slight understrength of the 7th Division, MacArthur, after conceiving the idea that South Korea might be called on to provide soldiers for American units, attached more than 8,000 Koreans to the division. On 11 August he directed General Walker to procure, screen, and ship to Japan for use in augmenting the 7th Division approximately 7,000 able-bodied male Koreans. Fortunately the ROK Government co-operated since no American commander had authority beyond merely requesting these men. As a commentary on the desperation out of which this measure was born, General Wright on 17 August talked to the chief of staff, GHQ, by telephone from Korea. He told him that about 7,000 Koreans were being shipped out of Pusan that day. "They are right out of the rice paddies," he said, "and have nothing but shorts and straw hats. I understand they have been inoculated, given a physical examination and have some kind of paper. I believe we should get busy on equipment." [37] These Korean men were brought to Japan, equipped and trained briefly, and then attached to the 7th Division. By 31 Au-

33 Memo, Gen Hickey, DCofS GHQ, for Gen Almond, 7 Aug 50, sub: Rpt of Visit to Korea, copy in OCMH.

34 Truman, *Memoirs*, II, 351.

35 Rad, CX 59802, CINCFE to DA, 10 Aug 50.

36 (1) Memo, CofS GHQ Reserve (Gen Ruffner) for CofS GHQ (Gen Almond), 28 Aug 50, sub: Strength of 7th Div, G–1 GHQ Daily Log, 28 Aug 50, Item 55. (2) Memo, G–1 GHQ for CofS GHQ, 8 Sep 50, sub: Status of 7th Inf Div Personnel, in CofS GHQ, UNC files.

37 (1) Memo, G–3 GHQ for CofS ROK, GHQ, 17 Aug 50, in CofS GHQ, UNC files. (2) Memo for Gen Beiderlinden, 8 Sep 50, sub: ROK Personnel With U.S. Units, in CofS GHQ, UNC files. (3) Rad, CX 59818, CINCFE to CG EUSAK, 11 Aug 50. (4) Rad, CX 60020, CINCFE to CG EUSAK, 13 Aug 50.

gust, 8,652 Koreans had joined the 7th Division.[38]

In a related action, General MacArthur ordered General Walker to strengthen each company and battery of American troops under his command by adding a hundred Koreans as rapidly as individual arms and equipment could be procured. The increase was to be made without regard to the present or future strength of the ROK Army. He authorized Walker to raise the ROK Army to any number he deemed practicable or advisable and to requisition equipment when the figure had been determined.[39] But, by the end of August, little progress had been made toward attaching Koreans to American units other than the 7th Division. The 1st Cavalry Division had 739 Koreans, the 2d Division had 234, the 24th Division had 949, and the 25th Division 240.[40]

Admiral Joy recommended to General MacArthur on 7 August that amphibious training of the 7th Division begin immediately even though the unit was then at less than half strength. He pointed out that the embarkation date for the prospective assault amphibious landing was 5 September and that training a RCT to conduct an opposed amphibious assault would delay it. He had already conferred with the commanding general of the 7th Division and had instructed him on the training objectives to be achieved before embarkation. These included proficiency in amphibious operations. General MacArthur ordered amphibious training for the 7th Division to begin as soon as possible, under the control and supervision of COMNAVFE.[41]

Airborne Units

MacArthur had no airborne troops when the fighting began in Korea. The 11th Airborne Division, which had served on occupation duties, had returned to the United States more than a year before. MacArthur now wanted airborne forces badly. The ability of such airborne troops to drop behind enemy lines, to sever lines of communications, and to disrupt rear-area activities had been proven during World War II. The increasing vulnerability of the North Korean Army to such tactics provided the perfect setting for airborne employment, particularly in conjunction with amphibious attack.

His early attempts to procure airborne troops included an effort on 8 July to have a complete regiment, with its equipment, flown to Japan. He apparently intended to use this airborne unit in Operation BLUEHEARTS. General Vandenberg, Air Force chief of staff, offered to fly the regiment and its equipment to Japan in C–119 aircraft if other members of the Joint Chiefs of Staff thought it necessary. But this emergency aerial movement would have required the diversion of Military Air Transport Service carriers and commercial planes which

[38] (1) The complete story of this unique experiment is contained in Mono, Maj Elva Stillwaugh, Personnel Policies in the Korean Conflict, copy in OCMH. (2) Telecon, TT 3708, DA and CINCFE, 30 Aug 50.

[39] Rad, CX 59709, CINCFE to CG Eighth Army, 9 Aug 50.

[40] Rpt, unsigned, 31 Aug 50, sub: Strength of South Koreans Attached to U.S. Divs as of 31 Aug, G–3 Opns Jnl, FEC and Pac Br, G–3, DA.

[41] (1) Rad, 0707027, COMNAVFE to CINCFE, 7 Aug 50. (2) Rad, CX 59636, CINCFE to CG Eighth Army and COMNAVFE, 8 Aug 50.

already were flying huge cargoes of men and matériel to MacArthur. If MacArthur's estimates were correct, these shipments were much more sorely needed than an airborne RCT, and should take precedence.

For this reason, and because no airborne RCT's, except for those of the 82d Airborne Division, were ready to fight immediately, the Joint Chiefs of Staff denied MacArthur's July request.[42] But they did take steps to ready an airborne unit for deployment as soon as possible. Whereas MacArthur actually had asked for an RCT from the 82d Airborne Division, the Joint Chiefs of Staff decided against weakening the only effective infantry division left in the United States and chose instead an RCT from the 11th Airborne Division. The commanding general of the 11th Airborne Division had been informed of the possible deployment on 7 July, but with the decision against air transport to Japan, no immediate action was taken. Planning continued, however, for possible movement by ship.

When General Collins learned during his conference in Tokyo that General MacArthur's plan for Inch'on included a role for the airborne RCT, he was somewhat concerned. He told General Almond, after hearing the latter describe the planned seizure of the north bank of the Han River by an airborne unit, that the Joint Chiefs of Staff would take a very personal interest in how General MacArthur employed the airborne troops. He assured General Almond that the Joint Chiefs of Staff would do their best to furnish planes to drop the vehicles and howitzers of the RCT, but cautioned against wasteful and improper employment of these specially trained troops. "Don't overestimate what one RCT can do," the Army Chief of Staff warned Almond. "Don't get too grandiose in your planned utilization of the limited troops available." [43]

When using the phrase "limited troops available," Collins was not exaggerating. The 11th Airborne Division had so few men that only one RCT, at less than half its authorized infantry strength, could be formed on 15 July. Since the beginning of July Army authorities had been assigning all officers and men completing the Army Parachute School at Fort Benning, Georgia, to the 11th, feeding in about 400 trained jumpers each week. General Bolté, investigating the readiness date for the airborne RCT, was told that by transferring trained jumpers from the 82d Airborne, the 11th Airborne RCT could be readied for shipment to MacArthur by 1 August. On the other hand, the current process of filling the RCT with graduates of the parachute school only would slow its departure until 20 September. The latter method did not disrupt the 82d Airborne, however, and was therefore the method most acceptable to General Bolté and General Collins. On 18 July, the Department of the Army told General MacArthur that the 11th Airborne RCT would be ready at home station by about 20 September. Asked to comment, he

[42] (1) Rad, C 57379, CINCFE to DA, 8 Jul 50. (2) Memo, G–3 DA for CofS, 8 Jul 50, sub: Troop Requirements Forwarded by General MacArthur to the DA for the JCS, in G–3, DA file 320.2 Pac, Case 21. (3) Rad, WAR 85328, DA to CINCFE, Collins (Personal) for MacArthur, 9 Jul 50.

[43] Memo, Col Dickson for Gen Bolté, sub: Record of Visit to FEC, 10–15 July 1950, in G–3, DA file 333 Pac, Case 3, Tab 6.

objected that his plans for the landing at Inch'on required these troops in his theater by 10 September and urged every effort to have them there on time.[44]

The brief description presented orally to General Collins during his visit apparently had not justified sufficiently the need for immediate deployment of the RCT. Whereupon, Washington asked General MacArthur for a more detailed explanation of the mission he would give the airborne RCT in the landing operation. On 23 July, General MacArthur replied that he planned to mount an airdrop from Japan, landing the airborne troops in the Inch'on objective area as soon after D-day as the situation warranted. They were to seize a key communication center immediately ahead of troops advancing out of the beachhead area.

At this time, when it was not at all certain that sufficient amphibious forces could be sent to MacArthur or that the landing at Inch'on would even be made, MacArthur's requirement for airborne troops appeared, to Army officials, secondary. The condition of the 11th Airborne Division, moreover, remained such that the Department of the Army deemed it impractical to send any of the division's regiments into combat in September. Army authorities informed General MacArthur in teleconference that the RCT would be operational in Japan by 23 October, but that he could not count upon using it in his landing operations. In turn, MacArthur remonstrated once again, asking that the Joint Chiefs of Staff expedite the arrival of the unit.[45]

Despite General MacArthur's protests, General Ridgway and General Haislip drew up a plan on 25 July to move the 187th RCT of the 11th Airborne Division to Japan with an operational readiness date in the Far East Command of 21 October. Infantry fillers would be transferred to the unit from the 82d Airborne if necessary. One hundred C–119 aircraft would arrive in the Far East Command in time to allow the RCT fifteen days of operational training prior to 21 October. On this basis, build-up of the 187th Airborne RCT went forward during July and most of August. By 19 August, the regiment had been built up to nearly 4,000 officers and men and was undergoing intensive training.[46] Arrangements progressed ahead of the original schedule and General MacArthur was told that the 187th RCT would be at the port of embarkation by 12 September. He again objected that in order to accomplish his planned operation he would have to have the unit and its required airlift in Japan by 10 September.

But General Ridgway, himself an airborne officer, opposed any stepped-up shipment of the airborne RCT. He advised General Collins, after studying General MacArthur's objections, ". . . I think the only justification for compli-

---

[44] (1) MFR, Col Dickson, G–3, DA, 17 Jul 50, sub: Readiness Date for the RCT of the 11th Abn Div. (2) Memo, Gen Ogden, Chief, Org and Training Div, G–3, DA, for Gen Bolté, G–3, DA, 15 Jul 50, same sub. Both in G–3, DA file 320.2, Case 6/5. (3) Rad, W 86323, DA to CINCFE, 18 Jul 50.

[45] (1) Rad, C 58473, CINCFE to DA, 23 Jul 50. (2) Telecon, TT 3573, DA with CINCFE, 24 Jul 50.

[46] (1) Memo, Gen Timberman, Opns Div, G–3, for Gen Bolté, G–3, DA, 25 Jul 50, sub: Movement of RCT of 11th Abn Div to FECOM, in G–3, DA file 320.2, Case 6/5. (2) Memo, Gen Bolté for Gen Ridgway, 19 Aug 50, sub: Movement of RCT of 11th Abn to FEC, with 1st Ind by Gen Ridgway (sgd F. F. Moorman) with Incls, same file.

ance would be a situation so desperate that the addition of an RCT as a straight infantry outfit was necessary to save the situation. It does not appear to me that such is the case." General MacArthur's objections were overruled and, in mid-August, he was told not to expect the airborne troops in time for his landing operation.[47]

General Collins, on a second visit to Tokyo late in August, found General MacArthur still insistent that the airborne RCT be sent in time to take part in Operation CHROMITE. Collins promised to do what he could and, upon returning to Washington, made a special effort to expedite arrangements. His investigation convinced him that his staff had been doing its best, and on 25 August he explained to General MacArthur that he had satisfied himself that an airborne RCT could not be sent by 10 September. He had even considered taking a regiment from the 82d Airborne instead of the 11th, but had found that this drastic action would have made no appreciable difference in the arrival date. For the delay was no longer caused by personnel shortages but by difficulties in procuring, assembling, and loading the specialized equipment required for airborne operations. General Collins felt that every reasonable and practicable measure had been taken to expedite the arrival of the RCT but that the unit would not be there for CHROMITE.

In his final word to General MacArthur on 28 August, he pointed out that by expediting to the maximum extent, the 187th Airborne RCT could reach Sasebo, Japan, on 21 September. The unit could then complete preparations for an airborne drop of the entire regiment by 29 September, but no earlier. "I strongly urge," General Collins said, "it not be committed prior to that date. The unit is presently capable of daylight operations only. However, I am confident that this unit will, in all respects, meet the high combat standards set by our airborne units in the last war." There appeared to be no appeal from these opinions of the Chief of Staff, and General MacArthur acquiesced, replying that his plans would be adjusted.[48]

The 187th Airborne RCT left Camp Stoneman, California, on 6 September and arrived in Japan on 20 September with a strength of about 4,400 men and officers.[49]

The Assault in Readiness

The U.S. X Corps, at its embarkation, numbered slightly less than 70,000 men. Included as its major units were the 1st Marine Division, the 7th Division, the 92d and 96th Field Artillery Battalions, the 56th Amphibious Tank and Tractor Battalion, the 19th Engineer Combat Group, and the 2d Engineer Special Brigade. The 1st Marine Division had a

[47] (1) Rad, CX 59999, CINCFE to DA, 13 Aug 50. (2) Rad, W 88966, DA to CINCFE, 16 Aug 50. (3) Memo, Gen Ridgway for CofS, 14 Aug 50, in CofS DA file 370, Case 11.

[48] (1) MFR, sgd Lt Col Herrick, 29 Aug 50, sub: Advancement of Date of Movement to FEC of 187th RCT of 11th Abn Div, in G-3, DA file 320.2, Case 6/4. (2) Rad, W 90063, DA (Collins) to CINCFE (MacArthur), 25 Aug 50. (3) Rad, WAR 89967, JCS (Collins) to CINCFE, 28 Aug 50. (4) Rad, C 71576, CINCFE (MacArthur) to DA (Collins), 30 Aug 50.

[49] (1) Interv, Capt Charles Thebaud with Maj C. M. Holland, 187th Abn RCT, Beppu, Japan, 19 Jan 52. (2) Ltr, Hq, 187th Abn RCT, to DA, 15 Dec 51, sub: Insignia. (3) War Diary, 187th Abn RCT, 1 Aug to 31 Oct 50.

strength of 25,040 men, including 2,760 attached Army troops and 2,786 Korean marines. The 7th Marines, which arrived on 21 September, added 4,000 men to the division strength.[50]

The echelon of command for CHROMITE progressed downward from General MacArthur through Admiral Joy, COMNAVFE, in the usual pattern established during World War II for amphibious operations. Admiral Struble, as Commander, Joint Task Force Seven, and Commander, Seventh Fleet, was actually in command of the amphibious phase of the operation. Under him, Admiral Doyle commanded the attack force (Amphibious Group One) which, in turn, controlled the landing force, composed of the 1st Marine Division. Command of the landing force was scheduled to pass to General Smith, Commanding General, 1st Marine Division, after the beachhead was secured and Smith had notified Doyle he was ready to assume command ashore. Command of the expeditionary troops, the U.S. X Corps, was to pass to General Almond from Admiral Struble after the corps had landed and Almond had indicated that he was ready to assume command.[51]

VICE ADMIRAL DEWEY STRUBLE

As D-day for Operation CHROMITE approached, the ports of Kobe, Sasebo, and Yokohama in Japan and Pusan in Korea became centers of intense activity. The 1st Marine Division, less the 5th Marines, loaded at Kobe, the 5th Marines at Pusan. The 7th Division loaded at Yokohama, and most of the escorting naval vessels, the Gunfire Support Group, and the command ships, at Sasebo. In order to reach Inch'on by 15 September, the landing ships, tank (LST's) had to leave Kobe by 10 September and the attack transports and cargo ships by 12 September. Only the assault elements were combat-loaded. The rest of the invasion force and the vast quantity of equipment and supplies were organization-loaded.[52]

General MacArthur, General Almond, and General Shepherd flew from Tokyo to Sasebo, joining naval commanders aboard the *Mt. McKinley* on the evening of 12 September. Some of the final arrangements for the landing were completed aboard the flagship.

[50] (1) 1st Marine Div SAR, vol. 1, Annex A, 5. (2) Hist Rpt, X Corps, G-3, Opn CHROMITE, p. 2.

[51] (1) Joint Task Force Seven, Inchon Rpt, Opn Plan. (2) X Corps Opn Order 1, Annex 1, 28 Aug 50.

[52] (1) 1st Marine Div SAR, 15 Sep-7 Oct 50, Annex D, p. 4. (2) War Diary, 7th Inf Div, Sep 50.

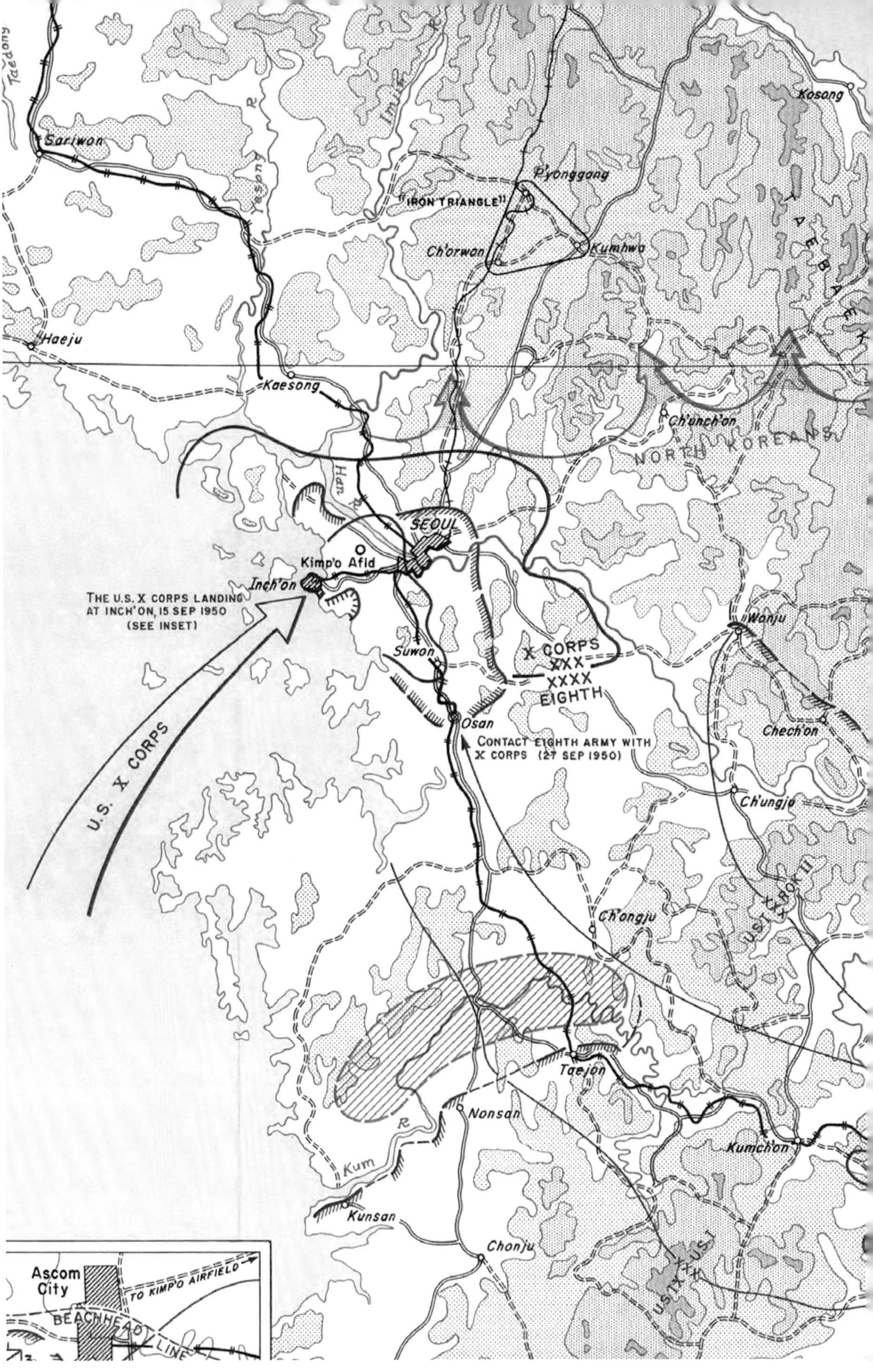

Taedong
Sariwon
Haeju
Kaesong
Yesong R
Imjin R
Pyonggang
"IRON TRIANGLE"
Ch'orwon
Kumhwa
Kosong
TAEBAEK
Ch'unch'on
NORTH KOREANS
Han R
SEOUL
Kimp'o Afld
Inch'on
THE U.S. X CORPS LANDING AT INCH'ON, 15 SEP 1950 (SEE INSET)
U.S. X CORPS
Suwon
X CORPS
XXX
XXXX
EIGHTH
Osan
CONTACT EIGHTH ARMY WITH X CORPS (27 SEP 1950)
Wonju
Chech'on
Ch'ungju
Ch'ongju
Taejon
Nonsan
Kum R
Kunsan
Kumch'on
Chonju
Ascom City
TO KIMP'O AIRFIELD
BEACHHEAD LINE

CHAPTER X

Crossing the Parallel: The Decision and the Plan

The Tide Turns

Events dramatically justified General MacArthur's firm confidence in Operation CHROMITE. American Marines, backed by devastating naval and air bombardment, assaulted Inch'on on 15 September and readily defeated the weak, stunned North Korean defenders. (*Map II*) On hand to see for himself the fruition of his plans, General MacArthur sent a cheering report from the scene to the Joint Chiefs of Staff: "First phase landing successful with losses slight. Surprise apparently complete. All goes well and on schedule." By mid-day, Marines had seized Wolmi-do, the fortress island dominating Inch'on harbor. By nightfall, more than a third of Inch'on had fallen into their hands. Obviously enjoying his first taste of victory in Korea, the U.N. commander again proudly reported to Washington, "Our losses are light. The clockwork coordination and cooperation between the Services was noteworthy. . . . The command distinguished itself. The whole operation is proceeding on schedule." [1]

Operation CHROMITE stayed on schedule. In the wake of the Marines, the 7th Division landed and struck south toward Suwon. Kimp'o Airfield fell to the Marines on 19 September, and on the 20th General MacArthur could tell the Joint Chiefs of Staff that his forces were pounding at the gates of Seoul.[2] So far, American forces had suffered only light casualties, while the North Koreans had lost heavily. At Inch'on, supplies were being unloaded at the rate of 4,000 tons daily; and Kimp'o Airfield

[1] (1) Rad, 142215Z, CINCUNC to JCS, 15 Sep 50. (2) Rad, C 63153, CINCUNC to CINCFE and JCS, 15 Sep 50.

[2] The Joint Chiefs of Staff were disturbed by newspaper reports that they had opposed the Inch'on landing and had not fully supported General MacArthur. One such dispatch said, "MacArthur sold the Joint Chiefs of Staff on the Inchon landing despite their unanimous objections to such an ambitious undertaking. . . . Sources close to General MacArthur said both General Collins and Admiral Sherman were opposed to the landing at Inchon." The Joint Chiefs notified General MacArthur that they were issuing a statement refuting these press reports and, to a limited extent, giving their own side of the background story. See Rad, W 91763, DA to CINCFE, 17 Sep 50.

A Russian-Made 76-mm. Gun *in fortified position on Wolmi Island.*

had swung into round-the-clock operation. When General Almond took command of all forces ashore in the Inch'on-Seoul area at 1800 on 21 September, he had almost 6,000 vehicles, 25,000 tons of equipment, and 50,000 troops.[3]

Fortunately, the success of MacArthur's plan did not depend upon an immediate juncture of the Eighth Army and X Corps. For, although MacArthur had ordered General Walker to attack out of the Pusan Perimeter beginning on the day after the X Corps landing, the North Koreans along the Naktong fought as fiercely on 16 September as they had on the 14th, and for nearly a week stood off all attempts by Eighth Army to punch through their defenses. The main body of the North Korean Army appeared unaware of the landing at Inch'on, approximately 180 air miles to its rear, and saw no reason to quit.

[3] (1) Rad, C 63187, CINCUNC to CINCFE and JCS, 20 Sep 50. (2) Rad, X 10042 IN, CG X Corps to CINCFE, 23 Sep 50. (3) Appleman, *South to the Naktong, North to the Yalu*, p. 519.

INCH'ON RAILROAD STATION *after bombardment by U.N. naval vessels.*

Eighth Army intelligence officers had predicted this kind of enemy reaction, pointing out that a success at Inch'on would not necessarily relieve the pressure on Eighth Army, since the enemy could still move men and supplies against the perimeter over alternate routes along the east coast.[4] Indeed, the Eighth Army G–3 had pessimistically speculated that the most likely enemy reaction to the landing would be an all-out drive to push the Eighth Army into the sea.

General Walker, who had never been convinced that he could break out on schedule, blamed equipment shortages for the delay. He complained to General Hickey on 21 September that he was ". . . ready to break loose if it weren't for the physical trouble." He could not get his armor across the Naktong, he pointed out, and, referring to the greater logistic support given the X Corps, noted, "We have been bastard children lately, and as far as our engineering equipment is concerned we are in pretty bad shape." He seemed anxious that General MacArthur's staff should appreciate his plight, telling

[4] Intelligence Annex (10 Sep 50), Eighth Army Opns Plan 10, 6 Sep 50.

LST's Unloading on Wolmi Island

Hickey, "I don't want you to think that I am dragging my heels, but I have a river across my whole front and the two bridges which I have don't make much."[5]

Walker's failure to keep to his schedule made General MacArthur somewhat doubtful that the Eighth Army would be able to break out of the Pusan Perimeter at all. He perhaps recalled earlier warnings by Eighth Army officers that Walker's divisions could not fight their way north even if the Inch'on landing were successful. At any rate, after three days of indecisive struggle along the perimeter, MacArthur ordered General Wright to implement the alternate plan for an amphibious landing at Kunsan, by using two of Walker's American divisions and one of his ROK divisions in the amphibious assault. Kunsan, on the west coast about one hundred air miles south of Inch'on, had originally been favored by General Collins as the primary objective area. A landing there now, MacArthur felt, would threaten the enemy's immediate rear and cause a

[5] Telecon, Gen Walker with Gen Hickey, 21 Sep 50, in CofS GHQ, UNC files.

North Korean collapse. When General Hickey discussed this plan with General Walker on 22 September, the latter objected to giving up any of his forces for a landing at Kunsan or anywhere else. But the argument ended there. For by this time, signs of an enemy collapse had appeared and MacArthur shelved the Kunsan plan. The signs proved correct and by the next day the North Korean Army, at last feeling the effects of its severed lines of communications and the presence of a formidable force in its rear, began a general withdrawal from the Pusan Perimeter. The withdrawal turned into a rout. During the next week, Eighth Army pursued the fleeing enemy. On the morning of 26 September, a task force from the 7th Cavalry Regiment, 1st Cavalry Division, of Eighth Army met elements of the 31st Infantry, 7th Division, of X Corps near Osan to mark the juncture of the two forces.[6]

General Almond's corps meanwhile had enlarged its holdings in the Inch'on-Seoul area. By 26 September, the Marine-Army team had wrested control of the South Korean capital from the enemy and North Korean resistance in the sector was dwindling rapidly.

The 38th Parallel—Genesis of the Decision

Two decisions in the third week of September 1950 were to rank among the most significant of the Korean War. The first of these, the decision to invade North Korea, stemmed in part from military expediency but the underlying issues were mainly political. The second decision, to use the X Corps in another amphibious operation, was completely military. General MacArthur figured to a large degree in the 38th Parallel decision and personally decided how the X Corps would be used. Both decisions were made as the recapture of Seoul became a certainty; and both were reached in the course of establishing a plan for operations in Korea that would best serve the interests of the United States and the rest of the free world.[7]

President Truman, of course, bore the full and final responsibility for choosing a course of action for Korea. But from his military and civilian advisers at several stations within the executive branch, he demanded and received the best advice available on all aspects of a problem, including the alternatives and consequences, before he took a stand.[8] Before the Korean War was three weeks old, and while American and ROK forces were falling back on Taejon, the President called on these advisers to tell him whether MacArthur should eventually send forces across the 38th Parallel. These advisers saw no need to test the legality of crossing the parallel. The basic authority under which the United

[6] (1) Opn Plan 100-C, JSPOG, GHQ, UNC files. (2) Rad, 063180, CINCUNC (Wright) to CINCFE (Hickey), 19 Sep 50. (3) Memo, Gen Hickey for Gen Wright, 23 Sep 50, JSPOG files. (4) For details of Eighth Army's breakout, see Appleman, *South to the Naktong, North to the Yalu,* Chapters XXVII and XXVIII.

[7] Except as otherwise indicated, this section is based on the 091 Korea file of G-3, Department of the Army, for 1950, Cases 14/14, 14/16; 14/17, 14/19; 14/20, 14/22; 14/28, 14/30; 14/31, and 79/3.

[8] Symbolic of his approach to decision-making, a small sign resting on President Truman's White House desk carried the reminder, "The Buck Stops Here."

States directed operations of the unified command in Korea lay in the U.N. Security Council's resolution of 7 July 1950; and within this resolution the United States had been called upon to direct United Nations forces so as "to assist the Republic of Korea in defending itself against armed attack and thus to restore international peace and security in the area." The United Nations' call for the restoration of peace and security in the area, was generally considered sufficient legal basis to enter North Korea.

The main concern was whether crossing the parallel would provoke an attack by the neighboring Chinese Communists or by Russia. Indeed, from the time of the President's first call for recommendations through the period of preparation for the Inch'on landing, American officials sought out the best ways to achieve military and political objectives without causing World War III. They tried, in particular, to determine a long-range policy toward Korea that would strengthen the United States' position in relation to that of the USSR. For they assumed that the USSR was America's chief antagonist in Korea and elsewhere, and that if the course chosen by the United States came too directly into conflict with Russian aims and interests, the United States might have to fight to hold that course.

Those authorities nearest the President concluded by 1 September 1950 that the United States was in no position to commit itself finally to any single course of action. There were too many unknowns, namely, what Russia or China might do and whether the United States could count on the United Nations, even on those members considered to be allies, to back up an American policy that might bring on a general war.

In searching for some flexible stand for the United States to take, Truman's top advisers became convinced that any crossing of the 38th Parallel by General MacArthur would evoke certain reactions from Russia. The Russians might encourage the Chinese to occupy North Korea, even to commit troops into battle in the hope of fomenting war between the United States and China. In the latter event, the American officials believed, U.N. forces should continue to fight as long as there was a reasonable chance of successfully resisting the Chinese; General MacArthur should be authorized to take appropriate air and naval action against Communist China; and the United States should take the matter to the U.N. Security Council in order to have the Chinese condemned as aggressors.

Or, as MacArthur's forces approached the parallel, the USSR itself could reoccupy North Korea and trump up an arrangement with the North Korean Government whereby the Russians would pledge to defend North Korean territory. If this proved the case, that is, if major Russians units entered the fighting either openly or covertly anywhere in Korea, the top advisory officials felt that General MacArthur should go on the defensive, make no move that would aggravate the situation, and report to Washington. Exactly what MacArthur would be told once he had reported to Washington was not yet decided. But it was definite that the United States did not want its resources tied up in Korea, an area regarded as of little strategic importance, if general war came.

In line with their own advice against commitment to any single course of action, these advisory officials recognized that certain military conditions could arise, such as an opportunity to destroy the North Korean Army completely which would, from a tactical point of view, justify military operations north of the parallel. But if the President, who alone had the authority and sufficient knowledge of all factors to make a decision on the crossing, did authorize a move above the parallel, there should be a clear understanding that no U.N. force would cross the northern boundary of Korea into Manchuria or the USSR, and that as a matter of policy only Korean units should operate in the border region. Further, if either Russian or Chinese forces had already entered Korea or had announced that they intended to enter, no matter how well the tactical situation might otherwise favor crossing the parallel at the time, General MacArthur should refrain from moving above the line. This did not mean, however, that he should discontinue air and naval operations in North Korea.

Truman's top advisers did not consider crossing the parallel to be a necessary ingredient of victory. They believed that the military situation eventually would be stabilized along the parallel and that the United Nations, instead of crossing, could offer surrender terms to the North Koreans as soon as a U.N. victory seemed assured.

The opinions of President Truman's closest advisers did not find favor among the Joint Chiefs of Staff or with General MacArthur. MacArthur, since mid-July, when he had received the United Nations 7 July resolution as a guide but no detailed instructions, held a directly opposing view. "I intend to destroy and not to drive back the North Korean forces," he told Generals Collins and Vandenberg at the time, adding that "I may need to occupy all of North Korea." [9] MacArthur continued to favor crossing the parallel even after his G–2, General Willoughby, reported on 31 August that ". . . sources have reported troop movements from Central China to Manchuria for sometime which suggest movements preliminary to entering the Korean theater." Willoughby placed the number of regular Chinese troops in Manchuria at about 246,000 men, organized into nine armies totaling thirty-seven divisions. Eighty thousand men were reported assembling near An-tung, just across the Yalu from Korea.[10]

The Joint Chiefs of Staff disagreed with the view that the Korean fighting would be stabilized along the 38th Parallel. While quite aware of the possibility of Russian or Chinese entry into the conflict, they did not believe that MacArthur should be held back from crossing the parallel if he wished to do so for tactical reasons. Any views and proposals to the contrary, the military chiefs told Secretary of Defense Johnson on 7 September, were unrealistic. They agreed with General MacArthur that the initial objective to be obtained was the destruction of North Korean forces. "We believe," they stated:

---

[9] Memo, Col Dickson for Gen Bolté, 15 Jul 50, sub: Rpt of Trip to FEC, 10–15 Jul 50, in G–3, DA file 333 Pac, case 3.

[10] DIS, GHQ, FEC, No. 2913, 31 Aug 50, p. 1–d.

George C. Marshall

that after the strength of the North Korean forces has been broken, which is anticipated will occur south of 38 degrees North, that subsequently operations must take place both north and south of the 38th Parallel. Such operations should be conducted by South Korean forces since it is assumed that the actions will be of a guerrilla character. General MacArthur has plans for increasing the strength of the South Korean forces so that they should be adequate at the time to cope with this situation.[11]

Touching next on the subject of the posthostilities period, the Joint Chiefs of Staff informed the Secretary of Defense that they and General MacArthur agreed that the occupation by U.N. forces should be limited to the principal cities south of the 38th Parallel and should be terminated as soon as possible. Further, U.S. troops should be taken out of Korea as early as safe to do so. The Joint Chiefs of Staff also pointed out that General MacArthur and President Rhee had agreed that the Government of the Republic of Korea should be re-established in Seoul as soon at it could be done. Rhee was willing, upon re-entry into the capital, to grant a general amnesty to all except war criminals and to call for a general election to set up a single government for all of Korea.

The final policy proposal sent to President Truman on 9 September included the views of the Joint Chiefs of Staff. Without making any changes, the President approved the proposal on 11 September.

In order that General MacArthur might have advance notice, the Joint Chiefs of Staff on 15 September sent him those provisions of the new national policy applicable to operations above the 38th Parallel and actions to be taken if Russia or Communist China intervened. The Joint Chiefs had not yet been told to work this new policy into a new directive for MacArthur, but were anticipating such instructions from the Secretary of Defense. General MacArthur had other things on his mind on the day he received this informative message (it was D-day for Operation Chromite), but he wanted to know more about the national policy on Korea. As soon as he could, he asked the Joint Chiefs to forward by courier the entire text of the approved policy paper. This the Joint Chiefs arranged by handing copies to an officer from the Far East Command who was returning after an official visit in Washington.

[11] Memo, JCS for Secy Defense, 7 Sep 50, sub: U.S. Courses of Action With Respect to Korea.

As of 18 September, the Secretary of Defense had not yet told the Joint Chiefs of Staff to prepare instructions for MacArthur based on the new policy. This inaction perhaps was occasioned in part by Secretary Johnson's resignation, which he had submitted on 12 September, and which President Truman had accepted and made effective as of 19 September. General of the Army George C. Marshall became the new Secretary of Defense on 21 September.

Meanwhile, hoping to lend impetus to the matter of new instructions to MacArthur, General Gruenther, Army Deputy Chief of Staff for Plans, proposed to draft a directive at Army level for submission to the Joint Chiefs of Staff. But the Joint Chiefs had anticipated Gruenther and had already worked out the new directive.

Ten days after American troops stormed Inch'on, the Joint Chiefs sent MacArthur's directive for future operations in Korea to Secretary Marshall. They told him that while they had dealt with military matters primarily, the implications of the directive affected other agencies of the United States Government; and they suggested that the Secretary obtain the concurrence of these other agencies. They had taken no action on aspects of the new national policy outside their purview, assuming that the responsible agencies would take care of these in directives of their own. The Joint Chiefs did ask, however, that they be allowed to comment from the military point of view on any directives prepared by other agencies.

Several days went by with no word on the directive and General Bolté became impatient. The reports from Korea, encouraging from the military viewpoint, were nevertheless disconcerting to the Army G–3, who knew that General MacArthur would soon reach the 38th Parallel and the limit of his current instructions. The advance information which had gone to MacArthur had made it plain that he would not cross the 38th Parallel without specific authority from the President. "In view of the rapidity with which military operations in Korea are approaching the 38th parallel," Bolté told the Chief of Staff on 27 September, "it is a matter of military urgency that the commander of the United Nations forces be given authority to cross this parallel to accomplish attainment of his military objective." [12] General Bolté was fearful that a delay in definite orders from Washington would cause U.N. forces to hesitate and break stride in their advance at the parallel thus enabling the North Korean Army to retreat in orderly fashion without being destroyed. He recommended that General Collins press the Secretary of Defense for approval of MacArthur's crossing of the parallel.

Actually, Secretary Marshall had been waiting for State Department concurrence in the directive before showing it to President Truman. The State Department approved the draft but added a paragraph of instructions on the return of Seoul to the Republic of Korea Government. Before General Bolté's objections had reached the Chief of

[12] Memo, Gen Bolté for CofS, 27 Sep 50, sub: U.S. Course of Action in Korea, with note by Gen Gruenther on original.

Staff, the President had approved the directive.[13]

The Joint Chiefs of Staff sent the directive to General MacArthur on 27 September, stipulating that it was being furnished to provide him with "amplifying instructions as to further military actions to be taken by you in Korea." They warned him, "These instructions, however, cannot be considered to be final since they may require modification in accordance with developments." Obviously wary of what the Russians or Chinese might do, they ordered MacArthur "to make special efforts to determine whether there is a Chinese Communist or Soviet threat to the attainment of your objective, which will be reported to the Joint Chiefs of Staff as a matter of urgency."[14]

For the first time MacArthur had a written directive to destroy North Korean forces.

> Your military objective is the destruction of the North Korean Armed Forces. In attaining this objective you are authorized to conduct military operations, including amphibious and airborne landings or ground operations north of the 38th Parallel in Korea, provided that at the time of such operation there has been no entry into North Korea by major Soviet or Chinese Communist Forces, no announcement of intended entry, nor a threat to counter our operations militarily in North Korea. Under no circumstances, however, will your forces cross the Manchurian or USSR borders of Korea and, as a matter of policy, no non-Korean Ground Forces will be used in the northeast provinces bordering the Soviet Union or in the area along the Manchurian border. Furthermore, support of your operations north or south of the 38th Parallel will not include Air or Naval action against Manchuria or against USSR territory.
>
> In the event of the open or covert employment of major Soviet units south of the 38th Parallel, you will assume the defense, make no move to aggravate the situation and report to Washington. You should take the same action in the event your forces are operating north of the 38th Parallel, and major Soviet units are openly employed. You will not discontinue Air and Naval operations north of the 38th Parallel merely because the presence of Soviet or Chinese Communist troops is detected in a target area, but if the Soviet Union or Chinese Communists should announce in advance their intention to reoccupy North Korea and give warning, either explicitly or implicitly, that their forces should not be attacked, you should refer the matter immediately to Washington.
>
> In the event of the open or covert employment of major Chinese Communist units south of the 38th Parallel, you should continue the action as long as action by your forces offers a reasonable chance of successful resistance. In the event of an attempt to employ small Soviet or Chinese Communist units covertly south of the 38th Parallel, you should continue the action.

MacArthur was directed to use all information media at his command to turn "the inevitable bitterness and resentment of the war-victimized Korean people" away from the United Nations and to direct it toward the Communists, Korean and Russian, and, "depending on the role they play," the Chinese Communists.

---

[13] The genesis of this directive is not clear in President Truman's memoirs. He states that he approved a statement of national policy on 11 September and that the JCS sent a "directive" based on this policy to MacArthur on 15 September. The JCS sent only the substance of the policy statement to MacArthur at that time, and did not send him the actual directive until 27 September. See Truman *Memoirs*, II, 59–60.

[14] Rad, JCS 92801, JCS (Personal) for MacArthur, 27 Sep 50. Because of its importance this directive will be quoted at length.

When organized armed resistance by North Korean forces has been brought substantially to an end, you should direct the ROK forces to take the lead in disarming remaining North Korean units and enforcing the terms of surrender. Guerrilla activities should be dealt with primarily by the forces of the Republic of Korea, with minimum participation by United Nations contingents.

Circumstances obtaining at the time will determine the character of and necessity for occupation of North Korea. Your plans for such occupation will be forwarded for approval to the Joint Chiefs of Staff. You will also submit your plan for future operations north of the 38th Parallel to the Joint Chiefs of Staff for approval.

MacArthur was advised that the United States was formulating instructions regarding "Armistice terms to be offered by you to the North Koreans in the event of sudden collapse of North Korean forces and Course of Action to be followed and activities to be undertaken during the post-hostilities period." The directive then continued:

As soon as the military situation permits, you should facilitate the restoration of the Government of the Republic of Korea with its capital in Seoul. Although the Government of the Republic of Korea has been generally recognized (except by the Soviet bloc) as the only legal government in Korea, its sovereignty north of the 38th Parallel has not been generally recognized. The Republic of Korea and its Armed Forces should be expected to cooperate in such military operations and military occupation as are conducted by United Nations forces north of the 38th Parallel, but political questions such as the formal extension of sovereignty over North Korea should await action by the United Nations to complete the unification of the country.

According to news reports appearing about the time the new directive reached MacArthur, General Walker had informed reporters that his forces were going to halt along the 38th Parallel for regrouping and, ostensibly, to await permission to cross. These reports, while unconfirmed, disturbed the Secretary of Defense to such an extent that he sent General MacArthur a personal message: "Announcement . . . may precipitate embarrassment in the United Nations where evident desire is not to be confronted with the necessity of a vote on passage of the 38th parallel." Secretary Marshall left no doubt, however, as to how he himself felt about the crossing when he said, "We want you to feel unhampered tactically and strategically to proceed north of the 38th parallel." [15]

General MacArthur had received no confirmation that General Walker had made a statement of this type and doubted that he had done so. But he took the precaution of warning Walker to make no comment on the 38th Parallel to anyone. "The matter is of such delicacy," he told the Eighth Army commander, "that all reference thereto will be made either from GHQ or direct from Washington." And in answer to the Secretary of Defense MacArthur replied that he had cautioned Walker against "involvement connected with nomenclature." "Unless and until the enemy capitulates," General MacArthur

[15] (1) Rad, JCS 92895, Secy Defense (Personal) to MacArthur, 29 Sep 50. (2) The President had been advised on 1 October that General MacArthur had informed the Joint Chiefs of Staff that he wished to issue a dramatic announcement when the 38th Parallel had been crossed. The Joint Chiefs of Staff had forbidden this, pointing out the unwisdom of such a statement. They had instructed him, instead, to go ahead with his operations but without calling special attention to the crossing of his forces into North Korea.

told General Marshall, "I regard all of Korea open for our military operations." [16]

The ROK Government Returns to Seoul

General MacArthur, before landing at Inch'on, had conferred with President Rhee and agreed informally that the government of the republic would be re-established in Seoul as early as possible. The two had also discussed arrangements for an election. In Washington, when the Joint Chiefs of Staff mentioned these dealings, great concern arose within the Department of State. That agency, then discussing means of a final settlement in Korea with other U.N. members, deplored any participation by the military commander in ROK governmental matters. Through the Secretary of Defense, the Department of State asked the Joint Chiefs of Staff to call upon MacArthur for a more complete accounting of his diplomatic activities. H. Freeman Matthews of the Department of State told the Secretary of Defense he did not wish to use diplomatic channels for this inquiry, believing, ". . . it would be extremely awkward for Sebald [Political Adviser to SCAP] to inquire into this matter, and equally awkward for Ambassador Muccio." [17]

When, acting on the request, the Joint Chiefs of Staff asked General MacArthur for complete details of his plans for restoring President Rhee's authority in Korea, MacArthur protested any thought of meddling in the Department of State's affairs. "I do not know precisely to what your message refers," he said,

> but I have no plans whatsoever except scrupulously to implement the directives which I have received. I plan to return President Rhee, his cabinet, senior members of the legislature, the United Nations commission, and perhaps others of similar official category to domicile in Seoul as soon as conditions there are sufficiently stable to permit reasonable security.

MacArthur pointed out that this involved no re-establishment of or change in government, since the ROK Government had never ceased to function and would merely resume control over its areas liberated from enemy control.[18]

Conditions in Seoul were not yet quite "sufficiently stable" for Rhee's return, for the X Corps had encountered exceptionally bitter resistance in and around the city. General Almond, under pressure from MacArthur, pushed his commanders to take the capital quickly. By 26 September, his troops had seized all key points within it, and the prize seemed almost within grasp. "On this basis," Almond said, "I advised General MacArthur that he might expect to enter Seoul on the 29th of September, that in my opinion the city would be perfectly safe to restore President Syngman Rhee to his rightful position at the Capital by that date." [19]

Almond also sent MacArthur a tenta-

---

16 (1) Rad, C 65035, CINCFE to CG Eighth Army, 30 Sep 50. (2) Rad, C 65034, CINCFE to DA for Secy Defense, 30 Sep 50.

17 Ltr, Mr. H. Freeman Matthews, Deputy Undersecy State, to Gen Burns, OSD, 18 Sep 50, in G–3, DA file 091 Korea, Case 14/26.

18 (1) Memo, Gen Bradley for Secy Defense, 7 Sep 50, sub: U.S. Courses of Action With Respect to Korea. (2) Ltr, Mr. Matthews to Gen Burns, 18 Sep 50. (3) Rad, JCS 92329, JCS to CINCFE, 22 Sep 50. (4) Rad, C 64159, CINCFE to JCS, 23 Sep 50.

19 Ltr, Gen Almond to Maj James F. Schnabel, 8 Jul 55.

tive program for the liberation ceremonies. But MacArthur replied:

> Arrangements suggested by you are not in accordance with those already set up by me. Following is the plan. Arrive Kimpo 0930. No honor guard or other ceremony there. Will proceed direct to capital building for informal conference with you and General Walker before arrival of Pusan party. Ceremony at 1200 hours. I will personally conduct the proceedings without being introduced. There will be no invocation or benediction necessary as the spiritual features are embodied in my own address. I will commence ceremony by five minute speech to be followed by speeches of similar duration by the Chairman UN COK, American ambassador and President Rhee, and I will conclude the proceedings.[20]

General MacArthur arrived in Seoul on the 29th as scheduled. In his address he told President Rhee:

> In behalf of the United Nations I am happy to restore to you, Mr. President, the seat of your Government, that from it you may better fulfill your constitutional responsibility. It is my fervent hope that a beneficent providence will give you and all of your public officials the wisdom and strength to meet your perplexing problems in a spirit of benevolence and justice, that from the travail of the past there may emerge a new and hopeful dawn for the people of Korea.

After leading his audience in the Lord's Prayer, MacArthur told Rhee, ". . . my officers and I will now resume our military duties and leave you and your Government to the discharge of civil responsibility." [21]

When MacArthur returned to Tokyo, he received protests from the Departments of State and Defense. Both departments noted with surprise and alarm that the American flag had been displayed with undue prominence over the ROK Capitol during the ceremonies, and complained that this placed too great an emphasis on the nature of the Korean War as a United States, rather than a United Nations, operation.[22] But congratulations also were in order. For, by the end of September, MacArthur had achieved the objectives of his landing, and the Eighth Army and the X Corps now controlled almost all of South Korea. Together, the two commands had routed the North Korean Army, had killed or captured huge numbers of its troops, and had destroyed or forced the abandonment of nearly all of its tanks, trucks, and artillery.

In congratulating MacArthur on 30 September, President Truman said, in part:

> No operations in military history can match either the delaying action where you traded space for time in which to build up your forces, or the brilliant maneuver which has now resulted in the liberation of Seoul. I am particularly impressed by the splendid cooperations of our Army, Navy, and Air Force and I wish you would extend my thanks and congratulations to the commanders of these services—Lt. Gen. Walton H. Walker, Vice Admiral Charles T. Joy, and Lt. Gen. George E. Stratemeyer. . . . I salute you all, and say to all, from all of us at home, 'Well and nobly done.'

The Joint Chiefs of Staff joined in the congratulations, praising MacArthur and his men for a ". . . transition from defensive to offensive operations [that]

[20] Rad, C 64724, CINCUNC to CG X Corps, 28 Sep 50.

[21] Text of message by General MacArthur on return of Government of Korea to Seoul, 29 September 1950, contained in *MacArthur Hearings*, page 3481.

[22] Rad, W 92972, DA to CINCFE, 30 Sep 50.

RHEE EXPRESSES APPRECIATION *to General of the Army MacArthur at liberation ceremonies at the Capitol Building, Seoul.*

was magnificently planned, timed and executed." [23]

General MacArthur passed along these compliments to all of his command, but they brought no particular joy to General Almond. For neither President Truman nor the Joint Chiefs of Staff had specifically credited the X Corps or Almond with any contribution to the success of the operations. Though the oversight presumably was unintentional, Almond complained that this absence of official recognition adversely affected the morale of his command.[24]

[23] Rad, ZC 18525, CINCFE to All Comdrs, 30 Sep 50.

[24] (1) *Ibid.* (2) Telecon, Gen Beiderlinden with Col Harrison, 2020–2100, 1 Oct 50, recorded in SGS GHQ, FEC 337 files, 1950.

The X Corps and General Almond were to have another opportunity for recognition as a result of the 27 September directive from the Joint Chiefs of Staff to MacArthur calling for the destruction of the North Korean armed forces. During the recent offensive large numbers of North Koreans had managed to slip away, particularly through the eastern mountains, into their home territory.

In connection with the assigned objective to destroy the North Korean armed forces, the Joint Chiefs of Staff authorized MacArthur to broadcast a surrender ultimatum to the North Korean Government. The broadcast also was to instruct the North Korean military leaders on how to handle prisoners of war, to assure them that on surrender their own forces would be fairly treated, to inform them that the Republic of Korea would be re-established with its capital in Seoul, and to point out that the question of the future of Korea was now before the United Nations. MacArthur, however, placed little confidence in a call to surrender. He doubted that the North Koreans would come to terms until he had beaten them so decisively as to leave them no alternatives but surrender or annihilation. He therefore concluded that he should try to crush the North Korean Army by a pursuit above the 38th Parallel. He, in fact, had made this decision before he received his newest directive from the Joint Chiefs of Staff, but he had seen the gist of the new policy underlying the directive and therefore was able to judge the latitude he would be allowed.[25]

New Operations Plans

Accordingly, on 26 September, MacArthur instructed General Wright and the JSPOG staff to plan another amphibious encirclement well above the 38th Parallel. The new landing was to be co-ordinated with a new overland attack. MacArthur wanted Wright to consider two conceptions of advance into North Korea. The first of these would send the Eighth Army in a main effort along the west coast in conjunction with an amphibious landing at Chinnamp'o or elsewhere. MacArthur's other idea provided for an overland attack to the east coast by the Eighth Army and a simultaneous amphibious landing at Wonsan, a city of some 150,000, also on the east coast.[26] The plan eventually used included features of both concepts.

General Wright furnished the hybrid plan, actually an up-to-date version of an alternate concept prepared earlier for Operation CHROMITE, on 27 September.[27] By this plan, the Eighth Army would make the main effort in the west to seize the North Korean capital, P'yongyang, and the X Corps would make an amphibious assault landing at Wonsan. Wright told General MacArthur that the amphibious landing could be staged within ten days of the order to load out if shipping was assembled early enough.[28]

Wonsan was an excellent choice for an amphibious landing. Besides being sufficiently deep into North Korea, it was the principal port on the east coast; it

[25] Rad, JCS 92762, JCS to CINCFE, 27 Sep 50.

[26] (1) Memo, Gen Hickey for JSPOG (Gen Wright), 26 Sep 50, sub: Plans for Future Opns, JSPOG, GHQ, UNC files. (2) See also, Appleman, *South to the Naktong, North to the Yalu*, pp. 609–14 and 618–21.

[27] General Wright, who carried out General MacArthur's planning directives during this period and supervised their conversion into concrete plans, felt that the method chosen for entering North Korea was a natural outgrowth of MacArthur's preoccupation since July 1950 with the possibility of a double amphibious envelopment. "Even while we were under the pressure of the Inchon planning," Wright has written, ". . . I had JSPOG concurrently assembling the data for a Wonsan operation." It was strictly the paucity of men and matériel that had led MacArthur to settle for a single envelopment at Inch'on in the first place, according to Wright. And he had kept the Wonsan operation in mind, for the time when he would have enough strength to mount it. "I think it can be inferred that he had rather definite plans for Wonsan immediately following the success of the Inchon operation." See Ltr, Gen Wright to Maj Schnabel, 14 Jun 55, copy in OCMH.

[28] (1) Memo, Gen Wright for CofS GHQ, 26 Sep 50, sub: Plans for Future Opns. (2) Interv, Col Appleman with Gen Wright, Feb 54.

was the eastern terminus of the easiest route across the narrow waist of the peninsula; and it was a road and rail communications center. Wonsan, in fact, was the principal port of entry for Russian supplies and military equipment received by sea from the Vladivostok area and a key point on the rail line from the same area. Moreover, from Wonsan a military force could move inland and west across the peninsula to P'yongyang or north to the Hamhung-Hungnam region, the most important industrial area in all Korea.[29]

General MacArthur readily accepted the plan tailored to his specifications. On 28 September he informed the Joint Chiefs of Staff: "If the North Korean Armed Forces do not surrender in accordance with my proclamation to be issued on 1 October 1950, dispositions will be made to accomplish the military objective of destroying them by entry into North Korea." He sketched his plan briefly. He would send the Eighth Army across the 38th Parallel through Kaesong and Sariwon to capture P'yongyang. Almond's X Corps would land amphibiously at Wonsan, thereafter "making juncture with Eighth Army." Presumably, this juncture would require the X Corps to attack west along the Wonsan-P'yongyang road.[30]

Mindful of the warning contained in his latest directive, General MacArthur promised Washington that he would use only ROK troops for operations above the line Ch'ungju-Yongwon-Hungnam. "Tentative date for the attack of Eighth Army," MacArthur reported, "will be not earlier than 15 October and not later than 30 October. You will be provided detailed plans later." Washington's concern over possible Chinese or Russian interference in the Korean fighting prompted General MacArthur to report also that there was no indication of "present entry into North Korea by major Soviet or Chinese Communist Forces." [31]

On the following day, just before he delivered his address in Seoul, MacArthur summoned General Walker, General Almond, Admiral Joy, and General Stratemeyer to a conference in a room on the second floor of the Capitol to tell them of his new plan. Although the Joint Chiefs of Staff had not yet approved the plan, he pointed out, approval was expected with no material change in the concept of the operation. He directed Almond to relinquish the Seoul area to Walker by 7 October, to plan on moving the 7th Division overland for embarkation at Pusan, and to embark corps troops and the 1st Marine Division from Inch'on. He tentatively set 20 October as the date for the Wonsan landing.[32]

The actual plan for destroying North Korean forces above the 38th Parallel was based on three assumptions. Two were correct, namely, that the bulk of the North Korean forces had been destroyed and that the United Nations Command would conduct operations

[29] (1) JANIS 75, ch. VIII (Korea—Cities and Towns), pp. 52–53. (2) GHQ FEC Terrain Study 6, North Korea, XIV, 26–27, and Map No. 760, Wonsan City Plan, Plate 12. (3) War Diary, X Corps, Oct 50, Opns, pp. 18–19, and Diary CG X Corps, 24 Oct 50.

[30] (1) Rad, C 64805, CINCFE to JCS, 28 Sep 50. (2) See also Douglas MacArthur, *Reminiscences* (New York: McGraw-Hill, 1964), pp. 357–60.

[31] Rad, C 64805, CINCFE to JCS, 28 Sep 50.

[32] Ltr, Gen Almond to Maj Schnabel, 8 Jul 55.

north of the 38th Parallel. The third, that there would be no outside interference, was less sound. The plan called on the forces of the Eighth Army and the X Corps to advance to and hold a line across Korea from Ch'ongju, through Yongwon, to Hamhung. The target date for the Eighth Army assault was set at twelve days after the Eighth Army had passed through the X Corps in the Inch'on-Seoul area. General Walker's ground attack might precede General Almond's amphibious assault by three to seven days. General Wright estimated that it would take six days to load the assault elements of the X Corps and four days to sail to Wonsan.[33]

Most of MacArthur's principal staff officers had assumed, before seeing the new plans, that the UNC commander intended to place the X Corps under General Walker after Seoul was returned to ROK control. MacArthur had created the X Corps specifically for the landing at Inch'on, had tailored it hurriedly, and had taken its key officers from his own staff. As the corps completed its mission in late September, it could logically be assumed that the combat elements of the corps would be assimilated by the Eighth Army and that the key officers would return to GHQ and their normal duties. Generals Hickey and Wright advised General MacArthur to follow this course; Maj. Gen. George L. Eberle, MacArthur's G–4, also strongly favored Walker's taking over the X Corps; and General Almond had always understood "that when the Inchon operation was completed that the X Corps troops would be absorbed by Eighth Army. . . ."[34] Subsequently, General MacArthur could not believe that these officers really disagreed with his decision.

> To the contrary, the decision to retain a function of GHQ command and coordination between the Eighth Army and the X Corps until such time as a juncture between the two forces had been effected was, so far as I know, based upon the unanimous thinking of the senior members of my staff. It but followed standard military practice in the handling and control of widely separated forces where lateral communications were difficult if not impossible.[35]

General Walker and the Eighth Army staff apparently felt very strongly that the X Corps should become part of the Eighth Army. Walker seems to have had two plans in mind for the possible employment of Almond's forces. In one of these, the X Corps would drive overland from Seoul to seize P'yongyang, and the rest of the Eighth Army, after coming up behind the X Corps, would then move laterally from P'yongyang to Wonsan on the east coast where it would join the ROK I Corps as the latter moved up the east coast. Such a maneuver might save a great deal of time, since the X Corps was already in position to advance on P'yongyang, and would establish a line across Korea at the narrow waist that could cut off a large number of North Koreans still trying to move northward through the central and eastern mountains. Meanwhile, the X Corps

---

[33] (1) Opn Plan 9–50, 29 Sep 50, in JSPOG, GHQ, UNC files. (2) Memo, Gen Hickey for JSPOG, Note 2, Gen Wright to CofS, GHQ, UNC, 26 Sep 50, sub: Plans for Future Opns.

[34] (1) Interv, Col Appleman with Gen Wright, Feb 54. (2) Interv, Col Appleman with Gen Eberle, 12 Jan 54. (3) Ltr, Gen Almond to Maj Schnabel, 8 Jul 55.

[35] Ltr, Gen MacArthur to Gen Snedeker, USMC, G–3, Hq USMC, Washington, D.C., 24 Feb 56, copy in OCMH.

could move on above P'yongyang toward the Yalu River. The operations of both the X Corps and the Eighth Army could be co-ordinated under Walker's command; and both could be supplied from Pusan and Inch'on until the Wonsan area fell, at which time the forces operating in the east could be supplied by sea through Wonsan and Hungnam, farther north.[36]

General Walker's second plan was to approach Wonsan by a more direct, diagonal route. Assuming that the X Corps became a part of the Eighth Army, Walker would, in this instance, send a corps to the east coast objective through the Seoul-Ch'orwon-Wonsan corridor.[37]

If these were the plans Walker had in mind, he did not ask authority to carry out either of them. Apparently unaware of what Almond's plans were he contented himself with asking General MacArthur discreetly that he be let in on what was going on: "To facilitate advance planning for the approaching juncture with the X Corps, request this headquarters be kept informed of the plans and progress of this Corps to the greatest extent practicable. To date the X Corps operations plans have not been received." [38]

General MacArthur told Walker that as soon as X Corps had completed its CHROMITE missions, he would place it in GHQ Reserve in the Inch'on-Seoul area and that he, MacArthur, would direct its future operations. These operations would be revealed to the Eighth Army commander at an early date.[39] MacArthur, in fact, consulted neither Walker nor Almond on the next operation until the plan was almost in final form.

MacArthur's guidance to his planners was tantamount to an order that they recommend another amphibious operation by the X Corps. While MacArthur did not specify that the X Corps would make the amphibious landing, no other element of the United Nations Command could have carried out the maneuver. Too, General MacArthur had been most favorably impressed by Almond's performance at Inch'on and by the over-all results of his operations. Furthermore, he saw amphibious maneuver as the best means of slashing deep into North Korea, of cutting off escape routes for thousands of fleeing enemy soldiers, and of seizing a major port to support his troops. This last-named purpose was perhaps uppermost in his thinking. Ammunition, food, gasoline, and most other supplies that kept the UNC divisions fighting in late September came into Korea through two ports, Pusan and Inch'on. As troops moved farther north, Pusan's value dwindled, since the rail lines and roads over which matériel had to be brought from the port to the combat units had been severely damaged in the earlier heavy fighting. The other port, Inch'on, had a limited capacity for receiving vessels and could scarcely have supported, with its facilities, all U.N. forces involved in the fighting.[40]

---

36 Interv, Col Appleman with Maj Gen Leven C. Allen, 15 Dec 53, copy in OCMH.

37 Ltr, Wright to Schnabel, 14 Jun 55.

38 Rad, G 25090 KGO, CG Eighth Army to CINCFE, 26 Sep 50.

39 Rad, CX 64610, CINCFE to CG Eighth Army, 27 Sep 50.

40 General Wright points out in this connection, "Inchon was not capable of fully supplying Eighth

General Wright, in later analyzing the decision and the planning for entering North Korea, said,

> Both General MacArthur and General Walker realized that any successful campaign in North Korea would need the full operation of an east coast port, preferably Wonsan or Hungnam. And I believe that their staffs were in full agreement. The point at issue was simply that of how to capture such a port and who should do it.[41]

Any campaign north of the P'yongyang-Wonsan corridor would certainly encounter a most difficult logistical problem. The northern Taebaek Range rose to rugged heights in the east central part of the peninsula, forming a nearly trackless mountain waste in the direction of the Manchurian border. Few roads or trails ran west and east. The principal lanes of travel were axial routes that followed the north and south trend of deep mountain valleys. The only reasonably good lateral road connected P'yongyang with Wonsan, where it joined the coastal road running northward to Hamhung and Hungnam. A rail line crossed the peninsula in the same general area between P'yongyang and Wonsan.

General MacArthur apparently decided that he could not supply both Eighth Army and X Corps through Pusan and Inch'on and over the crippled road and rail system in a campaign that he wanted to end quickly so that his forces would not have to fight during North Korea's severe winter weather. Weeks of concentrated work by all the available engineer troops would be needed before even the main lines of communication could be repaired as far as the 38th Parallel, not to mention the area to the north where the next phase of the campaign would be fought. But with the addition of the Wonsan port facilities, MacArthur reasoned, two separate forces, co-ordinated and supported from Japan, could operate in Korea without impairing the effectiveness of either.[42] Of the two methods by which he could seize Wonsan, amphibious encirclement took precedence over ground advance. The means were at hand in the X Corps, his directives specifically authorized amphibious operations in North Korea, and he apparently hoped the waterborne movement would be as successful as the one at Inch'on.

The Joint Chiefs of Staff, having already established the principle that MacArthur could carry the fight into North Korea, did not quibble over MacArthur's methods. They passed the plan on to the Secretary of Defense for final approval, asking that he act with great speed since "certain ROK Army Forces may even now be crossing the 38th Parallel." President Truman and General Marshall agreed to the plan at once, and the Joint Chiefs of Staff told General MacArthur to carry out his plan on schedule.[43]

Army and I think a logistical check will show that, temporary handicap to Eighth Army as it was, the movement out of X Corps enabled Eighth Army to provide itself with the logistic capability to perform its advance to the Pyongyang area." See Ltr, Wright to Schnabel, 14 Jun 55.

[41] Ltr, Wright to Schnabel, 14 Jun 55.

[42] Interv, Col Appleman with Gen Ruffner, formerly CofS X Corps, 20 Aug 51.

[43] (1) Rad, JCS 92975, JCS to CINCFE, 29 Sep 50. (2) Memo, Gen Bradley for Secy Defense, 29 Sep 50, sub: Future Korean Opns. (3) To later critics who noted that ROK troops captured Wonsan on 11 October before American units were even disembarked and that MacArthur had noticed this, General Wright pointed out that General MacArthur

had indeed noticed and was impressed by the remarkable advance of ROK soldiers up the east coast of Korea where, by late September, they had driven almost to the parallel. But those same ROK troops had, only weeks before, shown themselves to be extremely vulnerable to pressure and counterattack. And there was every good chance that these troops would run into guerrilla forces, reinforced by retreating North Korean survivors, when they reached the mountainous area west of Kaesong and Kojo. Too, MacArthur did not feel that he had sufficient control of ROK troops. While technically under his command, their subordination to him was based merely on an understanding between himself and President Rhee of the Republic of Korea. This fact, according to General Wright, made their conduct under certain conditions problematical, and had to be considered in any planning for a major operation. In other words, any plan which hinged on ROK troops to any degree (i.e., to seize and hold Wonsan) was felt to be leaning on a weak reed. See Ltr, Gen Wright to Maj Schnabel, 14 Jun 55.

CHAPTER XI

The Invasion of North Korea

At noon on 1 October, General MacArthur broadcast from Tokyo a call to the North Korean commander in chief, demanding his surrender. The call went unanswered.[1] Hence there appeared to be no alternative to sending UNC forces into North Korea if the remainder of the North Korean Army and the Communist regime were to be destroyed. But since the United Nations had not ordered or even clearly authorized the entry into and occupation of North Korea, American authorities were careful not to make public any plans for occupying the northern half of Korea while they worked to achieve some definite form of United Nations approval.

The United Nations Resolution of 7 October

The Departments of State and Defense agreed that if North Korea collapsed and its Russian and Chinese neighbors kept hands off, MacArthur should occupy North Korea under the auspices of the United Nations. Some officials favored a unilateral occupation by the United States if the United Nations took no new steps authorizing occupation, "even at the expense of some disagreement with friendly United Nations nations."[2] But this was decidedly a minority view.

In late September, the Joint Chiefs of Staff had sent MacArthur a State Department opinion which held, "It will be necessary to consult with and obtain the approval of United Nations members before the United Nations commander can be authorized to undertake the occupation of North Korea." The State Department proposed that MacArthur send mainly South Korean and other Asian troops to occupy only key points in North Korea. U.S. troops would leave Korea as early as possible. There would be no revenge or reprisal in the occupation. "The general posture of United Nations forces should be one of liberation rather than retaliation," State Department authorities believed. General MacArthur agreed and told the Joint Chiefs, "The suggested program from the standpoint of the field commander seems entirely feasible and practicable."[3]

[1] Text, Broadcast, CINCUNC to CINC NKPA, 1 Oct 50, in State Dept *Bulletin*, 9 Oct 50.

[2] Memo, Chief, Plans Div G–3 (Col Johnson) for DCS for Plans (Gen Gruenther), 21 Sep 50, sub: Program for Bringing Korean Hostilities to an End, in G–3, DA file 091 Korea, Case 99/3.

[3] (1) Rad, JCS 92608, JCS to CINCFE, 26 Sep 50. (2) Rad, CM–IN 15683, CINCFE to JCS, 26 Sep 50.

Secretary of Defense Marshall also wanted United States troops to stay out of the picture during any occupation of North Korea. "I wish to state," Marshall told Secretary of State Acheson on 3 October:

that the Department of Defense continues to believe that as few United States troops as possible should engage in the physical occupation and pacification of areas north of the 38th Parallel, once organized military hostilities have ended. It remains important, therefore, to increase the number of other United Nations troops sent to Korea, particularly from countries in Asia.[4]

General Marshall deplored the lack of an organized United Nations agency, other than military, to handle "the tremendous problems that will follow hostilities." He reminded Secretary Acheson that the United Nations Commission in Korea (UNCOK) was neither staffed nor equipped to meet the problems that would face it if the United Nations occupied North Korea. He called upon Acheson to sponsor the formation by the United Nations of one combined or three separate agencies to handle the three major problems—relief and reconstruction, political unification, and security.[5]

The Department of State had already drawn up a resolution for the United Nations to consider. This resolution supported the political objectives of the United Nations in Korea, including means for carrying them out through occupation if necessary. State Department officials talked informally with representatives of friendly member nations in the United Nations and solicited their support for the passage of the resolution. The United States could not work through the Security Council as in earlier days, since the USSR delegate to the council had returned to his seat in August, bringing a veto power likely to be used against any American-inspired resolution. Consequently, the American delegation moved the Korean question before the General Assembly where the USSR had no veto power and where American greatly outweighed Russian influence.

On 7 October the General Assembly passed the resolution. It did not clearly call for the conquest and occupation of North Korea but gave implicit assent. The General Assembly recommended:

(a) All appropriate steps be taken to ensure conditions of stability throughout Korea; and, (b) All constituent acts be taken, including the holding of elections, under the auspices of the United Nations, for the establishment of a unified, independent and democratic Government in the sovereign State of Korea. . . .

This resolution also established the United Nations Commission for the Unification and Rehabilitation of Korea (UNCURK) which replaced the old United Nations Commission in Korea.[6] The Joint Chiefs of Staff had already sent a draft copy of the resolution to General MacArthur, at the same time informing him that the United States Government considered it as supporting operations north of the 38th Parallel.[7]

MacArthur's Plans Change

"All appropriate steps" to "ensure conditions of stability throughout Korea"

[4] Ltr, Secy Defense (Marshall) to Secy State (Acheson), 3 Oct 50, Incl to JCS 1776/129, 3 Oct 50, in G–3, DA file 091 Korea, Case 99/11.

[5] *Ibid.*

[6] State Dept *Bulletin*, XXIII (23 Oct 50), 648–49.

[7] Rad, JCS 93555, JCS to CINCFE, 6 Oct 50.

meant only one thing to General MacArthur, particularly since the enemy refused to answer his surrender demand. He went ahead with his preparations for destroying the North Korean Army on its own ground. His original plans for doing so were scarcely recognizable by the time they went into effect. In order to keep pace with the swift advances in the east and west, General MacArthur had to change his scheme of late September. Other deviations from the prepared plans became necessary because of unexpected conditions encountered at Wonsan.

In late September, Walker's troops on the west and central fronts, although poised for the assault on North Korea, had to be held in check temporarily. Walker's divisions were delayed, not by reluctance or enemy opposition, but simply by a lack of sufficient food, fuel, and munitions for sustained operations in North Korea. All supplies had to come forward on badly damaged overland routes; incoming cargo jammed limited port facilities; and all available air transport was busy rushing supplies into the few usable airfields so that MacArthur's troops might attack as soon as possible.

As the Eighth Army and X Corps pressed into the crowded maneuver area along the border of west and central Korea, MacArthur established a boundary between them. He made Eighth Army responsible for establishment and publication of complete bombline locations for all of Korea. Boundary control points between areas of responsibility of the two major commands were selected by General Walker who then notified General Almond. Walker received permission to use roads through Almond's areas in carrying out the necessary surveying work. Direct communication between the two commanders was authorized. The U.N. commander, when it appeared that the enemy's lines of communication and other facilities would soon be under his control and would be needed in the advance into North Korea, changed policy and forbade any further unnecessary destruction of railroad facilities and equipment, bridges, and enemy airfields.[8]

The North Korean Army seemed to have melted away, so rapidly did it retreat. Even as MacArthur called for surrender and while American divisions waited in the west, the ROK 3d Division of the ROK I Corps on the east coast crossed almost unopposed into North Korea on 1 October. MacArthur reported the crossing to the Joint Chiefs the next day: "Probings by elements of the ROK Army are now well across the 38th Parallel. Advances on the extreme right are between ten and thirty miles in the coast sector with practically no resistance." These ROK troops were under Walker's command.[9]

General MacArthur foresaw that he might not need to use X Corps amphibiously, if successes in the east continued. "It is possible," he told the Joint Chiefs on 2 October, "if the enemy's weakness is pronounced that immediate exploitation may be put into effect before or in substitution for my prepared plans."[10] Yet he sent no more troops into the coastal operation in support of the ROK

[8] (1) Rad, C 64621, CINCFE to CG Eighth Army and CG X Corps, 27 Sep 50. (2) Rad, CX 65139, CINCFE to All Comds, 1 Oct 50.

[9] Rad, C 65252, CINCFE to DA (JCS), 2 Oct 50.

[10] *Ibid.*

drive, and on the same day issued orders for an overland attack north along the Kaesong–Sariwon–P'yongyang axis and an east coast amphibious landing at Wonsan to encircle and destroy North Korean forces south of the Ch'ungju–Kunu-ri – Yongwon – Hamhung – Hungnam line. The Eighth Army was to make the main ground attack on P'yongyang, and the X Corps was to perform the amphibious movement. After the Eighth Army had seized P'yongyang and the X Corps had invested Wonsan, each was to attack toward the other along an east-west axis, join up, and cut off all enemy escape routes. On the ground, only ROK troops would operate north of the line Ch'ongju–Kunu-ri–Yongwon–Hamhung–Hungnam, except on MacArthur's direct order. He ordered Admiral Joy, COMNAVFE, to outload X Corps. The assault force from the 1st Marine Division was to load at Inch'on; the remainder of the corps, principally the 7th Division, was to embark from Pusan. These orders were completely within the authority granted General MacArthur on 27 September.[11]

General Walker would command all United Nations ground forces in Korea with the exception of X Corps and the 187th Airborne RCT; X Corps, under General Almond, would revert to GHQ Reserve when passed through by Eighth Army and remain under the direct command of General MacArthur. Upon embarkation for the assault and while on the water, X Corps would be controlled by Admiral Joy. The commander of Joint Task Force Seven would command all forces in the amphibious assault until General Almond had landed and indicated his readiness to assume responsibility for further operations ashore.

On 3 October, General MacArthur canceled his previous delineation of the Inch'on-Seoul area as a X Corps objective and took direct control of the 187th Airborne RCT, which had entered Korea on 23 September and been operating in the Kimp'o area. General Walker relieved General Almond of responsibility for the Inch'on-Seoul area at noon on 7 October.[12]

By 1 October, the total ground force strength within the United Nations Command in Korea, divided among the Eighth Army, X Corps, and service units, amounted to more than a third of a million men. Far East Air Forces under General Stratemeyer, on the same date, totaled 36,677, and U.S. Naval Forces, Far East, under Admiral Joy numbered 59,438.[13]

The Likelihood of Chinese Intervention

Political Signs

From the very beginning of U.N. operations in Korea the United States and its allies had kept close watch on the political and military reactions of Korea's giant neighbor, Communist China. Possessed of a powerful army and led by men fanatically dedicated to communism, China could have interfered with serious effect during July, August, and

---

[11] GHQ, UNC Opns Order No. 2, 2 Oct 50.

[12] (1) Rad, CX 65371, CINCFE to All Comdrs, 3 Oct 50. (2) Rad, X 10665, CG X Corps to CINC FE, 8 Oct 50.

[13] (1) Rpt, ROK and U.N. Ground Forces Strength in Korea, 31 July 1950–31 July 1953, DA COA, 7 Oct 54. (2) Appleman, *South to the Naktong, North to the Yalu*, p. 605.

September. The fighting in Korea, however, received far less attention in Chinese newspapers and in policy statements than did Formosa, which China seemed to consider more important to her immediate interests. Too, the relationship between China, the USSR, and North Korea did not emerge clearly at first. The Chinese appeared content to allow Russian propagandists and officials to champion the North Korean cause in July and August.

On 13 July, Prime Minister Jawaharlal Nehru of India suggested to Premier Stalin and Secretary of State Acheson that Communist China, more formally, the Peoples' Republic of China, be admitted to the U.N. Security Council and that the United States, USSR, and China, "with the help and cooperation of other peace-loving nations," informally explore means to end the Korean War. Stalin promptly accepted, but the United States rejected the offer on 18 July. Chinese leaders made no immediate official comment.[14]

On 4 August, Jacob Malik, USSR representative to the United Nations, proposed that the "internal civil war" in Korea be discussed with Chinese Communist representation in the United Nations and that all foreign troops be withdrawn from Korea. On 22 August, Malik warned that any continuation of the Korean War would lead inevitably to a widening of the conflict. This statement seemed to signal a turning point for Chinese propagandists who, in public journals and official statements, began to hint darkly that if necessary the Chinese people would defend North Korea against its enemies. On 25 August, China formally charged the United States with strafing its territory across the Yalu. On 6 September, the U.N. Security Council voted down Malik's 4 August proposal and on 11 September, defeated his move to have Chinese Communists come to the United Nations tò consider Chinese charges of border violation by the United States.[15]

MacArthur's successful landing at Inch'on brought no actual intervention, as feared by some, but it did trigger a barrage of threatening pronouncements from high Chinese officials. On 22 September, the Chinese Foreign Office declared that China would always stand on the side of the "Korean people," and on 30 September, the Chinese Foreign Minister Chou En-lai publicly warned, "The Chinese people absolutely will not tolerate foreign aggression, nor will they supinely tolerate seeing their neighbors being savagely invaded by the imperialists." [16]

Late on 3 October, Chou En-lai called in the Indian Ambassador to Peiping, Dr. K. M. Pannikar, and, obviously expecting that his message would be conveyed to the U.S. Government, informed him that if United Nations troops entered North Korea, China would send in its forces from Manchuria. China would not interfere, however, if only South Koreans crossed the parallel. Next day, Pannikar communicated Chou's message to the United States through the British Minister at Peiping.[17]

---

[14] Allen S. Whiting, *China Crosses the Yalu* (New York: The Macmillan Company, 1960), p. 60.

[15] *Ibid.*, pp. 69–70, 92–94.

[16] *Ibid.*, p. 108.

[17] K. M. Pannikar, *In Two Chinas: Memoirs of a Diplomat* (London: Allen and Unwin, 1955), pp. 109–11.

CHOU EN-LAI

Because earlier threats had not materialized, and because of the assumption that the Chinese, if they really were serious, would not give away their intentions, Chou's warning caused no change in MacArthur's orders. The fact that the message came from Pannikar also raised doubts that the warning was genuine. For Pannikar had shown distinct Communist leanings and anti-American feelings in the past. Only a few days earlier, moreover, when the United States asked India to advise Communist China that it would be in the latter's best interest not to interfere in Korea,[18] Pannikar reported that China did not intend to enter Korea.[19] For these reasons, American intelligence officials discredited the newest message from Pannikar.

Still another factor which detracted from the validity of Chou's warning was a resolution pending before one of the committees of the General Assembly of the United Nations. The key vote on the resolution was to take place on 4 October. President Truman felt that the Chinese threat could well be a blatant attempt to blackmail the United Nations.[20]

Military Signs

Military indications of Chinese plans to invade North Korea were hard to come by. MacArthur on 29 June 1950 had been warned to stay well clear of Manchurian and Soviet borders. This order forced him to rely almost entirely upon outside sources for information on the strength and disposition of the Chinese Communist forces in Manchuria. Using these sources, General Willoughby, MacArthur's intelligence chief, reported on 3 July that the Chinese had stationed two cavalry divisions and four armies in Manchuria. A Chinese army normally possessed about 30,000 men but this figure varied.[21]

Other reports, often conflicting and of doubtful credence, told of troops of Korean ancestry being sent into North Korea by the Chinese. Throughout July and August 1950, the Department of the Army received a mass of second- and

[18] Since the United States did not recognize the Peiping government, it did not deal directly with the Communist Chinese.

[19] Memo, G-2 DA for DCofS for Plans, DA, 25 Sep 50, sub: Chinese Communist Attitude Toward Korean Hostilities, in CofS, DA file 000.1, Case 1.

[20] Truman, *Memoirs*, II, 362.

[21] MS, Col Bruce W. Bidwell, History of the War Department Intelligence Division, Part VII, ch. V.

third-hand reports that more Chinese troops were moving from south China to Manchuria. Willoughby estimated by the end of August that the Chinese had moved nine armies totaling 246,000 men to Manchuria.[22]

Indications that the Chinese Communists possibly intended to enter the fighting continued to be reported to the Department of the Army by the G–2 Section of the Far East Command. In daily teleconferences between officers at the Department of the Army and MacArthur's headquarters in Tokyo, General Willoughby, or his officers, relayed the latest information of Communist Chinese military activities. Each day, also, the United Nation Command's Daily Intelligence Summary (DIS) went to the Department of the Army by courier, arriving several days later in Washington. This summary carried all reports received from intelligence sources on the Chinese Communists and made an effort to evaluate these reports. At the top intelligence level, the Central Intelligence Agency combined reports from its own sources with those of the United Nations Command and then analyzed the actions and intentions of the Chinese for high-level governmental agencies.

To determine through outward manifestations alone whether the Chinese intended to intervene was virtually impossible. But by using such indications as movements of troops and supplies, American intelligence agencies could gauge this intention with some hope of accuracy. Penetration of Communist China to ascertain these movements was an almost impossible task. But certain agencies, particularly those allied with the Chinese Nationalist Government on Formosa and others operating out of Hong Kong, relayed reports of Chinese Communist military movements.

Although there were no definite military indications after Inch'on that the Chinese meant to enter the fighting in Korea, General Willoughby speculated that 450,000 Chinese troops were massed in Manchuria. The nation's planners had given full consideration to Chinese strength and the possibility of its employment in Korea when they drew up their blueprint for national policy in September.[23] While the primary concern of these authorities continued to be the possibility of intervention by the USSR, much attention had also been given to whether or not Chinese forces would come into Korea, and if so, what course should be followed by the United Nations Command. On 27 September, the Joint Chiefs of Staff directed General MacArthur to make a special effort to determine if the Chinese intended entering the war.[24] On the next day, General MacArthur assured them that there was no present indication of the entry into North Korea by Chinese Communist forces.[25]

On the day of Chou's warning, 3 October, the UNC intelligence staff reported some evidence that twenty Chinese Communist divisions were in North Korea and had been there since 10 September. They also commented on the reported warning from the Chinese Foreign Minister and other recent public statements that "Even though the utterances . . .

---

[22] *Ibid.*

[23] DIS, GHQ, FEC, 2934, 21 Sep 50.

[24] Rad, JCS 92801, JCS to CINCFE, 27 Sep 50.

[25] Rad, C 64805, CINCFE to JCS, 28 Sep 50.

are a form of propaganda they cannot be fully ignored since they emit from presumably responsible leaders in the Chinese and North Korean Communist Governments. The enemy retains a potential of reinforcement by CCF troops." [26]

On 5 October, noting the purported entry into North Korea of nine Chinese divisions, GHQ intelligence officers observed that recent reports were taking on a "sinister connotation" and concluded that the potential "exists for Chinese Communist forces to openly intervene in the Korean War if United Nations forces cross the 38th Parallel." [27] General Willoughby told Washington officials that the USSR "would find it both convenient and economical to stay out of the conflict and let the idle millions of Communist China perform the task as part of the master plan to drain United States resources into geographical rat holes of the Orient." He informed them that a build-up of Chinese forces along the Korean-Manchurian border had been reported by many of his sources and that "while exaggerations and canards are always evident, the potential of massing at the Antung and other Manchurian crossings appears conclusive." According to his computations, between nine and eighteen of the thirty-eight Chinese divisions believed to be in Manchuria were massing at the border crossings. Yet, MacArthur's intelligence chief did not, as far as is known, attempt to dissuade General MacArthur from crossing the parallel. Moreover, continuing reports of Chinese Communist troops crossing into Korea in early October were discounted by the Far East Command intelligence officers since "no conclusive evidence" existed; and the recent Chinese threat to enter North Korea if American forces crossed the 38th Parallel was characterized as "probably in a category of diplomatic blackmail." [28]

Nevertheless, the possibility that the Chinese Communists might actually intervene caused President Truman to direct the Joint Chiefs of Staff to give General MacArthur instructions covering such an eventuality. On 9 October, the Joint Chiefs of Staff directed MacArthur that ". . . in the event of the open or covert employment anywhere in Korea of major Chinese Communist units, without prior announcement, you should continue the action as long as, in your judgment, action by forces now under your control offers a reasonable chance of success. In any case you will obtain authorization from Washington prior to taking any military action against objectives in Chinese territory." [29]

One day earlier, the delicate balance of international relations received a substantial jolt when two of MacArthur's jet

[26] DIS, GHQ, UNC, 2946, 3 Oct, and 2947, 4 Oct 50.

[27] DIS, GHQ, UNC, 2948, 5 Oct 50.

[28] (1) DIS, GHQ, FEC, Nos. 2951, 2952, 2957, 8, 9, 14 Oct 50. (2) The Indian Ambassador to China asserts that on 9 October, Ernest Bevin, U.K. Foreign Minister, sent him a message to be transmitted to Chou En-lai personally and which was ". . . friendly in tone and contained vague assurances . . . that the Korean Commission would give the Chinese views their most careful consideration." Dr. Pannikar sent along this message which, in his viewpoint, added insult to injury since the Korean Commission consisted of such countries as the Philippines and Siam. "In any case," Pannikar notes, "Bevin's approach was too late, for the Chinese armies were already in Korea." See Pannikar, *In Two Chinas: Memoirs of a Diplomat*, pp. 111–12.

[29] (1) Rad, JCS to CINCFE, 9 Oct 50. (2) Truman, *Memoirs*, II, 362.

fighters attacked a Soviet airfield in the Soviet maritime provinces near Sukhaya Rechk. This incident was tailor-made for the USSR to use as an excuse to intervene in the Far East, especially since it occurred at almost the same time that American divisions moved above the 38th Parallel for the first time.

The United States informed the Soviet Union that the pilots had made a navigational error and had used poor judgment, that the commander of the Air Force Group responsible had been relieved, and that disciplinary action had been taken against the two pilots. The United States also expressed deep regret and offered to pay for all damages to Soviet property which were, it was reported, considerable. But the Russians did not acknowledge this offer.[30]

One intelligence report reaching President Truman on 12 October stated that Chinese military forces, while lacking the necessary air and naval support, could intervene effectively but not necessarily decisively. Further, in spite of statements by Chou En-lai and troop movements to Manchuria, there were no convincing indications of Chinese Communist intentions to resort to full-scale intervention in Korea. The general conclusion of the report was that the Chinese were not expected to enter North Korea to oppose the United Nations Command, at least not in the foreseeable future. Several reasons were given for this conclusion: The Chinese Communists undoubtedly feared the consequences of war with the United States. Anti-Communist forces would be encouraged and the regime's very existence would be endangered. The Chinese Communists also would hesitate to endanger their chances for a seat in the United Nations. Moreover, in the unlikely event that the Chinese entered the war without the benefit of Soviet naval and air support, they were bound to suffer costly losses. On the other hand, acceptance of Soviet aid, if forthcoming, would make China more dependent on Russia and would increase Russian control in Manchuria. This report agreed with many others that from a military standpoint, the most favorable time for intervention had passed. For all of these reasons, U.S. intelligence officials concluded that while full-scale Communist intervention in Korea had to be regarded as a continuing possibility, such action,

SECRETARY OF THE ARMY FRANK PACE, JR., *arrives at Seoul airbase for an inspection tour and is greeted by General Milburn.*

[30] (1) Rad, JCS 93885, JCS to CINCFE, 11 Oct 50. (2) Rad, No. 412, Secy State (Acheson) to USUN, N.Y., 18 Oct 50.

General Allen

barring a Soviet decision for global war, was not probable in 1950. This optimistic forecast was bolstered by a report from the Far East Command on 14 October implying that China and the USSR, "in spite of their continued interest and some blatant public statements," had decided against "further expensive investment in support of a lost cause." [31]

Eighth Army Enters North Korea

By 3 October, the ROK I Corps was well inside North Korea on the east coast. But until the second week of October, Walker's divisions in the west continued to occupy the Seoul-Inch'on area and to prepare for their drive on P'yongyang.

Supply shortages still plagued Walker's forces. Lt. Gen. Frank W. Milburn, now commanding the U.S. I Corps which was slated to lead the Eighth Army attack, was uneasy about these shortages, and especially wanted at least 3,000 tons of ammunition in forward supply points near Kaesong to support his divisions in the attack. But it was physically impossible to raise the forward supply levels. The Army simply had outrun its logistic support (I Corps, for instance, was 200 miles north of its railhead at Waegwan); and Inch'on helped hardly at all since unloading almost halted during the first half of October when its port facilities were diverted to the outloading of X Corps.[32]

Walker nonetheless was convinced by 7 October that it was time to move. Since MacArthur's order for the attack to the north had not designated a beginning date for the Eighth Army advance and since Walker had had no word since the initial order on 2 October, he directed his chief of staff, Maj. Gen. Leven C. Allen, to get in touch with Tokyo and find out what was wanted. Allen immediately called General Hickey, acting chief of staff, FEC GHQ, for an answer. "Your A-Day will be at such time as you see it ready," Hickey replied. Allen asked for and received immediate confirmation of this by radio. Two days later, on 9 October, Walker notified MacArthur that he had ordered his commanders to strike out for P'yongyang without delay.[33]

---

[31] (1) Rpt in CofS, DA file 323.3, 12 Oct 50. (2) DIS, GHQ, FEC, 2957, 14 Oct 50.

[32] (1) 3d Log Comd Hist Rpt, Oct 50. (2) EUSAK War Diary, G-4 Sec Rpt, 10 Oct 50. (3) Interv, Col Appleman with Gen Eberle, GHQ, FEC, UNC G-4, 12 Jan 54.

[33] (1) Telecon, Gen Hickey (Tokyo) with Gen Allen (Korea), 1130, 7 Oct 50. (2) Rad, CX 65711, CINCFE to CG Eighth Army, 7 Oct 50.

TANKS AND INFANTRYMEN *of the 1st Cavalry Division pursue Communist-led North Korean forces approximately 14 miles north of Kaesong.*

Also on 9 October, basing his action on the new U.N. Security Council resolution, General MacArthur made a second attempt to persuade the North Koreans to surrender. "In order that the decisions of the United Nations may be carried out with a minimum of further loss of life and destruction of property," he told enemy leaders by radio, "I, as the United Nations Commander-in-Chief, for the last time call upon you and the forces under your command, in whatever part of Korea situated, to lay down your arms and cease hostilities." He assured the enemy that the people of North Korea would be treated fairly and that the United Nations would rehabilitate their devastated country as part of a unified Korea. But he warned that unless he got an immediate agreement from the North Korean Government, "I shall at once proceed to take such military ac-

U.S. TROOPS *pass supply wagons abandoned by the North Koreans in their haste to escape from P'yongyang.*

tions as may be necessary to enforce the decrees of the United Nations." [34]

Kim Il Sung, the North Korean Premier, rejected this demand out of hand. He knew that, even as MacArthur's message reached him, Walker's divisions in the west were entering North Korea while in the east the ROK I Corps was fast approaching Wonsan.

Information on North Korean activities north of the parallel had already convinced the Eighth Army that hard fighting awaited on the road to P'yongyang. ROK intelligence agents described extensive North Korean fortifications and other defensive preparations, including the moving up of new units of fresh troops who had not fought in South Korea.[35]

In the U.S. I Corps zone, patrols from the U.S. 1st Cavalry Division crossed the 38th Parallel on 7 and 8 October; and on 9 October, the full division struck across the boundary north of Kaesong. The

[34] Radio Broadcast, CINCUNC to CINC NKPA, 9 Oct 50.

[35] EUSAK PIR Nos. 82, 89, and 90, dated 2, 9, and 10 Oct 50.

Toksil-li
Mup'yong-ni
PUJON (FUSEN) RESERVOIR
Kojang
Koin-dong
CHANGJIN (CHOSIN) RESERVOIR
30 OCT
29 OCT
Hagaru-ri
CCF 38
CCF 40
Huich'on
Pukch'ong
25 OCT
Onjong
Unsan
Ch'ongch'on R
Hongwon
Hamhung
Sinup
Hungnam
Kuryong R
Yongwon
Kunu-ri
U.S. I
ROK II
Sinanju
Pukchang-ni
Yanghung
20 OCT
Sunch'on
Sukch'on
EIGHTH
U.S. X
1 MAR
26-31 OCT
Yangdok
Wonsan
11 OCT
Kojo
P'YONGYANG
Taedong R
Sin'gye
14 OCT
11 OCT
P'yonggang
Kumsong
Kumhwa
Ch'orwon
Imjin R
Kumch'on
9-14 OCT
Hwach'on
Yesong R
Munsan-ni
Uijongbu

British 27th Brigade, the ROK 1st Division, and the U.S. 24th Division also took part in the drive. The attackers encountered prolonged and fierce resistance at Kumch'on, but on 14 October they seized that battered town and by 16 October the enemy front lines ceased to exist. American, British, and ROK troops then raced toward P'yongyang. *(Map III)*

The progress of the two divisions of the ROK I Corps along the east coast of Korea was even faster, and at times spectacular. Although enemy resistance appears to have been lighter in the area, the ROK advance nevertheless reflected a creditable offensive spirit.[36] The speed with which these South Korean soldiers pursued their adversaries up the peninsula made inevitable the bypassing of comparatively large numbers of enemy troops in the east coast mountains. These troops later turned to guerrilla warfare and proved an annoying, even dangerous, thorn in the side of U.N. forces.

Because General Walker was not sure of how much control he held over ROK units, General Allen, when he talked to General Hickey on 7 October, had asked for guidance. Referring to the attack order from GHQ, Allen said, "In the order you notice, there is a line up beyond which certain people [ROK] go. . . . Is KMAG under our control and logistic support? We would like to know if we can organize the ROK Army itself." Hickey was not able to provide an immediate answer, but called Allen back fifteen minutes later saying, "Red, I've got the confirmation on the way to you by wire regarding those elements you mentioned. They are to be considered as members of the team and working with the team in whatever area they may be employed." [37]

ROK units on 11 October captured Wonsan, the objective area for the pending X Corps assault. General Walker flew into the city on the day of its capture. He was so impressed by the ROK's successes that he tacitly established his own plan for cutting a line across Korea from P'yongyang to Wonsan. By taking Wonsan before X Corps arrived the ROK units had changed considerably the tactical picture existing at the time of the issuance of Operations Order No. 2, nine days before. The ROK forces seemed to be in a position to carry out the original mission assigned to X Corps, advancing along the Wonsan-P'yongyang axis to link up with other Eighth Army forces and sealing off Korea to that line.

The success scored by the ROK I Corps and mounting evidence of landing problems at Wonsan had already caused General MacArthur to think of changing the employment of X Corps. He had directed his planners to modify plans for Almond's landing and to prepare for a possible landing by the Marines at Hungnam instead of Wonsan. The 7th Division would land administratively a few miles north of Wonsan, then strike out overland for P'yongyang. The Marines, in the meantime, from their base at Hungnam would head toward P'yongyang also. On 8 October, General Wright presented General MacArthur with such a plan. This plan pointed

[36] General MacArthur paid tribute to the ROK forces engaged in this operation by stating that "In . . . the exploitive pursuit they are unequaled." See *MacArthur Hearings*, p. 4.

[37] Telecon, Gen Hickey and Gen Allen, 1115, 7 Oct 50, and 1130, 7 Oct 50.

out that the Hungnam area was a feasible location for an amphibious assault operation. After reviewing the plan, General MacArthur called in Admiral Joy, pointed out that ROK units were even then approaching Wonsan, and told him that he was considering this alternative method of landing. Joy strongly opposed the change. He pointed to the great disadvantages of splitting the two forces, the lack of time for detailed naval planning, and the impracticability of clearing both Wonsan and Hungnam harbors of mines in the short time left before the landing was to take place. General MacArthur accepted Joy's views, gave up the idea of changing landing places, and on 10 October ordered all major commanders to carry out the original plan as scheduled.[38]

General Walker, on the next day, reported to General MacArthur, "The I ROK Corps has entered Wonsan and is now mopping up enemy resistance. The II ROK Corps [is] advancing north on the Wonsan area from the vicinity of Chorwon-Kumhwa-Kumsong." Then, apparently believing that this welcome news gave him sufficient license, General Walker announced some plans of his own:

In order to support the planned operations of the ROK Army in securing the Wonsan area and advance to the west to Pyongyang in conjunction with the advance of the U.S. I Corps from the south and southeast, it is vital to provide for the supply of five divisions of the ROK Army through the port of Wonsan. Request that the harbor be swept clear of GF mines as soon as possible.[39]

General MacArthur had no intention of leaving X Corps out of the operations. He made this very clear to Walker, removing any delusions that Eighth Army was going to expand its mission. "Wonsan port facilities will be secured and utilized for operations of X Corps in accordance with the United Nations Command Operations Order No. 2," he instructed General Walker. He told him that the Navy would continue its sweeping operations to remove mines from Wonsan Harbor and would maintain its gunfire and air support of ROK divisions. But no additional LST's for carrying supplies to the ROK troops could be furnished until after X Corps troops had landed. MacArthur also told Walker that the Eighth Army would lose the ROK forces in the Wonsan area when X Corps came in. "I now plan to place X Corps in operational control of I ROK Corps. . . ." [40]

The X Corps Prepares

At the close of September, at X Corps headquarters in Ascom City near Inch'on, General Almond briefed his division commanders and principal staff officers on the coming amphibious operation. General Smith, commanding the 1st Marine Division, viewed the plan skeptically, especially the concept of marching westward across the peninsula to contact Eighth Army. "It involved a movement

[38] Karig, *Battle Report, The War in Korea*, ch. 25, pp. 301–02. (2) Rad, C 55002, CINCFE to All Cmdrs, 10 Oct 50. (3) Opn Plan CINCFE 9–50 (Alternate).

[39] (1) War Diary X Corps 8 Oct 50, Wonsan-Iwon Landings. (2) Rad, GX 25744, CG EUSAK to CINCFE, 11 Oct 50.

[40] Rad, CX 66169, CINCFE to CG Eighth Army, 11 Oct 50.

of 125 miles across the rugged central mountain chain of Korea," he wrote later. "There were many defiles and many stretches of the road were one-way. The Eighth Army in its rapid drive north from the Pusan Perimeter had bypassed thousands of North Korean troops. These enemy troops had faded into the central mountains and were making their way north to a sanctuary somewhere in North Korea. In a drive across the central mountain range, the protection of the MSR would present a serious problem, as the drift of the North Koreans would be across the MSR." The matter was not for Smith to decide, however, and the division officers began planning for the new operation at once.[41]

The 1st Marine Division, scheduled to assault the Wonsan beaches, began assembling in the Inch'on area on 4 October. By 7 October, the division and a regiment of South Korean marines moved into staging areas at Inch'on and on 9 October began boarding ship for the 830-mile sea voyage to Wonsan.

The other major component of the X Corps, the U.S. 7th Division, started moving south to Pusan by road and rail on 5 October. Several times during the long trip, groups of bypassed enemy soldiers attacked the column, but were beaten off. The leading regiment of the 7th Division reached Pusan on 10 October. By the 12th, all units were in their Pusan assembly areas; and on 16 and 17 October, the division boarded ship.

By General MacArthur's direction, the Eighth Army was responsible for the logistical support of all United Nations forces in Korea. Thus, General Walker was responsible for supplying the X Corps without having any control over the corps' operations. This arrangement added confusion and misunderstanding to an already unusual relationship between the two major commands. MacArthur may have felt that Almond's extremely tight time schedule in preparing for the amphibious move, the general dislocation of Almond's forces during the transfer of divisions, and the weaknesses inherent in corps logistical facilities as compared to an army, justified saddling Walker with this additional responsibility. Too, there was reason to believe that the Wonsan operation would be completed within a matter of weeks, thus rendering Walker's obligation a temporary measure of short duration.

Colonel Smith, X Corps G–4, had decided views on the effect of MacArthur's decision to make Eighth Army responsible for X Corps supply. "The preparation for the East Coast landing," Smith commented:

> was further complicated from a logistic viewpoint by action taken by GHQ to revise channels during the out-loading so that the Eighth Army became responsible for logistic support of the Corps instead of Corps dealing directly with logistical agencies in Japan. Through direct contact of X Corps staff with JLCOM [Japan Logistical Command] Agencies, detailed supply plans had been completed. . . . The introduction of Eighth Army into channels interrupted these arrangements at a critical time.
>
> Although Eighth Army made every effort to assist the Corps in out-loading in conformance with the plan, the loss of direct contact with JLCOM resulted in resupply difficulties during the unloading phases. Rations arrived on large ships, bulk loaded. In order to assemble logical menus for issue to troops, almost the entire ship had to be unloaded before a balanced meal could be

[41] General Smith's Chronicles, p. 371.

provided. This required emergency airlift of rations into the Corps area. Had the original plan for shallow-draft ships with cargo prepared for selective discharge been followed, it would have been possible to have met the troop requirements from day to day.

A similar problem occurred in the outloading of Signal supplies. At this time the Corps was utilizing three ports for unloading. Instead of distributing Signal supply items to permit this discharge at each area, all items were placed on one ship and unloaded at Iwon with the Seventh Division. Lack of rail facilities and limited truck transportation delayed redistribution of these supplies to other units.

POL intended for resupply of Seventh Division was never outloaded by JLCOM due to a misunderstanding based on a cancellation of what they thought to be a duplication. This necessitated an emergency shipment by LST to meet an urgent requirement for the Seventh Division.

It is believed that the above and many similar problems were created primarily by the change of channels at a time when all staffs were over-worked and involved in a very complicated operation requiring the closest of liaison and direct coordination. Many of these difficulties would not have arisen had X Corps continued to receive logistic support direct from JLCOM at least until the initial landings had been established on the East Coast.[42]

A New Obstacle

While the loading of X Corps ran its course, other developments in the objective area threatened even more directly than supply and shipping problems to wreck the entire landing operation. The enemy had mined Wonsan Harbor and all its approaches.

The U.S. Navy had discovered enemy mines in Korean waters as early as 4 September. Operations at Inch'on had been somewhat hampered by contact mines laid in the entrance channel. (Magnetic mines had also been discovered but, fortunately, ashore.) Between 26 September and 2 October five U.N. ships had struck mines. Intelligence reports confirmed that the enemy had mined the approaches to Wonsan Harbor. But the depth and thoroughness of the enemy's mining operations along the coast of northeast Korea did not become apparent until Almond's troops had already begun loading at Pusan and Inch'on.[43]

The Navy attack order, issued on 1 October, had called for minesweeping operations to begin five days before the landing date. But reports of mines at Wonsan and the possibility of bad weather or influence mines making the clearing a longer operation, prompted Admiral Struble to advance the beginning date to 10 October.[44]

When on the 10th ships of Joint Task Force Seven turned toward Wonsan to begin their minesweeping, they encountered mine patterns as concentrated and effective as any in the history of naval warfare. At least 2,000 mines of all types—contact inertia, contact chemical, pressure, and electronic—lay in the path of any invasion fleet. Intelligence reports later disclosed that Russian technicians and advisers had assembled the mines, planned the minefields, and su-

---

[42] Blumenson, Miscellaneous Problems and Their Solution, p. 52.

[43] (1) Comd and Hist Rpt, COMNAVFE, Sep–Nov 50, Annex to Rpt of GHQ, FEC, UNC. (2) Field, *History of United States Naval Operations, Korea*, pp. 229–42.

[44] Field, *History of United States Naval Operations, Korea*, p. 233.

pervised their laying. Civilians impressed from Wonsan had laid the mines simply and economically by rolling them off towed barges. The Russians had meant to lay 4,000 mines, but had not finished when ROK forces drove them out of Wonsan.[45]

Although 20 October had been established as the corps landing date, Admiral Doyle, commanding Task Force 90, directed that the landing be delayed until the transport and landing areas were positively clear of all mines. The command ship for the operation, USS *Mt. McKinley,* with Generals Almond and Smith aboard, proceeded to the Wonsan area to await developments; but the remainder of the assault shipping was ordered to delay arrival at Wonsan by alternately sailing north for twelve hours then south for twelve hours and while awaiting orders from Admiral Doyle to proceed to Wonsan.[46]

Since the ten American minesweepers in the theater were not enough to sweep the harbor in time for a 20 October landing, and with no time to bring more sweepers from the United States, Admiral Joy had petitioned General MacArthur for permission to use Japanese minesweepers. Necessity overcame any political objections, and General MacArthur granted permission. He stipulated that Japanese crew members must be volunteers and that they receive double pay. Subsequently, 19 minesweepers—10 American, 8 Japanese, and 1 ROK—concentrated for the sweeping operations which began on 10 October.[47]

The use of Japanese contract vessels and crews introduced problems stemming from misunderstanding. The Japanese had been informed that they would not be used in sweeping operations north of the 38th Parallel; they did not know how to communicate with the Japanese Maritime Safety Agency without breaking radio silence; a question arose as to whether the double pay feature was applicable to base pay and allowances or merely to base pay; they felt that they were inadequately supplied; they were sweeping with 3.2-meter draft ships while the mines were planted only three meters below the surface; and they were conducting the first sweep, the combat sweep, whereas they had been promised that they should perform only the second sweep. They registered their complaints but to little avail.[48]

The situation hardly improved on 17 October when one of the Japanese-manned vessels struck a mine and sank. The Department of State hurriedly cabled General MacArthur and cautioned him not to release any information on the sinking because of the great propaganda advantage that the Communists could gain from the fact of Japanese participation in Korean operations. MacArthur assured Washington that he was keeping the news under wraps, but insisted that his use of Jap-

[45] (1) Comd Rpt, NAVFE, Sep–Nov 50. (2) Field, *History of United States Naval Operations, Korea,* p. 237.

[46] General Smith's Chronicles, pp. 404–09.

[47] For additional details of minesweeping operations in Korea, see Field, *History of United States Naval Operations, Korea,* pages 229–42.

[48] (1) Rad, 012344Z, COMNAVFE to CINCFE, 1 Oct 50. (2) Rad, C 65564, CINCFE to COMNAVFE, 6 Oct 50. (3) Action Rpt, Joint Task Force Seven, Wonsan Opn, 1–C–2. (4) Rad, 230310Z, COMNAVFE to Comdr, Task Force Seven, Info CINCFE, 24 Oct 50.

MINESWEEPER *blows up in Wonsan Harbor from a Russian-type contact mine.*

anese minesweepers was perfectly legitimate. "These vessels," he asserted, "were hired and employed, not for combat, but humanitarian purposes involved in neutralizing infractions of the accepted rules of warfare." The infractions MacArthur mentioned referred to the use of free-floating mines by the North Koreans. Whether classified as a combat operation or a humanitarian effort, the sweeping continued with Japanese participation.[49]

[49] (1) Rad, No. 650, Secy State to SCAP, 19 Oct 50. (2) Rad, CX 67066, CINCFE to DA, 21 Oct 50. (3) Comd and Hist Rpt, NAVFE, Sep–Nov 50.

The Wake Island Conference

While Eighth Army troops pressed forward into North Korea and X Corps prepared to land at Wonsan, President Truman called General MacArthur to a conference at Wake Island. On 10 October, the President announced:

> General MacArthur and I are making a quick trip over the coming weekend to meet in the Pacific. . . . I shall discuss with him the final phase of United Nations action in Korea. . . . We should like to get our armed forces out and back to their other duties at the earliest moment consistent with the fulfillment of our obligations as a

RADFORD, MACARTHUR, PACE, AND BRADLEY *at Wake Island Conference.*

member of the United Nations. Naturally, I shall take advantage of this opportunity to discuss with General MacArthur other matters within his responsibility.[50]

President Truman had intended to take all of the Joint Chiefs of Staff with him. But after being advised of the danger of taking them all from Washington under conditions then existing in Korea and in other potential trouble spots, the President took only the chairman, General Bradley, and Secretary of the Army Pace as his military advisers.[51] In addition to his two military experts,

[50] Rad, 101910Z, DA to SCAP, 11 Oct 50.

[51] For the details of the Wake Island Conference, the author has relied upon a compilation of notes by General Bradley. These notes, which were kept by the Washington conferees, were augmented by shorthand recordings taken by a secretary who listened to the meeting from an adjacent room. General MacArthur later objected that he had no knowledge that a verbatim transcript was being taken, or that, indeed, any record of the conference was kept. General Bradley states that five copies of this material were forwarded to General MacArthur on 19 October 1950 and that one of General MacArthur's aides signed for them on 27 October 1950. According to his own testimony, General MacArthur

President Truman was accompanied to Wake Island by Ambassador-at-large Philip C. Jessup, W. Averell Harriman, and Assistant Secretary of State Dean Rusk. General MacArthur arrived with Maj. Gen. Courtney Whitney of his staff. Admiral Radford, Commander in Chief, Pacific, also attended the main conference. All conferees arrived on 15 October.

The President and General MacArthur first conferred privately for approximately an hour. Afterward, Mr. Truman opened the general conference by asking the United Nations commander to give his views on the problems facing the United States in rehabilitating Korea. In reply MacArthur was extremely optimistic, stating that he believed that formal resistance by the enemy would end by Thanksgiving. The North Korean Army was pursuing a forlorn hope in resisting the United Nations forces then attacking it. The enemy, according to MacArthur, had only about 100,000 men left and these were poorly trained, led, and equipped. They were fighting obstinately, but only to save face. "Orientals," General MacArthur pointed out, "prefer to die rather than to lose face."

General MacArthur described his tactical plan in broad outline, saying that he was landing X Corps at captured Wonsan from which this corps could cut across the peninsula to P'yongyang in one week. He compared this planned maneuver to the Inch'on operation and noted that the North Koreans had once again erred fatally in not deploying in depth. "When the gap is closed, the same thing will happen in the north as happened in the south."

Eighth Army, if things went according to General MacArthur's schedule, would be withdrawn to Japan by Christmas. The 2d and 3d Divisions and certain U.N. units of smaller size would remain in Korea under the X Corps to carry out security missions and to support the United Nations Commission for the Rehabilitation and Unification of Korea. He hoped that elections could be held before the first of the year, thus avoiding a military occupation. "All occupations are failures," General MacArthur commented.

General Bradley, who was quite concerned over the shortage of American forces in Europe and who saw the end of the Korean War as an opportunity to get another division into Europe in a hurry, asked General MacArthur if the 2d or 3d Division could be made available for shipment to Europe by January. MacArthur responded with a promise to make either division ready for shipment by that time, but recommended that the 2d Division be sent since it was a battle-proven organization and better trained than the newly arrived 3d Division.

Before their 1-day conference ended, President Truman asked MacArthur what chance there was of Chinese interference. The United Nations com-

---

did not bother to look at the copies of the record furnished him. He said, however, "I have no doubt that in general they are an accurate report of what took place."

Immediately after his return from Wake Island, Secretary Pace reported to the Army Policy Council on the meetings. His report, made at a time when the record was not the political issue it later became, is identical in content to General Bradley's notes. See also Committee on Armed Services and Committee on Foreign Relations, U.S. Senate, *Substance of Statements Made at Wake Island Conference on October 15, 1950* (Washington, 1951).

President Truman

mander replied, "Very little." He felt that the Red Chinese had lost their chance to intervene effectively. He credited the Chinese with having 300,000 men in Manchuria, with between 100,000–125,000 men along the Yalu, and estimated that 50,000–60,000 could be brought across the Yalu. But the Chinese had no air force, according to General MacArthur; hence, in view of U.N. air bases in Korea, "if the Chinese tried to get down to Pyongyang there would be the greatest slaughter." This broad assurance from MacArthur must have done much to allay any fears entertained by Mr. Truman and the other top authorities that China meant to intervene.[52]

[52] See Appleman, *South to the Naktong, North to the Yalu*, pp. 760–61.

CHAPTER XII

Signs of Victory

New Plans of Advance

Upon his return from Wake Island, General MacArthur moved vigorously to validate his prediction of imminent victory. Although enemy mines had temporarily stymied the landing of the X Corps, MacArthur did have troops in North Korea in the east as well as on the west and west central front. As a first step in exploiting the situation in the east, he placed all units in the Wonsan area under Almond's command. On 16 October, MacArthur told General Walker:

> In order to exploit to the maximum all forces under CINCUNC and to implement the full concept of operations . . . X Corps, operating as an independent Corps of GHQ Reserve, will, effective at 1200 hours, 20 October 1950, and until further orders, assume operational control of all UN and ROK ground forces operating north of 39 degrees and 10 minutes north.[1]

Strengthening Almond's command still more, MacArthur ordered the 3d Division—two regiments of which were then in Japan and the other, the 65th Infantry, in Korea—readied by 2 November for shipment to Wonsan.[2]

Turning to Walker's zone, MacArthur alerted the 187th Airborne RCT which had gone into GHQ Reserve around Kimp'o Airfield, for an airdrop across the two main arterial routes north of P'yongyang near the towns of Sukch'on and Sunch'on. He directed the 187th to be ready to drop on 21 October to stop enemy withdrawals to the north, to cut off enemy reinforcements, and to disrupt enemy communications. He hoped, also, to capture important North Korean officials and to rescue U.N. POW's before the enemy could move them northward. MacArthur apparently still believed that the X Corps would be operating ashore within a few days. For he directed that when the corps had landed and driven west to establish contact with the 187th north of P'yongyang, General Almond was to assume operational control of the airborne unit. The commanding general, Far East Air Forces, became responsible for lifting the 187th and for furnishing

[1] Rad, C 66549, CINCFE to CG Army Eight, 16 Oct 50.

[2] Rad, C 66553, CINCFE to COMNAVFE and CG X Corps, 15 Oct 50.

tactical air support to the airborne troops throughout the operation.[3]

On 17 October, as Walker's troops approached P'yongyang, MacArthur issued new orders designed to wring every advantage out of the favorable battlefield situation. He set up a boundary between the Eighth Army and the X Corps to become effective on his further order. MacArthur wanted the Eighth Army to advance on the left of the new boundary to the general line Sonch'on–Chongsanjangsi–Koin-dong–Pyongwon. The X Corps would advance to the eastern extension of this line, Toksil-li–Pungsan–Songjin.[4] For this general advance, MacArthur removed the restrictions against using any but ROK forces north of the line Ch'ongju–Kunu-ri–Yongwon–Hamhung. But he directed that only ROK troops would operate north of the new objective line. His new concept, of course, also canceled his previous plan to cross and seal off the peninsula between P'yongyang and Wonsan. For the new objective line ranged from 80 to 130 miles north of the P'yongyang-Wonsan road and approached within 40 miles of the Manchurian border.[5]

P'yongyang fell on 19 October, whereupon MacArthur started the all-out drive toward the new objective line. He stepped up the airborne operation by one day, sending the 187th Airborne into the Sukch'on-Sunch'on area in the first airdrop of the Korean campaign on 20 October. The Eighth Army assumed operational control of the 187th RCT after it hit the ground.

From his own plane, General MacArthur, accompanied by Generals Stratemeyer, Wright, and Whitney, watched the parachute troops land and assemble. He then flew to P'yongyang, where he commented to reporters that the airborne landings seemed to have completely surprised the North Koreans and that "This closes the trap on the enemy." When he returned to Tokyo on 21 October, he predicted that "the war is very definitely coming to an end shortly." [6]

On the east coast, General Almond went ashore at Wonsan by helicopter on 19 October to take charge of the ROK I Corps which was still moving rapidly to the north. By that date, the 1st Marine Air Wing had been flying out of Wonsan Airfield for five days; and shore parties, engineers, and advance billeting parties were in Wonsan preparing for the landing of the 1st Marine Division.[7]

Although to Admiral Struble, commander of the Seventh Fleet, Almond expressed unhappiness over the delayed landing of his remaining forces, Struble would not authorize a landing until he, himself, was satisfied with mine clearance operations. Hence, transports arriving

[3] (1) UNC Opn Order No. 3, 16 Oct 50. (2) Rad, CX 66591, CINCUNC to All Comds, 16 Oct 50.

[4] The new boundary ran westward along the 39th Parallel from the Sea of Japan to a point at 126 degrees 45 minutes east longitude, then generally to the north and northeast through Sinup at 39 degrees, 28 minutes north, 127 degrees, 8 minutes east, and Changjin at 40 degrees, 54 minutes north and 127 degrees, 10 minutes east, to Huch'ang at 41 degrees, 24 minutes north and 127 degrees, 4 minutes east.

[5] UNC Opns Order No. 4, 17 Oct 50.

[6] (1) *Pacific Stars and Stripes*, October 21, 1950, p. 1, col. 6. (2) New York *Times*, October 21, 1950, p. 1, col. 8, Parrott dispatch; also p. 3, col. 1 and p. 16, col. 2. (3) EUSAK War Diary, 22 Oct 50, Daily News Bulletin.

[7] General Smith's Chronicles, pp. 404–10.

EQUIPMENT AIR-DROPPED *in the Sunch'on drop zone.*

off Wonsan with the 1st Marine Division on 20 October steamed back and forth outside the harbor, a maneuver that the Marines promptly dubbed "Operation Yo-yo." Meanwhile, the 7th Division remained idly afloat in Pusan Harbor awaiting word to proceed to the objective area.

General Almond did not easily give up his efforts to get Marines ashore before the minefields were completely cleared. On 21 October, he asked that a battalion of the 1st Marine Division be landed at Kojo, about thirty-nine miles south of Wonsan, to relieve ROK soldiers guarding a supply dump there. Since Japanese and ROK LST's had put into Kojo even though the adjacent waters had not been swept, Almond felt that Navy LST's could do the same. But Admiral Doyle did not consider a Kojo landing urgent enough to justify jeopardizing the troops and ships.

On 22 October, Admiral Doyle estimated that the Wonsan landing would not be feasible for another two or three days. A longer delay appeared possible on 23 October when a new row of magnetic mines was encountered. But on the 24th, the results of sweeping operations indicated that the magnetic mines were about cleared out, and that the troop landings could therefore soon be made.

Another Change

On the same date, General MacArthur in an extraordinary order commanded Walker and Almond to drive forward with all possible speed using all forces at their command. The objective line he had set up only a week before was merely to be an initial objective; and the restriction he had placed on using other than ROK forces was removed since, as he reminded his commanders, the prohibition had been established only in view of a possible enemy surrender.[8]

This order conflicted with the instructions the Joint Chiefs of Staff had sent MacArthur on 27 September wherein they had told him that "as a matter of policy no non-Korean ground forces will be used in the northeast province bordering the Soviet Union or in the area along the Manchurian border." The Joint Chiefs, upon learning of MacArthur's new order, objected in the form of an inquiry. "While the Joint Chiefs of Staff realize," they told him, "that you undoubtedly had sound reasons for issuing these instructions they would like to be informed of them, as your action is a matter of some concern here." [9]

MacArthur defended his action with characteristic vigor. He held that his order had been prompted by military necessity since his ROK forces had neither sufficient strength nor enough skilled leadership to take and hold the border areas of North Korea. As to the legality of his decision, MacArthur pointed out that the Joint Chiefs had told him that the directive of 27 September was not final, that it might require modification in accordance with developments. For additional justification, General MacArthur emphasized that the Joint Chiefs had not actually banned the use of other than ROK forces but had merely stated that it should not be done as a matter of policy. Finally, in his mind, the instructions from the Secretary of Defense on 30 September, which had assured him, "We want you to feel unhampered tactically and strategically to proceed north of the parallel," had certainly modified any prior instructions from the Joint Chiefs and he had proceeded to issue his orders on that basis. He made no move to placate his superiors. While he assured them that he understood their concern, he also hinted of dire developments if he took any other course and clinched his argument by claiming that "This entire subject was covered in my conference at Wake Island." [10] Thereafter, the Joint Chiefs allowed General MacArthur's order to stand.

[8] Rad, CX 67291, CINCUNC to All Comdrs, 24 Oct 50.

[9] (1) Rad, JCS 94933, JCS (Personal) for MacArthur, 24 Oct 50. (2) General Collins later pointed to this instance as an example of violation of orders by the U.N. commander in *MacArthur Hearings*, pp. 1240, 1301–02.

[10] Rad, C 67397, CINCFE for JCS, 25 Oct 50. None of the other persons attending the conference on 15 October recorded any mention of the use of non-ROK soldiers along the Manchurian and Soviet borders.

X Corps Lands

By 1500 on 25 October, ships began coming into the swept area of Wonsan Harbor preparatory to the landing of troops and equipment. That evening, five LST's containing advance parties hit the beaches and on the following day, the main troop landings began.

General Almond meanwhile made almost daily reconnaissance flights up the east coast, checking the progress of the ROK troops and searching for a suitable alternate landing area for the 7th Division. In view of the recent change of plans, he chose not to set the 7th Division ashore in the wake of the Marines, but to land it deeper in North Korea and thus accelerate carrying out MacArthur's new instructions for a drive to the border. Following close aerial inspection of the coast, he decided to land the Army division at the small port of Iwon, 105 miles northeast of Wonsan. After he completed arrangements for landing at the new site with Admirals Struble and Doyle, the 7th Division sailed north from Pusan on 27 October and began landing at Iwon two days later.[11]

Occupation Plans

As military prospects brightened during October, American planners turned their efforts to devising a system of military government for North Korea. In Washington, the Army staff prepared a detailed directive for military government, and on 10 October forwarded it to General MacArthur for comment.

Under this directive, the occupation of North Korea would have three phases. In the first phase, which would last until internal security had been restored, General MacArthur would act as supreme authority in North Korea, subject to the control of the United Nations and the United States Government. During the second phase, which would last until national elections had been held throughout Korea, MacArthur would retain complete authority, but a United Nations commission would furnish advice and recommendations which he would honor within the bounds of security of his forces. The third phase, from the completion of national elections to the withdrawal of all non-Korean United Nations forces, would see a gradual release of control to the elected government of Korea. The military commander, in this final phase, would be responsible only for such missions as might be assigned to him.[12]

MacArthur's primary duties during the occupation would be to establish public order, to rebuild the nation's wrecked economy, and to prepare the people for unification. But while he would definitely dissolve the Communist government of North Korea, he would not replace it with the ROK Government of President Rhee. In fact, he would create no central government for North Korea other than as part of his occupational control machinery. This would be deferred until free, Korea-

[11] (1) X Corps WD, Oct 50. (2) General Smith's Chronicles, pp. 404–10.

[12] Rad, W 93721, DA to CINCFE, 10 Oct 50.

wide elections had been held under jurisdiction of the United Nations.[13]

Neither the United States nor the United Nations intended that Rhee's government should automatically assume control of liberated areas of North Korea. General MacArthur had been so advised by the Joint Chiefs on 27 September when they had told him, "political questions such as the formal extension of sovereignty [ROK sovereignty] over North Korea should await action by the United Nations to complete unification of the country." On 9 October, he was reminded that the authority of the Republic of Korea over North Korea had not been recognized and that as United Nations commander he would not recognize any such authority.[14]

A fuller explanation of this potentially explosive issue came on 12 October when the Interim Committee of the General Assembly of the United Nations formally resolved that the United Nations recognize no government as having "legal and effective control" over all of Korea. The committee asked that the unified command (U.S. Government) assume provisionally all responsibilities for the government and civil administration of all parts of North Korea coming under control of the U.N. forces "pending consideration by the United Nations Commission for the Unification and Rehabilitation of Korea of the administration of these territories." [15]

At the Wake Island Conference, General MacArthur had expressed fears that the committee's action might have a bad effect on U.N.-ROK relations. "I have," he told Mr. Truman, "been shaking in my boots ever since I saw the United Nations resolution which would treat them [South Koreans] exactly on the same basis as the North Koreans." President Truman sided with MacArthur and stated that the United States would continue to support the ROK Government. But the President announced no decision to place the Rhee government in control of captured North Korean territory.[16]

When President Rhee learned of the U.N. resolution, he protested hotly to General MacArthur. Never temperate in his approach to the unification of Korea, the veteran statesman accused the U.N. committee of reviving and protecting communism by its resolution, and asserted that his government was "taking over the civilian administration whenever hostilities cease." MacArthur passed Rhee's protest along to President Truman.[17]

The President attributed Rhee's reaction to an incomplete understanding of the U.N. committee's resolution. The American Government continued to support the U.N. position that the jurisdiction of the Republic of Korea did not automatically extend over North Korea and that the U.N. commission would arrange for elections and other constituent acts required to unify the country.[18] On 29 October, this policy

[13] *Ibid.*

[14] (1) Rad, JCS 92801, JCS to CINCFE (Personal) for MacArthur, 27 Sep 50. (2) Rad, WAR 93721, DA to CINCFE, 9 Oct 50.

[15] Rad, W 94093, DA to CINCFE, 12 Oct 50.

[16] Substance of Statements, Wake Island Conference, 15 Oct 50.

[17] Rad, CX 66554, CINCFE to CG USARPAC, MacArthur (Personal) for Truman, 16 Oct 50.

[18] (1) Rad, JCS 94710, JCS to CINCFE, transmitting message from President Truman, 20 Oct 50. (2) Rad, WAR 94472, DA to CINCUNC, 18 Oct 50.

was affirmed and General MacArthur was directed not to recognize the authority of Rhee's government in North Korea but to consult with that government "in matters of national scope" through Ambassador Muccio.[19]

General MacArthur apparently interpreted this directive too literally to suit authorities in Washington. When it appeared to them that MacArthur was prohibiting South Koreans from participating in civil affairs matters in North Korea, they explained to him on 2 November:

> It is not intended that pertinent directives or any other instructions prohibit the use of ROK administrators, police, military forces, or any other ROK asset in North Korea as long as it is clearly and publicly understood that such resources are not under control of ROK but rather are designated as UN instrumentalities, and are placed under CINCUNC's control and at CINCUNC's disposal.[20]

The Department of State was anxious to release to the Interim Committee of the United Nations and to send to the press the text of the directive for conducting civil affairs in North Korea. But when General MacArthur's opinion of such a move was solicited, he replied, "I believe that the public release of the directive would be premature, at least until major hostilities have come to an end. The effect upon troops engaged in serious combat of this type of advance planning based upon assumptions not yet realized, cannot be overestimated." [21]

[19] Rad, JCS 95238, JCS to CINCFE, 29 Oct 50.

[20] Rad, WAR 95715, JCS to CINCUNC, 2 Nov 50.

[21] Rad, C 67920, CINCFE to DA, 31 Oct 50.

The Mirage of Victory

In some respects, the critical situation in Korea during the first four months of fighting had proved beneficial to the United States Army. Moved by the obvious need for greater combat strength in Korea and aware of the increased danger of Communist aggression elsewhere in the world, American leaders had relaxed the rigid controls over military appropriations prevalent from 1945 to mid-1950. As one result, the Army had expanded its strength and facilities. Some of the Army expansions during these months had little direct relationship to Korea's needs, but were aimed instead at placing the Army in a more favorable position to meet any general emergency. On 27 September, the Secretary of Defense had authorized the Army to increase its strength during Fiscal Year 1951 to 1,263,000.[22] Keeping step with the increase in manpower, substantial increases in logistical support for the Army were authorized and steps were taken to transmute these authorizations into matériel.

In connection with this growth, Secretary of the Army Pace pointed out to the Army Policy Council on 5 October that many of the measures taken by the Army as a result of the Korean fighting had not received the specific approval of the Congress. More important, the Army, in its efforts to build up rapidly, had obligated and spent funds that had not yet been appropriated by the Congress. He anticipated less concern among the nation's law-makers over the

[22] (1) Memo, Secy Defense for Secy Army, 27 Sep 50. (2) Ltr, DA to All Cmdrs, 15 Sep 50, sub: Recall of Additional Reserve Officers to Active Duty.

use of funds than over the question of whether the Army had violated a principle and usurped authority rightfully belonging to the Congress. He directed a careful and thorough re-examination of all Army programs under the assumption that the United States requirements in Korea would be much less than originally contemplated.[23]

Signs of Retrenchment

Throughout October, optimism grew that the fighting would soon end. Consequently, the Department of the Army's planners began to anticipate a need to curtail much of the support developed for Korea. The efforts of responsible officials had started a dynamic flow of men and matériel to the Far East Command which, if allowed to continue at its current rate after the fighting stopped, would flood that theater with unneeded forces and supplies. Millions of dollars would be wasted because of poor storage facilities in Japan and Korea and the great cost of returning men and supplies to the United States.

General MacArthur, after returning from Wake Island, ordered the JSPOG staff to prepare detailed plans for the withdrawal of forces from Korea and for keeping certain units as occupation troops. He based his instructions on agreements and understandings that he felt had been reached on this general topic at Wake Island. On 20 October, JSPOG issued CINCFE Operations Plan No. 202 outlining procedures to be followed after combat operations had dwindled, so that some U.N. forces could be withdrawn from Korea. The plan assumed that the fighting would end in the destruction of organized enemy forces, but that North Korean guerrillas would still resist in the mountains. The plan also assumed that neither Soviet nor Chinese Communist forces would intervene.[24]

General MacArthur, as Commander in Chief, United Nations Command, would carry out such missions and functions as became necessary to bring things to a satisfactory conclusion in Korea. The X Corps was named to stay in Korea for occupation duty. It would have one American division, plus all other U.N. units in Korea, ROK Army units, and KMAG. The U.N. units would be withdrawn progressively, with European units leaving first. Insofar as possible, no forces other than ROK would be stationed in South Korea.

The Eighth Army headquarters, along with its original four American divisions, would return to Japan; and the 5th RCT would go back to Hawaii. Within Japan, after the return of the Eighth Army, the Northern and Southwestern Commands would be dissolved and their functions assumed by the Eighth Army. The Japan Logistical Command would be retained to perform all army logistic functions in Japan.[25]

---

23 Min, Army Policy Council, 2 Oct 50 and 5 Oct 50, in CofS DA file 334 APC.

24 On the day the plan was issued, the GHQ Daily Intelligence Summary carried what it termed a "reliable report" that 400,000 Chinese Communist soldiers were in border-crossing areas, alerted to cross into North Korea. To detect any such crossings, the U.N. Command ordered daily air reconnaissance flights over the border area.

25 (1) CINCFE Opn Plan No. 202, 20 Oct 50, in JSPOG, GHQ, UNC files. (2) General Smith, the Marine division commander, gives some hint of the effect of this planning and what he calls the "end of the war atmosphere." He recalls, "On 21 October

This plan had scarcely reached the interested staff members of General MacArthur's headquarters when word from Washington disrupted its entire concept. On 21 October, the Joint Chiefs of Staff told General MacArthur that demands for American troops in other parts of the world were forcing them to withdraw the 2d and 3d U.S. Divisions from the Far East Command as soon as possible after fighting ended in Korea. Consequently, the forces to guard Japan and also to occupy Korea would have to come from the four divisions originally based in Japan.[26]

MacArthur had left Wake Island under the impression that the 3d Division would be kept in Korea as part of the occupation force. For General Bradley had asked only that one division, either the 2d or 3d, be made available for Europe by January 1951, and MacArthur had recommended the 2d. He therefore objected when he learned that the Joint Chiefs planned to take both divisions from him. He explained his understanding of the arrangements agreed upon at Wake Island, saying he was ". . . under the impression that this proposal had received approval of all concerned. I resubmit it at this time for your consideration."[27]

The Joint Chiefs of Staff had not hastily reached their decision to take both divisions. The American force in western Europe badly needed strengthening; and on the basis of estimates from General MacArthur and other sources, they had concluded that the Far East Command would soon revert to its pre-Korean War status. Since their planning was global and took in the needs of the American defense system as a whole, they undoubtedly felt that their decision should not be unduly influenced by the opinions of a theater commander who was understandably biased in favor of his own needs. On the other hand, the Korean fighting had not yet ceased and the Far East Command was still on a wartime basis. So they made no final decision for the moment.

When General Boltè went to the Far East in late October, he discussed the matter with General MacArthur. He sent a full report of these talks to General Collins, sketching the current status of the 3d Division and recommending that the Joint Chiefs modify their previous decision slightly. General MacArthur had ordered the 3d Division to embark for Korea from Japan on 9 November to operate initially in the southern sector of North Korea and later to serve as the only U.S. occupation division. "When I departed Washington,"

I received a dispatch from COMNAVFE stating that upon the conclusion of hostilities it was the intention to recommend to CINCFE that the 1st Marine Division, less one RCT, be returned to the United States. The RCT not returned to the United States would be billeted at Otsu, Japan. I was requested to comment. On 24 October, we learned that X Corps had received a document, for planning purposes only, to the effect that the present Corps commander would become Commander of the Occupation Forces. One American division, probably the 3d Infantry Division, would remain in Korea as part of the Occupation Forçes. Under this plan the Eighth Army would return to Japan. The receipt of information such as that cited above could not help but spread the impression that the war was about over. There was a noticeable let-down. However, it was only a matter of days until the operations at Kojo brought home to us forcefully the fact that the war was not over. Talk of redeployment was never heard again." See General Smith's Chronicles, p. 403.

[26] Rad, JCS 94651, JCS to CINCFE, 21 Oct 50.

[27] Rad, C 67065, CINCFE to JCS, 21 Oct 50.

General Bolté recalled, "the final disposition of the 3d Division was unresolved." MacArthur had apparently raised some very convincing arguments, for Bolté told General Collins, "From what I have learned here regarding the many administrative and organizational problems in reconstituting the FECOM balanced force coupled with the requirement for a division now to accomplish operational tasks in North Korea, I recommend that FECOM be authorized to retain the 3d Division until elections in North Korea are held but not later than 1 May 1951." He concluded by pointing out that if later required in western Europe, the 3d Division could be placed in an effective status of training and combat readiness as rapidly in the Far East Command as elsewhere.[28] The Joint Chiefs, probably on the basis of General Bolté's recommendation, suspended the provisions previously set up for the 2d and 3d Divisions, but notified General MacArthur to be ready to move the 3d Division from the theater within sixty days of receipt of orders. The 3d Division could be kept in the theater until 1 May 1951.[29]

In the same vein, Washington proposed a cutback of personnel support to General MacArthur. He was told on 25 October:

> Reduction of the scale of operations in Korea compels immediate reconsideration of the number of service enlisted fillers and replacements previously scheduled for shipment to FECOM. To reduce this number to the minimum, Department of the Army proposes to cancel shipment of enlisted reserve corps personnel presently scheduled for October and November except 17,000 NCO grades.

Normal rotation of foreign service tours would resume in January 1951.[30]

Some of the other nations committed to support the United Nations in Korea read the handwriting on the wall and found it encouraging. The United Kingdom had offered the 29th Infantry Brigade for service in Korea and its first echelon had sailed from England on 4 October. The heartening news from Korea led the British General Staff to conclude that requirements had diminished and that some remaining elements of the 29th Infantry Brigade need not be sent. They proposed to withdraw the armored regiment, the 8th Hussars, from the brigade since "the possibility of future battles in Korea in which heavy armor will be required seems remote" and "in the event of a general war the 8th Hussars would be required elsewhere." But General Bolté objected. He considered the future course of the war in Korea, for the moment and for some time to come, too uncertain, at least until all of North Korea was cleared of enemy and the intentions of the USSR and the Chinese Communists could be determined. Too, the probability of a general war in the near future was not great enough to justify withholding forces from the Korean effort.[31]

General Collins took the matter up informally with the other Joint Chiefs on 4 October and was instructed to advise the British representative in Washington that the United States felt the

28 Rad, C 677985, Bolté (Personal) for Collins, 1 Nov 50.

29 Rad, W 95625, DA to CINCFE, 3 Nov 50.

30 Rad, S 94985, DA to CINCFE, 25 Oct 50.

31 Memo, G–3 (Bolté) for CofS (Gruenther), 4 Oct 50, in CofS DA file CX 370, Item 21.

fighting in Korea was far from over and that if the British engaged the enemy, General MacArthur had no surplus tanks available for supporting them. The British Government pursued the topic no further and the 8th Hussars accompanied the 29th Brigade to Korea.[32]

Canada during this period also proposed to cancel the shipment of a special fighting force which had been slated for Korea. This proposal, together with the earlier suggestion from the British Government, brought the need for a policy on continuing United Nations support into sharp focus. General Bolté took the stand that all U.N. forces in Korea and scheduled to arrive there might be required to conclude the fighting successfully and would certainly be needed to provide an adequate occupation force. He voiced concern to General MacArthur over the fact that recent operational successes were fostering a growing tendency among other United Nations members to consider additional forces unnecessary. He suggested that MacArthur include in his next report to the United Nations a special statement to encourage additional contributions from these other nations. General Bolté forcefully recommended to General Collins that, as a matter of policy, the Department of the Army should oppose the release of foreign U.N. military units previously accepted for service in Korea.[33]

By late October, troops of five nations, totaling about 9,000 men and officers, were serving alongside U.S. and ROK troops in Korea. Two British units, the 2-battalion infantry brigade from Hong Kong and a marine commando unit, and a 5,000-man Turkish infantry brigade were already in Korea. Additional U.N. units, totaling 27,000 men, were either en route to the battlefront or preparing for departure from their home countries; and infantry brigades from Great Britain, Canada, and Greece, infantry battalions from Belgium, the Netherlands, France, Australia, and Thailand, and an artillery battalion from New Zealand had been offered to the unified command by the respective countries and these offers accepted. But the developments in Korea, and particularly the favorable prognosis brought back from the Wake Island Conference by the returning conferees, had a profound effect on General Bolté's attitude and on that of all concerned with deciding what should be done about curtailing the shipment of forces to the battlefront. On 23 October, General Bolté advised the Chief of Staff that he felt a total of 36,000 U.N. troops other than U.S. and ROK soldiers was too great and that the time was now ripe for reducing the current and projected strength of such troops to about 15,000. "In view of the state of operations in North Korea," Bolté said:

> and in general consonance with the Wake Island conversations, it is considered desirable to review current and projected plans for utilizing other United Nations in Korea. The problem is to reduce logistic burdens on the United States and at the same time

---

[32] MFR, sgd Gen Collins, 5 Oct 50, CofS DA file CS 370, Item 21.

[33] (1) Rad, WAR 93605, DA (Bolté) to CINCFE, 6 Oct 50. (2) Memo, Gen Bolté for Gen Collins, 13 Oct 50, sub: Continuing Need for U.N. Forces in Korea, in G-3, DA file CS 091, Case 25.

CANADIAN TROOPS CROSS THE IMJIN RIVER IN LIGHT ASSAULT BOATS

retain the political advantages of multinational United Nations representation.[34]

General Collins agreed with his G-3 and presented his views to the Joint Chiefs of Staff at once. He told his fellow service heads that the United Nations contingents were a logistical burden on the United States, although not so pronounced in the case of British units that used their own arms and equipment. From the viewpoint of operational usefulness there was also a wide divergence in the quality of the various contingents. Turning to fiscal matters, General Collins noted that "the degree of ultimate reimbursement to the United States may prove problematical in many cases." He did see the political advantages to be gained by having a wide representation of United Nations contingents even if only token forces were involved. He felt, as did General Bolté, that the Korean operations were approaching a

[34] Memo, G-3 (Bolté) for CofS (Collins), 23 Oct 50, sub: Cutback in U.N. Ground Forces in Korea (other than U.S. and ROK).

TROOP REPLACEMENTS *from Thailand arrive at Pusan.*

point where some of the U.N. units could be dispensed with. Basing his final judgment on all the factors—logistic, political, and operational—General Collins then recommended changes in the U.N. lineup.

After considering Collins' views, the Joint Chiefs of Staff agreed that the request for the Belgian battalion, the Netherlands battalion, the French battalion, the New Zealand artillery battalion, and the additional Australian battalion should be canceled and the countries concerned asked not to send these units. They decided also to ask Canada and Greece to reduce their respective brigades to battalion size and for the redeployment of the Philippine battalion and the British marine commando unit to their own countries. Finally, the U.K. 27th Brigade would be returned to British control as soon as the U.K. 29th Brigade arrived from Hong Kong to replace it. General MacArthur was notified of these decisions at once, and the Secretary of Defense was asked to co-ordinate with the Secretary of State in obtaining the concurrence of the countries concerned. On 2 November, Secretary Marshall wrote the Secretary of State asking that he make the proper arrangements.[35]

Equally as pronounced a reaction to the success of United Nations operations came from those persons charged with logistic support planning for the Korean fighting. During a visit to the Far East Command in late September, Lt. Gen. Thomas B. Larkin, the Department of the Army G–4, conferred with General MacArthur on the matter of cutting back the wartime flow of supplies to his theater. He asked particularly that MacArthur take steps to reduce substantially his requisitions for ammunition and other items.[36] The Army's supply experts were very worried because the bulk of the Army's supplies and equipment were being drawn to the Far East. Because of the rapid developments in Korea since July, the Far East Command had not been able to get its inventory and stock reports in good condition and there was a tendency to draw out of the United States larger amounts of supply and equipment than were needed.[37]

Upon his return from Japan, General Larkin expressed concern that no real plans to reduce the logistic support of the Far East Command to conform with posthostilities requirements had yet been made. "Present indications," Larkin said, "are that hostilities in Korea may end at an early date (VK–day)." He ordered his staff to begin planning on a priority basis for the cutback and diversion of excess supplies en route, for the roll-up of supplies in excess of FEC needs, and for the movement and housing of forces diverted and redeployed from the Far East. But in order to plan, his staff had to know how many troops were going to be stationed in the Far East and Korea after the war ended, and what levels of supply and special reserves would be needed for the U.S. Army, the ROK Army, and the other U.N. forces. He asked General Bolté to provide that information as soon as possible, "due to the possibility of early suspension of major hostilities in Korea and the probable necessity of diverting cargo ships on the high seas." [38] Bolté saw the problem but had no ready answer and turned to the Chief of Staff for advice. Bolté pointed out that the Army could not make plans until the Joint Chiefs had determined what forces were going to be kept in the Far East when the war was over and how much support was going to be given the ROK and Japanese security forces. He insisted that the matter should receive the highest possible priority.[39] General Collins on 3 October asked the Joint Chiefs to start considering the problem.[40] Bolté, giving Larkin such information as he could without waiting for a decision by

---

[35] (1) *Ibid.* (2) JCS 1776/144 and 2d corrigendum thereto, 25 Oct 50. (3) Memo, Secy JCS (Adm Lalor) for CofS USA (Collins), 27 Oct 50, sub: Cutback in Ground Forces in Korea. (4) Ltr, Secy Defense (Marshall) to Secy State (Acheson), 2 Nov 50. All in G–3, DA file 091 Korea, Case 111–111/6.

[36] Min, 43d mtg Army Policy Council, 19 Oct 50, CS 334.

[37] Min, 42d mtg Army Policy Council, 5 Oct 50, CS 334.

[38] Memo, DA G–4 (Larkin) for DA G–3 (Bolté), 26 Sep 50, sub: Post-Korean Hostilities Basis for Logistic Support, in G–3, DA file 091 Korea, Case 99.

[39] Memo, Gen Bolté (G–3) for CofS, 27 Sep 50, sub: Responsibilities for Supply in FEC, Posthostilities in Korea, in G–3, DA file 091 Korea, Case 99/6.

[40] JCS 1776/124, 3 Oct 50.

the Joint Chiefs, told him on 3 October that levels of supply in FECOM after the fighting ceased should be those in effect before the fighting started. But, in addition, MacArthur's theater should be given a 90-day reserve at combat rates for a 4-division balanced force which, it was contemplated, would constitute the posthostilities troop basis. The Army planned to continue moving those troop units and casuals which were en route or scheduled for shipment for the Far East as of that date. This procedure would permit the use of the incoming filler personnel to complete the organizational structure of units already in the theater. Furthermore, the few service troops who were alerted for the Far East would be needed to support the revised 4-division troop basis for the FEC. Future shipments of units and soldiers not already alerted for the theater would be carefully screened to prevent unnecessary redeployment. Two days later Bolté furnished Larkin with a list of American units tentatively planned to garrison the Far East Command after the fighting ceased. He accompanied this with an Army plan for the deployment of troops from the Far East Command.[41]

While the Joint Chiefs of Staff considered the problem of the posthostilities troop basis and supply needs in the Far East Command, and before they could arrive at a final decision, the problem was overtaken by events. Nevertheless, the Department of the Army took its planning seriously, and on 15 October informed General MacArthur of the projected size of his command once the campaigns in Korea were over. In a definite move toward retrenchment, the Department of the Army also asked him to cancel immediately all outstanding requisitions for supplies from the ZI and to resubmit his requests on the basis of the anticipated posthostilities force.[42]

Several logistic actions taken within the Far East Command indicate that the Department of the Army officials were not alone in foreseeing an early end to the fighting. On 22 October, General Walker requested authority from General MacArthur to divert to Japan all bulk-loaded ammunition ships arriving in Korea after 20 October. Ammunition already in Korea, Walker believed, could take care of the North Koreans and still leave a balance for posthostilities requirements. MacArthur granted this request and ordered Japan Logistical Command to take the necessary actions. In the same vein General Weible, commander of Japan Logistical Command, asked MacArthur to authorize the return to the United States of six ships loaded with 105-mm. and 155-mm. artillery ammunition and aerial bombs. Both General MacArthur and General Stratemeyer agreed that this ammunition was in excess of the needs of the Korean theater in view of the existing tactical situation. MacArthur, on the other hand, felt it would be highly desirable to have a reserve stock of ammunition placed at Hawaii for use in the event of another emergency and asked the Department of the Army at least to consider diverting these ammunition ships to Hawaii before ordering them back to the United

---

[41] Memo, Bolté for Larkin, 3 Oct 50 and 5 Oct 50, with Incls, sub: Post-Korean Hostilities Basis for Logistic Support, in G–3, DA file 091 Korea, Case 99/6.

[42] Rad, W 94227, DA to CINCFE, 15 Oct 50.

States. General Weible also requested the San Francisco Port of Embarkation to cancel all outstanding requisitions for ammunition and to unload any ammunition-carrying vessels that had not left port.[43]

A Minority View—Just in Case

Regardless of the general feeling of optimism and the retrenchment moves, a disturbing, low-key concern over a possible Chinese intervention still remained. Consequently, while planning for the mundane aftermath of victory in Korea, the Army staff gave some attention to what might happen if the Chinese suddenly moved against MacArthur. On 11 October, a survey of MacArthur's resources for meeting such an attack was completed and sent to General Bolté by Brig. Gen. Ridgely Gaither, Chief of the Operations Division, G–3, Department of the Army. Gaither's survey showed that U.N. members other than the United States were scheduled to furnish additional troops to the U.N. Command to bring the total up to 29,700 troops by March 1951; General MacArthur had indicated his intention to activate five more ROK divisions, one each month beginning in November, with each division having an approximate strength of 11,000; 60,000 fillers for American units were scheduled to be furnished between 10 November and late December; and by 31 December, American strength in the Far East Command would total 6 Army divisions, 1 marine division, 1 infantry RCT, 1 airborne RCT, and 1 infantry regiment, all at full strength.[44]

If the Chinese intervened and further reinforcements were required, the situation would be a little tight. The most readily available source of nondivisional unit reinforcement were 108 units previously allocated for shipment to the FEC. The nondivisional artillery and Engineer units on this list would be operationally ready by 1 January and could materially strengthen the combat capability of the United Nations Command. The four National Guard divisions federalized in September would be trained and ready about 1 June 1951. The 82d Airborne Division was ready at any time and could be committed within thirty days if necessary. An additional marine division, to be ready by June 1951, had been organized and was now in training. The rest of the General Reserve in the United States, including the 2d Armored Division, 11th Airborne Division, 14th RCT, 196th RCT, and 278th RCT, was just beginning a program of reconstitution but could be ready by June 1951.[45]

Underscoring the Chinese threats of intervention early in October, another warning was sounded, if faintly, on 18 October when American reconnaissance planes flying close to the Yalu found almost one hundred Russian-built fighters lined up on An-tung Airfield across the river in Manchuria. MacArthur's air commander, General Stratemeyer, minimized this ominous discovery by telling General Vandenberg in Washington that the planes were probably there

[43] (1) Rad, CX 7506, CINCFE to CG Army Eight and CG JLCOM, 26 Oct 50, with MFR on file copy, SGS files, GHQ. (2) Rad, CX 67702, CINCFE to DA, 28 Oct 50, and Memo for Record on file copy, SGS, GHQ.

[44] Memo, Gen Gaither (Opn Div G–3) for Gen Bolté (G–3), 11 Oct 50, sub: Reinforcement of U.N. Forces in Korea, in G–3, DA file 091 Korea, Case 109.

[45] *Ibid.*

purposely to lend "color and credence to menacing statements and threats of Chinese Communist leaders, who probably felt that this display of strength involved no risk in view of our apparent desire to avoid border incidents." Stratemeyer certainly did not believe that the Chinese meant to use these fighters to attack his planes since they had not done so when the observation aircraft, an easy target, had come close. "I believe it especially significant," he told Vandenberg, "that, if deployment for possible action in Korea were under way, it would be highly unlikely that aircraft would have been positioned to attract attention from south of the border." [46]

Almost at the same time that Chinese fighters appeared on the border, the Department of State suggested to the Joint Chiefs that General MacArthur be instructed to disavow publicly any intention of destroying certain hydroelectric power facilities along the Manchurian border. This suggestion stemmed from some evidence that the Chinese intended to move down into Korea to protect the Suiho Hydroelectric Power Plant and the installations along the Yalu. An announcement by MacArthur would have the dual purpose of allaying Chinese Communist fears of trespassing into Manchuria by the United Nations Command and of showing the rest of the world that his expedition into North Korea was not primarily destructive in purpose. The State Department asked also that General MacArthur's announcement contain a statement that the U.N. Commission for the Unification and Rehabilitation of Korea would consult with all interested parties on this and other problems which might come before it. Since Communist China qualified as an interested party to any operation along the Manchurian border, this could be construed as a willingness by the commission to deal with Communist China on matters involving Korea. The Joint Chiefs felt such an announcement would be militarily undesirable. They were, however, directed by President Truman to send the suggestion to MacArthur. They did so, telling him he could issue the text of the announcement if he wished. "It is considered desirable," they concluded, "that President Rhee be advised with regard to this action if it is to be taken." [47]

General MacArthur did not feel that the time was propitious for such an announcement, especially since the Suiho Hydroelectric Power Plant at Sinuiju was not under United Nations control and no determination could be made at long range of how much power was being turned out or where it was going. MacArthur did not propose, if he could avoid it, to tie his own hands with a commitment that he might later find militarily inconvenient.

> There would certainly be no intent on the part of this command to disturb any peaceful and reasonable application of this power supply, and it would be repugnant to destructively interrupt any constructive uses to which it is being applied. If, however, this power is being utilized in furtherance of potentially hostile military purposes through the manufacture of munitions of war or there is a diversion of it from the minimum peaceful requirements of the Korean people, most serious doubts would at

[46] Rad, A 25438 INT-IE, CG, FEAF to CS USAF, 20 Oct 50.

[47] Rad, JCS 94799, JCS to CINCFE, 21 Oct 50.

once arise as to our justification for maintaining status quo.

Nor did MacArthur feel that he should speak for the United Nations Commission for the Unification and Rehabilitation of Korea by predicting any of its future policies or decisions. The Joint Chiefs did not press the matter and the announcement was never made.[48]

Despite Mao Tse-tung's early October threats and the midmonth discovery of fighter aircraft just over the Manchurian border, apprehensions subsided as the end of the month approached without Chinese intervention. Likewise, the difficulties of supplying the Eighth Army and the unexpected hitch in landing the X Corps at Wonsan became mere annoyances as Walker's forces moved deep into western North Korea and as the force that Almond so far commanded ashore, the ROK I Corps, moved far up the east coast. The Wake Island Conference truly had been the catalyst that generated wide confidence in MacArthur's march to victory.

[48] Rad, C 67154, CINCFE to JCS, 22 Oct 50.

CHAPTER XIII

The Chinese Take a Hand

The Chinese Move

Unbeknown to MacArthur or to Washington authorities, the Chinese Communist government had begun sending infantry divisions into North Korea on 14 October. Between 14 October and 1 November, it appears, some 180,000 troops from the *Chinese Fourth Field Army* crossed the Yalu into Korea. More than two-thirds of this force had been in Manchuria, near the border, since July 1950. Three days before these crossings, the Chinese Ministry of Foreign Affairs prefaced this invasion with a blunt public warning: "Now that the American forces are attempting to cross the thirty-eighth parallel on a large scale, the Chinese people cannot stand idly by with regard to such a serious situation created by the invasion of Korea. . . ." Ominously, the Chinese Foreign Office described MacArthur's operations as ". . . a serious menace to the security of China. . . ." [1]

The first real proof that Chinese soldiers had entered the fighting came about noon on 25 October, when the ROK 1st Division in western Korea clashed with a Chinese force and captured the first Chinese Communist soldiers taken in the Korean War. The ROK 1st Division commander, General Paik Sun Yup, found many dead Chinese on the battlefield and so informed General Milburn, the I Corps commander. General Paik also took prisoners, who, by revealing that they belonged to organized units and that Chinese soldiers were in Korea in large numbers, gave ample proof of intervention. In eastern Korea, Almond's troops met Chinese units and took their first Chinese prisoners on the same day as did the Eighth Army. By 31 October they had captured a total of twenty-five Chinese and General Almond had personally interrogated some of them.[2]

The evidence, however, was not accepted at face value in any quarter that counted. General Willoughby, after reporting to Washington that Chinese troops had been captured and that he believed organized Chinese units to be in Korea, said, on 28 October:

> From a tactical standpoint, with victorious United States divisions in full deployment, it would appear that the auspicious time

[1] (1) Whiting, *China Crosses the Yalu*, p. 115. (2) Appleman, *South to the Naktong, North to the Yalu*, pp. 687 and 767.

[2] Appleman, *South to the Naktong, North to the Yalu*, pp. 675–708.

General Paik Sun Yup

for intervention has long since passed; it is difficult to believe that such a move, if planned, would have been postponed to a time when remnant North Korean forces have been reduced to a low point of effectiveness.[3]

Indeed, American ears were attuned to victory and the ominous harbinger of military disaster wrought no change in General MacArthur's plans to advance to the border with all speed and with all forces.

Despite the assurances they had received, the Joint Chiefs of Staff showed apprehension over the indication of Chinese Communist intervention. General Bradley noted on 31 October that the intervention was conforming to none of the patterns envisaged by the Joint Chiefs in their studies and in their directives to General MacArthur. Chinese actions were "halfway between" and left some doubt as to what the specific countermoves by MacArthur should be. Bradley based his analysis on information which showed that elements of five Chinese Communist divisions had been identified south of the Yalu, the largest element being a regiment. On the same day, General Collins told the Army Policy Council that these reported crossings of the Yalu River might reflect a face-saving effort since Chou En-lai had said that his government would not stand idly by and watch the North Koreans go down in defeat. Collins did not think that the Chinese would cross the river in sufficient numbers to risk a serious beating by MacArthur's forces. Nevertheless, when asked if the Chinese could become a real threat to the United Nations Command, Collins replied that they definitely could in spite of their lack of airpower and their weakness in artillery.[4]

Eighth Army Grows Cautious

Despite the Chinese threat, General Walker attempted to carry out his current instructions from General MacArthur. He sent two of his corps plunging toward the border in the west. In the coastal sector, the I Corps fanned out in a 3-pronged drive, crossing the Ch'ongch'on River and rolling up hard-fighting North Korean units making a last-ditch stand. Farther east, the ROK II Corps pushed columns northward along main roads. One of these columns, a regiment from the ROK 6th Division,

[3] DIS, GHQ, UNC, 2971, 28 Oct 50.

[4] (1) Memo, CVRS for Maj Gen Robinson E. Duff, 31 Oct 50, in G-3, DA file 091 Korea, Case 14/38. (2) Min, Army Policy Council, 31 Oct 50, in CofS, DA file 334 (APC), 31 Oct 50.

stepped far ahead of the others and on 26 October pushed its reconnaissance troops all the way to the Yalu at Ch'osan. There was little elation over this feat, for almost simultaneously other ROK divisions of the ROK II Corps ran head on into very strong Chinese units. Not only was the ROK 1st Division badly mauled, but the regiment of the ROK 6th Division on and near the border also was cut off by the Chinese and nearly destroyed. The climax of this early Chinese intervention came on the night of 1–2 November when the 8th Cavalry Regiment, 1st Cavalry Division, was attacked by a full Chinese division in positions near Unsan and very roughly handled.[5]

As soon as Walker saw what was happening, he ordered the advance halted and withdrew his forces back across the Ch'ongch'on, holding only a shallow bridgehead above the river. When General MacArthur learned that Walker had stopped driving toward the border, he directed his acting chief of staff, General Hickey, to telephone Walker in Korea and find out why. Hickey reached Walker's chief of staff, General Allen, who furnished an interim explanation on 1 November. Walker a few days later followed this with a fuller and more detailed explanation in a letter to General MacArthur in which he said:

On 26 October Eighth Army was advancing on a broad front in widely separated columns in pursuit of defeated North Korean Forces. The advance north of Pyongyang was based upon a calculated logistical risk involving supply almost entirely by airlift. Supplies available were sufficient for bare maintenance of combat operations of one reinforced American division and four ROK divisions against light opposition with no possibility of accumulating reserves to meet heavier opposition. An ambush and surprise attack by fresh, well-organized and well trained units, some of which were Chinese Communist Forces, began a sequence of events leading to complete collapse and disintegration of ROK II Corps of three divisions. Contributing factors were intense, psychological fear of Chinese intervention and previous complacency and overconfidence in all ROK ranks. There were no indications that Chinese troops had entered Korea prior to contact. The presence of Chinese troops increased materially the will to fight of remaining and reconstituted North Korean units. The ROK Corps retreated in confusion to a position in the vicinity of KUNURI, 13 miles from the only crossing area into the I Corps combat zone, before some semblance of order could be restored. Losses in equipment and personnel were large. The collapse on the east flank together with heavy attack on the 1st ROK Division and 8th Cav RCT on the east flank of the I US Corps seriously threatened the only road supplying the I Corps and dictated temporary withdrawal of exposed columns of 24th Inf Div on the west, a regrouping of forces, an active defense, a build-up of supplies pending resumption of offensive and advance to the border. By intense effort progress is being made in reorganization and stabilization of II ROK Corps, however, it is at most only fifty percent effective at present. The 2d US Division has been brought up in a position to take over in the event of collapse by ROK forces.

There has never been and there is now no intention for this Army to take up or remain on a passive perimeter or any other type of defense. Every effort is being made to retain an adequate bridgehead to facilitate the resumption of the attack as soon as conditions permit. All units continue to execute local attacks to restore or improve lines. Plans have been prepared for re-

[5] See Appleman, *South to the Naktong, North to the Yalu*, pages 689–708, for detailed account of these actions.

sumption of the offensive employing all forces available to the Army to meet the new factor of organized Chinese Communist Forces. These plans will be put into execution at the earliest possible moment and are dependent only upon the security of the right flank, the marshalling of the attack troops and the restoration of vital supplies. In this connection there now exists in the forward areas only one day of fire. Opening of Port of Chinnampo and extension of railroad to Pyongyang is essential to movement of supplies and troops.[6]

Meanwhile, traces of further Chinese participation in the west faded quickly as the new enemy pulled back into the mountain masses from which he had come.

X Corps and the New Enemy

Across the Taebaek mountain range far to the east of the Eighth Army and completely out of physical contact with Walker's forces, General Almond meantime had landed the rest of his American forces and was preparing to carry out his part of General MacArthur's directive to advance with all possible speed to the border. His plan of advance called for the ROK I Corps to drive up the east coastal road to the northeastern border of Korea. He ordered the U.S. 7th Division, which had landed at Iwon, to drive up the corridor from Pukch'ong through Pungsan and Kapsan to the Yalu River at Hyesanjin. To the U.S. 1st Marine Division, Almond assigned the task of pushing up from Hamhung to the Changjin Reservoir, from where it could either drive north to the border or shift its attack to the west and then north again. The rear areas around Wonsan-Hamhung were to be guarded by the 3d Division, which was en route from Japan after having completed a very brief training period.[7] Until the 3d Division arrived, the 1st Marine Division had the responsibility for securing the Wonsan-Hamhung area and therefore would not be entirely free to concentrate for its advance to the Changjin Reservoir.

It was while moving toward this reservoir that the ROK 26th Regiment had on 25 October found its way blocked by strong Chinese forces at the small village of Sudong. Prisoners taken in the heavy fighting identified the unit opposing the ROK regiment as the CCF *124th Division.* The ROK regiment fell back and for the next few days made no headway. But the 7th Marines, U.S. 1st Marine Division, coming up from Hamhung, took over from the ROK unit on 2 November and in a fierce running encounter, marked by the first and last appearance of Chinese tanks in the area, virtually destroyed this Chinese division. The *124th* was barely accounted for, however, before the Marines picked up prisoners from a fresh division, the *126th.*

It was now quite clear that the eventuality so long discussed by American planners, Communist China's entry on the side of North Korea, was no longer hypothetical. Yet there was great reluctance at Eighth Army and X Corps headquarters, at GHQ in Tokyo, and in Washington to accept this intervention at face value.

---

6 Memo, Gen Walker for Gen MacArthur, 6 Nov 50.

7 War Diary, X Corps, 4 Nov 50, G–1 Rpt, Notes on Conference Between CG X Corps and Gen Partridge, CG 5th AF, 4 Nov 50.

Immediate Effects

The open employment of Chinese units against the United Nations Command in late October and early November caused a swift reversal of plans to reduce support to that command. General Bolté, Department of the Army G–3, had flown to Tokyo on 31 October to confer with MacArthur. On the next day he discussed with General MacArthur, General Wright, and the JSPOG staff the cancellation and reduction of forces to be sent to Korea and the plans for redeployment of forces already there. Even as these officers were talking, the Chinese struck Walker's advancing divisions and hurled them back. From Tokyo Bolté flew to Korea where he conferred with Walker, all corps commanders, and some division commanders. As a result of his observations, General Bolté cabled Washington that he was ". . . convinced that any deferment, cutback, or cancellation of requested units, individuals, or material would be premature." He urged that full support of MacArthur be continued. He pointed out that even the psychological impact of withdrawal or diminution would now adversely affect the command in the field. "Recent reverses—in Eighth Army," Bolté concluded, "require assurances of full support, which I have given them." [8]

Army officials in Washington had anticipated General Bolté's call for reappraisal. Maj. Gen. Robinson E. Duff, acting G–3 during Bolté's absence, immediately replied that action had been started on 3 November to halt the cutback trend. He had recommended to General Collins that all units originally intended for shipment to FEC, but held in the United States when it appeared they would not be needed to finish the job in Korea, be sent at once. He also asked that the planned reduction in United Nations units be called off.[9]

Robert A. Lovett, Deputy Secretary of Defense, also noted the new developments with some alarm. He was particularly disturbed by an intelligence report on 1 November which estimated that the Chinese Communists had decided to establish a *cordon sanitaire* south of the Yalu border. This report, coupled with what Lovett described as "the renewed vigor of the enemy attacks in the border area," caused him to counsel the Joint Chiefs of Staff to reconsider their plans to reduce the number of United Nations units to be sent to Korea. The acting G–3, General Duff, seconded this suggestion, pointing out that "the possible utilization of the enormous manpower of Communist Asia against the United States makes it mandatory at this time to provide General MacArthur with the largest possible United Nations force until the overall situation is better clarified." Duff held that it was now necessary, politically and militarily, to set a pattern for the development and use of the manpower of other friendly world areas, particularly those where manpower was the chief military resource. The United States, Duff felt, could not continue to provide the bulk of anti-Communist manpower for military operations, but ". . . must preserve its technical and scientific resources, its

[8] Rad, Gen Bolté to Gen Duff, 5 Nov 50, in G–3, DA files.

[9] Rad, WAR 95938, DA to CG RYCOM (Pass to Bolté), 6 Nov 50.

productive capacity, and its trained military leadership." [10]

On 6 November, the Joint Chiefs of Staff agreed that the time was no longer propitious for reducing contributions from other nations and set out to amend the steps they had already taken in that direction. They recommended to the Secretary of Defense that, "due to the fluid situation in Korea," action on cutting back United Nations forces be deferred. The Secretary of Defense subsequently passed this recommendation to the Department of State and the move to retrench was halted.[11]

From the Far East Command, the G–1, General Beiderlinden, again demanded replacements. "This theater has been operating throughout hostilities seriously understrength," he charged on 5 November. "The shortage of essential replacements is continuing and reaching critical stage." The current situation made it mandatory that replacement schedules be revised and that substantial numbers of men be shipped at once. General Beiderlinden admitted that it would be difficult to estimate what the losses would be in November and December, but he maintained that the trend was upward to a marked degree. Department of the Army officials were unable to understand, since there was no accompanying explanation, on what basis Beiderlinden was predicting increased battle casualties. When General Brooks, the Department of Army G–1, asked Beiderlinden for an explanation at the next teleconference, it became apparent from the reply that Beiderlinden had accepted Chinese Communist intervention as a proven fact and that he expected the worst. He pointed out that battle casualties had risen from 40 per day in October to 326 per day in the first week of November and that this upward swing was no flash in the pan. He based his theory on a number of disturbing considerations, and felt that the United Nations Command faced a situation as dangerous as that of the Pusan Perimeter. He noted that the new enemy would be better trained and equipped than the North Koreans, and that the fighting would be carried on in bitter winter weather. Furthermore, the Eighth Army was already experiencing transportation, supply, and evacuation difficulties which would multiply; combat divisions were understrength; periods which the individual soldier must spend in combat would probably increase; and there had already been a great drop in morale among combat men at the prospect of continued heavy fighting.[12]

At a higher level, General MacArthur on 7 November appealed to the Joint Chiefs of Staff for more combat strength. He told them that the appearance of the Chinese Communist forces in strength had completely changed the over-all situation and asked that all previous plans for sending men and units to his command be put into full effect immediately. Holding that it was essential for the replacement flow to his theater to be re-

[10] (1) Memo, Robert A. Lovett, Dep Secy Defense for JCS, 4 Nov 50, sub: Reduction in Forces to be Deployed in Korea. (2) Memo, G–3 (Duff) for CofS, USA, 6 Nov 50, sub: Utilization of U.N. Ground Forces Contingents, in G–3, DA file 091 Korea, Cases 111/8, 111/10.

[11] (1) Decision on JCS 1776/152, 6 Nov 50. (2) Ltr, Secy Defense to Secy State, 16 Nov 50, in G–3, DA file 091 Korea, Cases 111/8, 111/2.

[12] (1) Rad, CX 68300, CINCFE to DA, 5 Nov 50. (2) Telecon, TT 3982, DA and GHQ (Brooks and Beiderlinden), 7 Nov 50.

sumed at full force, MacArthur asked also that all Army combat and service units previously requested be sent without delay. He could not say at that date whether more Army, Air Force, and Navy units than had already been asked for would be required, but he was certain "that the full requirement for balanced forces as stated during the earlier phases of the campaign must now be met with possible appreciable augmentation thereof." "The alternatives," General MacArthur warned, "are either a stalemate or the prospect of losing all that has thus far been gained." [13]

Department of Army officials had already taken action to send 40,000 replacements to the Far East in November and December, and it was estimated that all units in the Far East would be up to strength by March 1951. With regard to the additional units, 112 combat and service units of various types had been recommended for shipment to the Far East, although shipment of 92 of these was contingent upon approval to send civilian component units overseas. On 16 November, the Chief of Staff told the United Nations commander that steps to fulfill his needs were being taken. "In view of the gravity of the current situation," Collins said, ". . . the flow of Army replacements has been resumed at an increased rate and you will be informed earliest of the combat and service type units alerted for movement to your command." [14]

One interesting outgrowth of the Chinese intervention was the effect on personnel policy in the Far East with respect to the ROK soldiers who had been integrated into American units in September. The Far East Command had established a policy during the period of October optimism of releasing South Korean soldiers to ROK Army control as American replacements arrived to take their places. As of 7 November, over 8,000 of these Korean soldiers had been released from Eighth Army's units. On that date, however, the practice was suspended because of the new threat, and more than 20,000 Koreans remained in American divisions.[15]

A Time for Reappraisal

One of the first intelligence reports on the Chinese intervention to reach Washington was that of General Willoughby, who revealed on 2 November that 16,500 Chinese Communist soldiers had entered North Korea. The Chinese Communist government reputedly was labeling these troops "volunteers." The Sinuiju radio had announced that these troops formed the "Volunteer Corps for the Protection of the Hydroelectric Zone" and had entered Korea expressly to prevent the destruction of hydroelectric facilities along the Yalu. General Willoughby admitted that the increasing resistance being met by MacArthur's forces had removed the problem of Chinese intervention from the realm of the academic and turned it into "a serious proximate threat." He was puzzled by the Chinese device of committing "volunteers" in

---

13 Rad, CINCFE to DA, 7 Nov 50.

14 (1) Memo, Bolté for CofS USA, 10 Nov 50, sub: Regarding CCF in Korea, in G–3, DA file 091 Korea, Case 14/42. (2) Rad, WAR 96672, DA (Collins) to CINCFE, 16 Nov 50.

15 Telecon, TT 3992, DA (Brooks) and GHQ (Beiderlinden), 9 Nov 50.

"special units" instead of in regular organized regiments of the Chinese Communist Army. He speculated that the Chinese, who he acknowledged were extremely subtle and obsessed with "saving face," might be doing this in order to have their cake and eat it too. By labeling their troops "volunteers" and claiming that no recognized units of their army were in Korea, the Chinese would avoid the appearance of intervention. Nor would they involve the prestige of the Chinese Communist Army if defeated. On the other hand, by furnishing troops to North Korea, China could claim credit for helping North Korea in its hour of need. MacArthur's intelligence chief concluded by warning:

> Although indications so far point to piecemeal commitment for ostensible limited purposes only, it is important not to lose sight of the maximum potential that is immediately available to the Chinese Communists. Should the high level decision for full intervention be made by the Chinese Communists, they could promptly commit 29 of their 44 divisions presently deployed along the Yalu and support a major attack with up to 150 aircraft.[16]

On the same day, 2 November, the American Consul General in Hong Kong sent Washington a report that in August, at a conference of top Sino-Soviet leaders, a joint decision had been made that Communist China would enter the Korean War. According to his report, the formal decision had come on 24 October at a meeting presided over by Premier Mao Tse-tung. An estimated twenty Chinese Communist armies had been sent to Manchuria.[17]

On the next day, Willoughby reported 316,000 regular Chinese ground forces and 274,000 Chinese irregulars, or security forces, in Manchuria. Most of the regulars were believed to be along the Yalu at numerous crossing sites.[18]

These disclosures had an extremely ominous ring and, coupled with the news of the withdrawal of Eighth Army before Chinese forces already in Korea, caused the Joint Chiefs of Staff to call on General MacArthur for an evaluation. They requested his earliest "interim appreciation of the situation in Korea and its implications in light of what appears to be overt intervention by Chinese Communist units." [19]

MacArthur's reply scarcely enlightened them. He told them, "It is impossible at this time to authoritatively appraise the actualities of Chinese Communist intervention in North Korea." MacArthur posed four courses of action which the Chinese Communists might be following. The first was open intervention with full force and without restraint; the second possibility, covert intervention concealed for diplomatic reasons; the third course might be the use of "volunteers" to keep a foothold in Korea; the fourth, Chinese forces might have entered Korea assuming they would meet only ROK units which they could defeat without great difficulty.[20]

Full intervention, according to General MacArthur, would represent a "momentous decision of the gravest international importance." "While it is a

---

16 Telecon, TT 3968, G–2 DA (Bolling) with G–2 FEC (Willoughby), 2 Nov 50.

17 Intelligence Rpt, 2 Nov 50, in G–2, DA files.

18 Telecon, TT 3971, DA and GHQ UNC, 3 Nov 50.

19 Rad, WAR 95790, CSUSA to CINCFE, 3 Nov 50.

20 Rad, C 68285, CINCFE to DA for CSUSA for JCS, 4 Nov 50.

distinct possibility," he told the JCS, "and many foreign experts predict such action, there are many fundamental logical reasons against it and sufficient evidence has not yet come to hand to warrant its immediate acceptance." Although he made no definite prediction, MacArthur felt that a combination of the last three courses of action by the Chinese were, at the moment, the most likely. In a cautious mood, he told the Joint Chiefs, "I recommend against hasty conclusions which might be premature and believe that a final appraisement should await a more complete accumulation of military facts." Nothing in the tone or content of General MacArthur's report implied that an emergency existed or that the situation even showed signs of getting out of hand. His report was, in a sense, reassuring.[21]

On the same day that General MacArthur sent this appraisal to Washington, the Chinese Communist government in an official statement charged that the United States was bent on conquering not only Korea but also China, as "the Japanese imperialists have done in the past." The statement, possibly made to prepare the Chinese people for further moves in Korea, claimed that in order to protect China, Chinese military forces must now assist North Korea.[22]

Chinese troops had crossed and were continuing to cross into North Korea over a number of international bridges leading in from Manchuria. By 3 November, General MacArthur's headquarters accepted the possibility that 34,000 Chinese had entered Korea and that 415,000 regular troops were located in Manchuria, ready to cross if ordered. Two days later, General Willoughby warned that the Chinese Communist forces had the potential to launch a large-scale counteroffensive at any time.[23]

The appearance of Chinese military formations in Korea, and evidence that these forces were being augmented rapidly, caused MacArthur to call for an all-out air effort to smash them. On 5 November, he directed General Stratemeyer to throw the full power of the Far East Air Forces into a 2-week effort to knock the North Koreans and their new allies out of the war. "Combat crews," he ordered, "are to be flown to exhaustion if necessary." He instructed Stratemeyer to destroy the Korean ends of all international bridges on the Manchurian border. From the Yalu southward, and excluding only Rashin, the Suiho Dam, and other hydroelectric plants, the Far East Air Forces would "destroy every means of communication and every installation, factory, city, and village." MacArthur warned that there must be no border violations and that all targets close to or on the border must be attacked only under visual bombing conditions.[24]

On 6 November, General MacArthur notified Army authorities that he intended to have his B–29's take out immediately the international bridges across the Yalu between Sinuiju and An-tung. He hoped, by destroying these bridges, to prevent or at least slow down

21 *Ibid.*

22 Telecon, TT 3975, DA and GHQ, UNC, 5 Nov 50.

23 DIS GHQ, FEC UNC, No. 2977, 3 Nov 50 and No. 2979, 5 Nov 50.

24 USAF Hist Study No. 72, United States Air Force Operations in the Korean Conflict, 1 November 1950–30 June 1952, ch. I, p. 22.

the flow of Chinese military strength into Korea. MacArthur conveyed this information to Washington in a routine manner during a teleconference with the Army staff at the Pentagon.[25]

Had this matter been handled routinely by the Army staff and merely reported through channels, the mission might have been well under way before the nation's leaders learned of MacArthur's intentions. However, General Stratemeyer, apparently feeling that his chief's decision held more than passing interest, sent to Air Force authorities in Washington a message describing his orders from MacArthur. Within minutes, Under Secretary of Defense Lovett had been informed and the fat was in the fire.[26]

Lovett doubted very seriously that the advantage of bombing the Sinuiju-An-tung bridges would offset the great danger of bombing Chinese territory. He went at once to discuss the problem with Secretary of State Acheson and with the Assistant Secretary of State for Far Eastern Affairs, Dean Rusk. The latter pointed out that the United States had promised the British Government not to take action which might involve attacks on Manchuria without consulting the British. Also, the United States was even then taking steps to have the Security Council pass a resolution calling on the Chinese to halt action in Korea, a resolution that surely would be jeopardized if bombs fell in Manchuria. Rusk was concerned, too, over possible Soviet reaction if China should invoke the mutual-assistance treaty with the Soviet Union.

Acheson and Lovett agreed that MacArthur's attack should be held up until the Korean situation became much clearer, particularly in view of Rusk's comments. Lovett then called Secretary Marshall and informed him of the details. Marshall agreed that unless a mass movement across the Yalu was threatening the security of MacArthur's forces, the planned bombing was unwise. Lovett then directed the Air Force Secretary Thomas K. Finletter, to tell the JCS of the feeling at State and Defense that the action by MacArthur should await a decision from the President himself. As a final step, Acheson called the President who was in Independence, Missouri. President Truman stated that he would approve this bombing only if there was an immediate and serious threat to MacArthur's forces.

Since MacArthur had reported no such threat and, indeed, only two days before had cautioned Washington against precipitate judgment and had recommended a wait-and-see attitude, the puzzlement of Mr. Truman and his chief advisers was natural. The President directed that the attack be put off and that MacArthur be asked to explain why he found this potentially dangerous action suddenly so necessary.

The Joint Chiefs of Staff, acting on the President's instructions, immediately directed MacArthur to call off until further orders any bombing of the international bridges. "Consideration is being urgently given to the Korean situation at the governmental level," they told him.

One factor is the present commitment not to take action affecting Manchuria without consulting the British. Until further orders

[25] Telecon, TT 3976, DA and GHQ, UNC, 6 Nov 50.

[26] Truman, *Memoirs*, II, 374.

postpone all bombing of targets within five miles of the Manchurian border. Urgently need your estimate of the situation and the reason for ordering bombing of Yalu River bridges as indicated.[27]

This order from Washington brought from General MacArthur an immediate protest couched in strong terms which portrayed the situation in Korea in the most pessimistic vein since July and August. He warned on 6 November that "men and materiel in large forces are pouring across all bridges over the Yalu from Manchuria," and, for the first time since Chinese entry had become evident, admitted that the situation was serious. "This movement not only jeopardizes but threatens the ultimate destruction of the forces under my command." He described for them how the Chinese were moving across the bridges under cover of darkness. Chinese troops could be committed without being attacked effectively by air because of the short distances from the river to the front lines. "The only way to stop this reinforcement of the enemy is the destruction of these bridges and the subjection of all installations in the north area supporting the enemy advance to the maximum of our air destruction," General MacArthur declared. "Every hour that this is postponed will be paid for dearly in American and other United Nations blood." He had intended hitting the main crossing at Sinuiju within the next few hours but in accordance with the Joint Chiefs's order had suspended the strike "under the gravest protest that I can make." He pointed out that his original order to bomb the bridges was, in his opinion, entirely within the scope of his directives, the rules of war, and the resolution made by the United Nations. It constituted to him no slightest act of belligerency against Chinese territory.

It is interesting to note MacArthur's reference to the resolution of the United Nations since he received his operating instructions and directives from the very quarter at which he was lodging his protest, the Joint Chiefs of Staff. The resolutions of the United Nations were merely guides which the United States Government, as the Unified Command under the United Nations, used in determining the specific policies for the United Nations Command in Korea. General MacArthur seemed sure that the Joint Chiefs did not realize the disastrous effect, both physical and psychological, that would result from the restrictions they were imposing. In an extraordinary request, he asked that President Truman be informed of the restriction, saying, "I believe that your instructions may well result in a calamity of major proportion for which I cannot accept the responsibility without his personal and direct understanding of the situation." He concluded by asking immediate reconsideration of the decision.[28]

---

[27] (1) Rad, JCS 95878, JCS (Personal) for MacArthur, 6 Nov 50. (2) This series of actions reveals clearly the speed with which important decisions could be taken and the "streamlining" of the normal policy-making methods. Stratemeyer's message had been received in Washington about three and one-half hours before his planes were scheduled to take off on their missions. In the interim every appropriate official within the Defense and State Departments had been consulted and the Presidential decision based on their advice had been reached. The JCS had sent out the order to MacArthur only an hour and twenty minutes before the B-29's were scheduled to take off from Japan. See Truman, *Memoirs*, II, 374-75.

[28] Rad, C 68396, CINCFE to DA (for JCS), 6 Nov 50. That MacArthur did not fully understand the

The sense of grim urgency conveyed by MacArthur's protest and his accompanying picture of a sudden, mammoth build-up of Chinese Communist forces in Korea surprised Washington. General Bradley called the President and read to him MacArthur's message. Still concerned over the dangers of bombing Manchuria by mistake, Mr. Truman nevertheless agreed to let MacArthur go ahead with his plans. President Truman, because MacArthur was on the scene and felt very strongly that this was of unusual urgency, told Bradley to give him the green light.[29]

Nevertheless, it was evident that both Truman and the military planners in Washington were gravely concerned by the tone of MacArthur's protest. The Joint Chiefs of Staff told MacArthur in an immediate reply that the situation he now depicted had changed considerably from that described in his last report of 4 November. They agreed that destruction of the bridges in question would probably alleviate the immediate problem but that the cure might be worse than the ailment. It might well bring increased Chinese Communist effort and even Soviet contributions in response to what the Communists might construe as an attack on Manchuria. Not only would this endanger MacArthur's forces, it would enlarge the area of conflict and American involvement to a dangerous degree.[30]

But in view of the apparent emergency, with men and matériel pouring across the Yalu bridges, the Joint Chiefs of Staff told MacArthur that he could bomb these bridges but with certain restrictions. ". . . you are authorized to go ahead with your planned bombing in Korea near the frontier including targets at Sinuiju and Korean end of the Yalu bridges." This did not mean, General MacArthur was cautioned, carte blanche to bomb any dams or power plants on the Yalu River. The Joint Chiefs expressed deep concern that careless action by the United Nations Command near the Yalu might trigger a crisis which would cause the fighting to spread. They specifically warned MacArthur on this, urging him to enforce extreme care to avoid Manchurian territory and airspace and to tell them promptly of any hostile action from Manchuria. They chided him obliquely for being lax in reporting new developments, prompted no doubt by the great discrepancy between his description of the situation on 4 November and that of 6 November. Certainly the routine and special reports from his command had not indicated so great a change in the Chinese Communist situation as appeared to have actually taken place. "It is essential," the Joint Chiefs maintained, "that we be kept informed of important changes in the situation as they occur and that your estimate as requested . . . be submitted as soon as possible." [31]

American intelligence agencies had been busy, meanwhile, preparing the best possible estimate of Chinese intentions based on the pooled information from all

mechanical procedures by which the instructions he received were evolved in Washington is evident. He seems to have assumed that the Joint Chiefs of Staff had ordered a halt to his planned attack without Presidential backing.

29 Truman, *Memoirs*, II, 376.

30 Rad, JCS 95949, JCS to CINCFE, 6 Nov 50.

31 *Ibid.*

their sources. This estimate was furnished all high-level planning and policy groups, including the Joint Chiefs of Staff, on the same day General MacArthur's pessimistic report arrived. The estimate concluded that between 30,000 and 40,000 Chinese were now in North Korea and that as many as 700,000 men, including 350,000 ground troops could be sent into Korea to fight against the United Nations forces. These Chinese forces would be capable of halting the United Nations advance by piecemeal commitment or, by a powerful all-out offensive, forcing the United Nations to withdraw to defensive positions farther south. The report concluded with a significant warning:

A likely and logical development of the present situation is that the opposing sides will build up their combat power in successive increments to checkmate the other until forces of major magnitude are involved. At any point the danger is present that the situation may get out of control and lead to a general war.[32]

This chilling prognosis was followed at once by another report from the United Nations commander. He confirmed that the Chinese threat was a real and developing one. That Chinese forces were engaging his troops was unquestionable although their exact strength was difficult for his commanders to determine. They were strong enough to have seized the initiative from Walker's forces in the west and to have materially slowed Almond's advances in the east. "The principle seems thoroughly established," General MacArthur declared, "that such forces will be used and augmented at will, probably without any formal declaration of hostilities." He emphasized that if the Chinese augmentation continued it could force the United Nations Command to perform a "movement in retrograde." But he affirmed his intentions to resume his advance in the west, possibly within ten days, and to try to seize the initiative, provided the enemy flow of reinforcements could be checked. In his first reference to what he later termed a "reconnaissance in force," General MacArthur told the Joint Chiefs of Staff, "Only through such an offensive effort can any accurate measure be taken of enemy strength."[33]

Once again he reiterated his conviction that the bridges had to be bombed "as the only resource left to me to prevent a potential build-up of enemy strength to a point threatening the safety of the command." This bombing was, in his eyes, so plainly defensive that he could hardly conceive of its causing increased intervention or provoking a general war, as the Joint Chiefs had intimated it might do. He promised that there would be no violation of the Manchurian or Siberian borders and that he would not destroy the hydroelectric installations along the Yalu.[34]

Twelve railroad and highway bridges spanned the Yalu and Tumen Rivers from Manchuria and Russia into Korea. The most important of these were the rail and highway bridges at Sinuiju and An-tung. These bridges were 3,000 feet long and very sturdy. The highway

[32] Intelligence Estimate, 6 Nov 50, sub: Chinese Communist Intervention in Korea, in G–2, DA files.

[33] Rad, C 68465, CINCFE to DA for JCS, 7 Nov 50.

[34] *Ibid.*

bridge at Sinuiju had been built in 1900 by the American Bridge Company and perhaps equaled in strength any in the world. Despite the swift current, winter ice, and spring floods, the builders had laid the foundations on bedrock. The Japanese had built an equally sturdy double-track rail bridge of twelve trusses in 1934, 350 yards north of the highway bridge. It was the largest rail bridge ever built by the Japanese.[35] Near Sakchu a double-track railway bridge spanned the Yalu, while at Manp'ojin both a rail and footbridge crossed the river. Other highway bridges were located at Ongondong, Ch'ongsongjin, Lin-chiang, Hyesanjin, Samanko, and Hoeryong.

As authorized by the President, MacArthur sent his bombers, starting on 8 November, against the bridges at Sinuiju, Sakchu, Ch'ongsongjin, Manp'ojin, and Hyesanjin. But bombing the Yalu River bridges involved almost insurmountable difficulties. Antiaircraft fire from Manchuria forced the bombers above 20,000 feet, and enemy jet fighters threatened them on their bomb runs. MacArthur's orders positively forbidding any violation of Manchurian airspace severely limited the possible axes of approach to the bridges and permitted enemy antiaircraft artillery to zero in on the flight path of the bombers. Also, the provision that the bridges could be attacked only under visual bombing conditions meant that any cloud cover at the target diverted the bombers to secondary or last-resort objectives.

On 12 November, carrier-based Navy bombers joined in the effort to destroy the bridges. All during November, the aerial attacks against the bridges continued but the results were disappointing. By the end of the month, the air effort had succeeded, at great cost, in cutting four of the international bridges and in damaging most of the others. But by this time the Yalu was frozen over in many places and enemy engineers were building ponton bridges across the Yalu at critical points. On 5 December, the bridge attacks were suspended.[36]

---

[35] (1) DA WIR No. 91, 17 Nov 50, p. 38, NK Border Crossings. (2) FEC Intelligence Digest, vol. I, No. 13, 16–30 June 1953, p. 26.

[36] (1) USAF Hist Study No. 72, U.S. Air Force Operations in the Korean Conflict, 1 November 50–30 June 1952, ch. 1, pp. 27–31. (2) Maj. Gen. Emmett "Rosie" O'Donnell, Commander, FEAF Bomber Command, during the period in question, testified before the Senate committee investigating General MacArthur's relief to the difficulty of destroying the Yalu bridges in November 1950. General O'Donnell said, "We were not, however, allowed to violate Manchurian territory, and by violation of territory I mean we were not allowed to fly over an inch of it. For instance, the Yalu has several very pronounced bends like most rivers before getting to the town of Antung, and the main bridges at Antung we had to attack in only one manner. There was only one manner you could attack the bridge and not violate Manchurian territory, and that was a course tangential to the southernmost bend of the river. So you draw a line from the southernmost bend of the river to the bridge and that is your course, and these people on the other side of the river knew that, and they put up their batteries right along the line and they peppered us right down the line all the way. . . . In addition to that, they had their fighters come up along side; while I didn't see them myself, the combat mission reports indicate that they would join our formation about 2 miles to the lee and fly along at the same speed on the other side of the river while we were making our approach, and just before we got to bombs-away position, they would veer off to the north and climb up to about 30,000 feet and then make a frontal quarter attack on the bombers just about at the time of bombs-away in a turn. So that they would be coming from Manchuria in a turn, swoop down, fire their cannon at the formation, and continue the turn back into sanctuary—and the boys didn't like it." See *MacArthur Hearings*, pp. 3069–70.

The Sanctuary and Hot Pursuit

On at least three separate occasions, American pilots, through error, had previously violated their instructions and attacked targets in Manchuria and Siberia. Although the U.S. Air Force attributed these incursions upon neutral territory to pilot and navigational error, these incidents, regardless of their cause, were serious matters. It was entirely possible that either China or Russia could have used the incidents as an excuse for expanding the war or for retaliating in other forms.[37] Indeed, after mid-August, Chinese antiaircraft batteries in Manchuria fired at U.N. aircraft flying south of the Yalu. By late October one American plane had been shot down and another damaged.[38]

The Department of State had been particularly apprehensive lest further such encroachments should provide the Russians or Chinese a semblance of justification for overt attacks against the United States. In a series of pointed questions addressed to the Secretary of Defense, the Secretary of State had asked whether it was necessary for American air and naval forces to operate along the North Korean border and whether the Joint Chiefs should not caution MacArthur against such operations. Secretary Marshall had requested advice from the Joint Chiefs. The Joint Chiefs, who had already instructed General MacArthur on the matter of border violations, felt that MacArthur was fully aware of the necessity for avoiding such incidents and that they could not curtail his mission. Consequently, on 1 November, the Joint Chiefs told the Secretary of Defense, "The need for air interdiction operations in areas contiguous to the international boundaries of Korea is sufficient justification for not further delimiting air operations." They pointed out that it appeared to them that all United Nations forces would be required to operate clear up to the international boundaries of Korea. "Therefore," they said, "it is not considered desirable from the military point of view, to deny these ground troops air and naval support in these areas, nor would acceptance of the loss of life entailed by such denial be justified." [39]

The efforts to bomb out the international bridges brought the question into sharp focus. The sorties against these bridges continued to be strongly opposed. Russian-built jet aircraft, later identified as MIG–15's presumably piloted by Chinese pilots, had been encountered by American pilots in the area since 1 November, when one such aircraft made a nonfiring pass at a U.N. plane; and when the U.N. air force undertook the bombing of the Yalu bridges on 8 November the enemy jet pilots attacked in earnest. One MIG was sent down in flames on the first day of the attacks.

The enemy jets did not stray far from the Manchurian border, and since American planes were forbidden to cross, enemy pilots enjoyed an almost insurmountable advantage. They could break off combat whenever things got

[37] USAF Hist Study No. 72, ch. 5, pp. 80–81.

[38] Rad, CX 67701, CINCFE to DA, 28 Oct 50.

[39] (1) Memo, Ch Intnl Br, G–3, for Gen Schuyler, sub: Delimitation of Air Opns Along the Northern Border of Korea, JSPC 853/60. (2) JCS 2150/9, Incl A, Memo, JCS for Secy Defense, 1 Nov 50, in G–3 DA file 091 Korea, Case 115.

too hot for them and dash across the border to safety. American Air Force commanders naturally complained to General MacArthur about the protection afforded enemy pilots by their Manchurian sanctuary.

MacArthur had already sought help. "Hostile planes are operating from bases west of the Yalu River against our forces in North Korea," General MacArthur informed the Joint Chiefs of Staff on 7 November. These planes were increasing in number; and the distance from the Yalu to the main line of contact was so short that it was almost impossible to deal effectively with the hit-and-run tactics that enemy pilots were employing. "The present restrictions imposed on my area of operation," MacArthur asserted, "provide a complete sanctuary for hostile air immediately upon their crossing the Manchurian–North Korean border. The effect of this abnormal condition upon the morale and combat efficiency of both air and ground troops is major." General MacArthur predicted that unless corrective measures were promptly taken the air problem could assume serious proportions, and asked for instructions for dealing with this new and threatening development.[40] He did not, it should be noted, ask specifically for permission to bomb Manchurian air bases or to follow enemy planes across the border.

The Joint Chiefs could not tell MacArthur to send his fighter planes into Manchuria after the fleeing Chinese pilots. All they could do was push the matter with their superiors, and in an immediate reply to the United Nations commander they told him that "urgent necessity for corrective measures" was being presented for highest United States-level consideration.[41]

Meanwhile, other member nations of the United Nations had noted the situation growing out of Chinese intervention and American border violations with mounting alarm. The French Government made two proposals designed to reassure the Chinese that the United Nations Command meant to respect their territory. The French first proposed that the United Nations General Assembly should publicly call upon the United Nations Command to refrain from bombing the Yalu River power installations "except the military necessity arises." The second proposal was in the form of a resolution to be passed by the General Assembly which would assure the Chinese that the United Nations Command considered the Chinese border "inviolate." The Joint Chiefs had no objection to the first French proposal since General MacArthur had already assured them he did not intend to bomb the power installations. Besides, the phrase "military necessity" was extremely elastic. But they considered the second French resolution wholly unacceptable because the term "inviolate" would convey an impression to the Chinese that the United Nations would not, under any conditions, trespass beyond the border, whereas there was no guarantee that the United States might not have to operate across the Chinese frontier even before the General As-

[40] Rad, CS 68411, CINCFE to DA for JCS, 7 Nov 50.

[41] Rad, CX 95978, JCS to CINCFE, 7 Nov 50.

sembly had a chance to adopt the French-sponsored resolution.[42]

On 10 November, the Secretary of Defense transmitted to the Secretary of State the views of the Joint Chiefs of Staff on the French proposals for resolutions by the United Nations General Assembly. Marshall agreed with the State Department view that some form of reassurances to the Chinese Communists was called for. "I believe it should be made clear," he told Mr. Acheson, "that a sanctuary for attacking Chinese aircraft is not explicitly or implicitly affirmed by an United Nations action." [43]

A draft resolution calling for withdrawal of the Chinese forces from Korea and sponsored by six nations of the United Nations, including France, the United Kingdom, and the United States, was placed before the Security Council of the United Nations on 10 November. This draft resolution assured the Chinese that ". . . it is the policy of the United Nations to hold the Chinese frontier with Korea inviolate and fully to protect legitimate Chinese and Korean interests in the frontier zone. . . ." (Department of Defense objections to the term "inviolate" obviously were unavailing.) But this measure was never passed by the Security Council since the representative of the Soviet Union exercised his power of veto against it.[44]

On 16 November, President Truman also attempted to reassure the Chinese Communist government that the United Nations Command had no designs on its borders and, further, that the United States desired no expansion of the war. In a public announcement, he took note of the resolution then under consideration by the Security Council, affirmed American support of this resolution, and declared that "Speaking for the United States Government and people, I can give assurance that we support and are acting within the limits of United Nations policy in Korea, and that we have never at any time entertained any intention to carry hostilities into China." [45]

Chinese aircraft operating out of bases in Manchuria meantime attacked MacArthur's planes with increasing intensity. By mid-November, large groups of Communist jet aircraft were ranging across the border to intercept U.N. fliers. The Joint Chiefs of Staff agreed with General MacArthur that this should be stopped and proposed corrective action. They favored the removal of as many restrictions on U.N. air operations as would allow MacArthur's airmen to pursue enemy attackers six or eight miles across the Manchurian border. This would greatly reduce the Communist fliers' advantage of being able to attack and escape without suffering effective retaliation. Secretary Marshall approved this scheme and later testified that his views were shared by Secretary Acheson and President Truman. But these American

[42] Memo, Actg ACofS G–3 (Duff) for CofS USA (Collins), 10 Nov 50, sub: Resolutions in the U.N. to Provide Certain Assurances to the Chinese Communists, JCS 1776/157, Incl to Decision on JCS 1776/157, in G–3, DA file 091 Korea, Case 14/39.

[43] *Ibid.*

[44] U.N. Doc S/1894, Joint Resolution of the Security Council, 10 Nov 50, contained in Department of State Publication 4263, *United States Policy in the Korean Conflict, July 1950–February 1951*, Doc 13.

[45] Statement, President Harry S. Truman, 16 Nov 50, quoted in Department of State Publication No. 4263, Doc 14.

authorities were reluctant for the United States to take the initiative in giving MacArthur permission to enter Manchurian airspace. To do so without consulting other nations whose forces were in Korea would be viewed by those nations, it was believed, as unilateral American action and might well cause a rift between the United States and its allies.[46]

Secretary Acheson, in mid-November, instructed his ambassadors in certain key nations to sound out their attitude toward "hot pursuit," as the issue was termed, of Chinese aircraft into Manchuria. The reaction was unanimously against any such action by the United Nations Command. Typical was the attitude of one nation whose official spokesman expressed the fear that "United States unilateral action in this regard would afford a basis to the Soviet charge that the United Nations is only a front for the United States." One American ambassador after interviewing the officials of the country to which he was accredited stated that he firmly believed that the Atlantic Pact nations would disassociate themselves from such American action as being unilateral and without United Nations endorsement. As a result, the United States shelved the idea of carrying the air war into Manchuria.[47]

[46] All of the following are in the *MacArthur Hearings*: Testimony of General Marshall, p. 329, 1912; Testimony of General Vandenberg, p. 1410; Testimony of Secretary Acheson, p. 1723.

[47] JCS 2150/10, Note by the Secys to the JCS, sub: Reactions to Proposal to Permit U.N. Aircraft to Pursue Attacking Enemy Aircraft into Manchuria, 4 Dec 50, in G–3, DA file 091 Korea, Case 115/5.

The Mission Is Re-examined

The Joint Chiefs of Staff meanwhile reacted cautiously to the mounting evidence of Chinese intervention. After examining intelligence from the theater and other sources, they informed General MacArthur that the eventuality anticipated in their instructions to him of 27 September, "entry into North Korea by major . . . Chinese forces," appeared to have arrived. At least the introduction of Chinese forces to the extent reported by him would so signify. "We believe therefore," they warned him, "that this new situation indicates your objectives as stated in that message, 'the destruction of the North Korean armed forces,' may have to be re-examined."[48] A change of mission in the face of Chinese pressure could mean abandoning the drive to the Yalu, going on the defensive, and consolidating the ground seized since Inch'on.

But MacArthur was of no mind to abandon his drive to the Yalu. Protesting to the Joint Chiefs against any re-examination of his mission, MacArthur pointed out that their instructions to him on 10 October had exactly defined his course of action in this present situation. They had told him, in the event of the open or covert employment anywhere in Korea of major Chinese Communist units without prior announcement, to continue the action as long as in his judgment his forces had a reasonable chance of success.[49]

MacArthur rejected completely any course of action short of his original intentions. "In my opinion it would

[48] Rad, JCS 96060, JCS to CINCFE, 8 Nov 50.

[49] Rad, C 68572, CINCFE to DA for JCS, sgd MacArthur, 9 Nov 50.

be fatal to weaken the fundamental and basic policy of the United Nations to destroy all resisting armed forces in Korea and bring that country into a united and free nation," he charged. General MacArthur proclaimed his faith in the effectiveness of air interdiction by telling the Joint Chiefs that he could, with his air power, keep the number of Chinese reinforcements crossing the Yalu low enough to enable him to destroy those Chinese already in Korea. He meant to launch his attack to destroy those forces about 15 November and to keep going until he reached the border. "Any program short of this," he explained:

> would completely destroy the morale of my forces and its psychological consequence would be inestimable. It would condemn us to an indefinite retention of our military forces along difficult defense lines in North Korea and would unquestionably arouse such resentment among the South Koreans that their forces would collapse or might even turn against us.

He charged that anyone who hoped that the Chinese, once they had succeeded in establishing themselves in North Korea, would abide by any agreement not to move southward would be indulging in wishful thinking at its very worst.[50]

The Joint Chiefs had told MacArthur that consultation with the British Government on any new course of action against China was an integral part of American policy. In an unusually vehement burst of impatience, MacArthur directed a scathing comment at what he termed, "The widely reported British desire to appease the Chinese Communist by giving them a strip of Northern Korea," and cited British action at Munich in 1938 as historic precedent for their present attitude.[51] He went further and referred to a State Department criticism of the British appeasement of Hitler to lend emphasis to his statement. He charged that any such appeasement of the Communists carried the germs of ultimate destruction for the United Nations. "To give up any portion of North Korea to the aggression of the Chinese Communists," General MacArthur declared, "would be the greatest defeat of the free world in recent times. Indeed, to yield to so immoral a proposition would bankrupt our leadership and influence in Asia and render untenable our position both politically and militarily." MacArthur asserted that by moving to halt his forces short of the Yalu River American authorities "would follow clearly in the footsteps of the British who by the appeasement of recognition lost the respect of all the rest of Asia without gaining that of the Chinese segment." [52]

[50] *Ibid.*

[51] A current news report had stated that Mr. Bevin favored a buffer zone south of the Yalu. Dr. Pannikar recalls that in mid-November Mr. Bevin sent a message through his minister in Peiping to be conveyed to Chou En-lai or the highest accessible Chinese official. Pannikar says of this message to which he apparently was given access, "It was a strange communication, an elucidation of the objectives of the United Nations in Korea, an assurance from Britain that Chinese boundaries would be respected. . . . when Hutchinson [the British Minister] discussed the matter with me I frankly told him that I doubted whether the Chinese would look at any proposal which did not include an offer of direct negotiations of the whole issue with them; and that I considered that the idea of Britain assuring China of the inviolability of her boundaries was patronizing, to say the least." See Pannikar, *In Two Chinas: The Memoirs of a Diplomat*, pp. 114–15.

[52] Rad, C 68572, CINCFE to DA for JCS, sgd MacArthur, 9 Nov 50.

After elaborating his point, General MacArthur said that he believed that the United States should press the United Nations for a resolution condemning the Chinese Communists and calling upon them to "withdraw forthwith to positions north of the international border on pain of military sanctions by the United Nations should they fail to do so." He ended his protest on a note of confidence as he told the Joint Chiefs of Staff that complete victory could be achieved if "our determination and indomitable will do not desert us." [53]

Despite the optimism implicit in General MacArthur's protest that his mission should remain unchanged, the Joint Chiefs of Staff were not cheered. His protest merely underscored the critical need for a firm course of action to meet the Chinese interference in Korea. President Truman directed the National Security Council to meet on 9 November to consider on an urgent basis what the national policy should be toward Chinese Communist participation. The Joint Chiefs had been instructed to furnish their views on what should be done. It will be recalled that the national policy agreed upon in September had provided that in the case of Chinese Communist intervention the United States should attempt to localize the action in Korea and thereby avoid a general war. A second position, but one that had only tentative approval for use as a planning guide, stated, in substance, that United Nations forces would continue the action so long as such action had a reasonable chance of success, and that the United Nations commander should be authorized to take appropriate air and naval action outside Korea against Communist China.[54]

As of 8 November, however, no firmly established set of instructions outlining detailed measures against Chinese Communist intervention, regardless of degree, had been agreed upon by the nation's leaders. That was the task which faced them as MacArthur, heartened by the dwindling evidence of Chinese participation after the first week of November, demanded to be allowed to continue his original line of action in Korea.

The Joint Chiefs of Staff on 9 November forwarded to the Secretary of Defense for the National Security Council a lengthy analysis containing their view on the significance of the Chinese intervention and some suggestions on what the United States should do about it. Without accepting the theory that the Chinese troops in Korea were volunteers, the Joint Chiefs expressed the opinion that such a view was feasible in the event that the Chinese merely wanted to gain time for the defeated and disorganized remnants of the North Korean Army. But they pointed out that intelligence reports did not back up this theory, since they showed that Chinese Communist soldiers were entering Korea both as individuals and in well-organized, well-led, and well-equipped units, probably of division size.[55]

Examining Chinese motives in send-

[53] *Ibid.*

[54] Memo, ACofS G–3 (Bolté) for CofS USA (Gruenther), 8 Nov 50, sub: U.S. Courses of Action With Respect to Korea, in G–3, DA file 091 Korea, Case 37.

[55] Memo, JCS (Bradley) for Secy Defense (Marshall), 9 Nov 50, sub: Chinese Communist Intervention in Korea.

ing military forces against the United Nations Command, the Joint Chiefs saw three possibilities, although none of these had as yet been made clear by Chinese actions either in Korea or in Manchuria. The Chinese might wish to protect the Yalu River and the Changjin-Pujon Reservoir power complexes and establish a *cordon sanitaire* in North Korea; they might wish to continue the active but undeclared war in Korea to drain American resources without expending too much of their own military strength; or they could be planning to drive the United Nations forces from Korea. If the Chinese Communists were prevented, through United Nations action, from obtaining electricity from the Yalu power systems, Manchuria's economy would suffer severely. Consequently, if the Chinese Communists had intervened in North Korea solely to protect the power plants, it might be well, the Joint Chiefs of Staff told the Secretary of Defense, to announce an unmistakably clear guarantee that the United Nations would not infringe on the sovereignty of Manchuria, would not damage the power plants, and would not interfere with their operation. If the Chinese Communists rejected such a guarantee, the United States could feel fairly certain that they had had some other objective in intervening.[56]

That the Chinese might be planning a limited war of attrition in Korea to tie down and dissipate United States strength was also a real possibility. As the Joint Chiefs pointed out, "Korea is at such a distance from the United States that it would be expensive for the United States in manpower, materiel, and money to conduct an undeclared war in that area over a long period." Conversely, the Chinese, being next door to Korea, would find it comparatively inexpensive, with their practically unlimited manpower and Soviet equipment, to carry on such a war indefinitely. The continued involvement of United States forces in Korea would, in the opinion of the Joint Chiefs of Staff, be in the interests of Russia and of world communism by imposing a heavy drain on U.S. military and economic strengths. They still considered Korea a "strategically unimportant area" and felt that, in the event of a global war, fighting in Korea would leave the United States off-balance while Russia completed its plans for global conquest. The Joint Chiefs could also visualize quite clearly a situation whereby the United States, through concentrating its strength to defeat the Chinese in Korea, might, "win the skirmish in Korea but lose the war against the USSR if global war eventuates." [57]

The Joint Chiefs did not truly believe that the Chinese Communists intended to drive the United Nations forces from all of Korea. While it was possible that the Chinese did have that intention, the Joint Chiefs felt they could not force MacArthur's men off the peninsula "without material assistance by Soviet naval and airpower." If Russia did intervene to that extent, it would be evident that World War III had begun and the United States should get its divisions out of Korea as fast as possible.[58]

If the Chinese intervened in full

[56] *Ibid.*

[57] *Ibid.*

[58] *Ibid.*

strength, the Joint Chiefs foresaw three possible courses of action for United Nations forces: to continue the action as planned; to set up a defensive line short of Korea's northern border; or to withdraw. In the first instance, some augmentation of United Nations military strength in Korea might be necessary if a drive to the Yalu were to succeed, even if no more Chinese troops entered the fighting. The second course, pause and dig in, was, in the eyes of the Joint Chiefs, perfectly feasible and, indeed, perhaps expedient in the face of unclarified military and political problems raised by Chinese entry. But they rejected withdrawal because "if conducted voluntarily it would so lower the world wide prestige of the United States that it would be totally unacceptable. . . ." If the United Nations forces were compelled to leave Korea involuntarily it "could only be accepted as the prelude to global war." With specific reference to global war, the Joint Chiefs maintained that current conditions did not conclusively indicate that global war was imminent, only that the risk of global war had been increased.[59]

One significant conclusion drawn by the Joint Chiefs of Staff from the evidence which they had was that the United States should, as a matter of urgency, make every effort to settle the problem of Chinese intervention by political means. They recommended that, through the United Nations, the Chinese be reassured concerning the intentions of the United Nations Command and, if necessary, that direct negotiations be carried on through the diplomatic channels of nations that had recognized Communist China and thus had some access to the leaders of its government. Insofar as General MacArthur's assigned mission in Korea was concerned, the Joint Chiefs were willing to await clarification of the Chinese Communists' military objectives before recommending a change in the plan to drive to the Yalu. But with respect to American preparedness elsewhere, they recommended that plans and preparations be made on the basis that the risk of global war had been substantially increased by the Chinese action.[60]

At the very important meeting of the National Security Council on 9 November in Washington, General Bradley presented views developed by the Joint Chiefs of Staff with regard to possible intentions of the Chinese Communists in Korea. Bradley ventured a personal opinion that U.N. forces could hold in the general area of their present positions but that the question of how much Chinese pressure these forces would have to take before being impelled to attack the Manchurian bases would become increasingly urgent. He pointed out, however, that the Joint Chiefs of Staff had agreed that any decision to attack Chinese territory would have to be a U.N. decision since under the terms of U.N. authority, which now was the guiding force behind MacArthur's directives, no such attack was authorized. Bradley told the assembled leaders that he did not agree with MacArthur that the bombing of the Yalu bridges would stop the Chinese from entering Korea in strength should they choose to continue

[59] *Ibid.*

[60] *Ibid.*

their incursions. General Smith added that within fifteen to thirty days the Yalu would be frozen anyway, rendering the entire question of bridge bombing academic.[61]

When Secretary of Defense Marshall questioned the disposition of X Corps, which he felt was in some danger because of its great dispersion and lack of depth, Bradley defended MacArthur's reasons. He pointed out that in deploying his troops MacArthur sought to carry out his directives to occupy all of North Korea and to hold elections. On this same point, Secretary Acheson pressed Bradley to tell them whether there was not a better line for MacArthur's forces to occupy in Korea. Bradley agreed that, from a purely military point of view, the chances of defending a line in Korea would increase as that line was moved south of the Yalu. But he noted also that any backward movement on MacArthur's part would reduce U.N. prestige and might adversely affect the will of the South Koreans to fight.[62]

Acheson then recommended consideration of a buffer zone twenty miles deep, ten miles on each side of the Yalu. He felt that the Russians were interested in such an arrangement. Insofar as the Chinese were concerned, he saw them as having two interests, first, to keep the United States involved and, secondly and less important, to protect the border and its power plants. The Chinese, of course, would, if such a buffer zone were proposed, insist on the departure of all foreign troops from Korea. This would have the effect of abandoning Korea to the Communists.[63]

After studying these various views and recommendations, the National Security Council recommended certain interim measures to the President of the United States. These recommendations represented the combined sentiments of the nation's policy-makers, and largely aimed at a possible political solution to the problem of Chinese Communist intervention and in keeping with the established policy of avoiding, by every honorable means, a general war.

The President later recalled that November 1950 found the United States mainly concerned with three moves with regard to Korea. The United States was attempting to reassure its European allies, especially England and France, that it did not intend to widen the conflict or to abandon Europe for new entanglements in Asia. Secondly, in the United Nations the United States was attempting to gain maximum support for resistance to Chinese Communist intervention, at the same time avoiding any United Nations move toward military sanctions against Peiping—which would have undoubtedly meant war. Third, the United States was making every effort to ascertain the strength and direction of the Chinese Communist effort.[64]

Although these moves may seem to have been inadequate in light of the problem that developed later, it should be remembered that the problem itself was then in the formative stage. The

---

61 Truman, *Memoirs*, II, 378–80.

62 *Ibid.*

63 *Ibid.*

64 *Ibid.*, 381

moves, at the time, did present a logical basis from which to proceed as the problem developed, and formed the framework of the policy that the United States pursued until changes were forced by the pressure of events.

CHAPTER XIV

The Threshold of Victory

The gauntlet cast down by the Chinese in late October and early November left American intelligence experts guessing. Were the Chinese merely saving face? Were they bluffing? Or did the Communist Chinese seriously mean to throw their vast armies into Korea to defeat MacArthur's United Nations forces? American and other intelligence analysts might disagree on Chinese motives and intentions. But all corroborated that Chinese armies had massed in great strength along the Yalu in Manchuria, disposed for early action in Korea if the signal came, and that an unknown number had entered Korea. It was indeed a time for careful treading and sober consideration.

Eighth Army's Plans and Problems

The temporary setbacks in early November did not alter MacArthur's plans. He continued to prepare for the northward advance in the face of proof that Chinese Communist forces had entered Korea. General Bolté had visited Korea just after, as he described it, "the Chinese had destroyed the 8th Cavalry Regimental Combat Team." He found General Walker apprehensive but confident over the ultimate outcome. Walker assured Bolté that he had no intention of going on the defensive and had withdrawn only as a temporary regrouping measure. Walker, at the time, was bringing up his IX Corps on the right of his I Corps in order to renew the attack in greater strength.[1]

Walker intended to advance three corps abreast, the U.S. I Corps on the west, the U.S. IX Corps in the center, and the ROK II Corps on the east. He had set D-day at 15 November and given his army the mission to "attack to the north destroying enemy forces, and advance to the northern border of Korea in zone." [2]

Walker's main concern in preparing for the attack lay in alleviating a shortage of supplies in his forward areas. Since moving above P'yongyang, the Eighth Army had been supplied mainly by airlift. General Milburn, commanding the I Corps, told General Bolté that his corps was operating with only one day of fire and one and one-half days of POL in reserve. General Walker felt that he

[1] Memo, G–3 (Bolté) for CofS USA, 14 Nov 50, sub: Visit of Gen Bolté and Party to the Pacific Area, in G–3, DA file Pac, Case 8/2.

[2] (1) EUSAK Opns Plan No. 14, 6 Nov 50. (2) War Diary, EUSAK G–3 Sec, 6 Nov 50.

could not improve this dangerous situation in the face of the limited transportation, the poor roads, and the long distances involved, unless the Chinnamp'o port was in full operation. General Bolté thought that the solution to these supply problems lay in greater effort by the Air Force. He pointed out that the Air Force was lifting 1,000 tons daily but could double this with more flight crews and better maintenance. "Cargo aircraft stand idle and supply is critical," Bolté complained to Washington, "Cannot this be remedied soonest? I emphatically recommend more help including triple crews immediately." [3]

In response to Bolté's question, General Vandenberg, Chief of Staff, United States Air Force, stepped in and asked General Stratemeyer, the FEAF commander, if his command was supporting the Eighth Army to its fullest capability. MacArthur's air chief replied indignantly that his planes could do no more to step up airlift tonnage because the capacity of Korean airfields simply would not permit doubling airlift while at the same time rendering combat tactical air support. "General Bolté's statement re the ground situation is quite correct," he asserted, "but his statements re Tunner's [CG FEAF Combat Cargo Command] are not quite so accurate. We could use much more airlift than is available, but Bolté's recommended solution of triple crews is an over-simplification." [4]

Reporting by teleconference to the Army chief of transportation on 9 November, the transportation officer of the Far East Command sketched the situation with regard to ports and lines of communication in Korea. Pusan was handling about 15,000 metric tons of supply daily, and Inch'on about 8,000 metric tons. Chinnamp'o, a vital port since it was much closer to the front, had been opened for partial operation but could handle only shallow-draft vessels. The port had not been completely mineswept, and the large tidal basin at the port had silted up considerably. Some LST's were being unloaded even though they rested on the bottom of the harbor at low tide.[5]

Rail lines were equally restrictive. Single-track bottlenecks and destroyed bridges materially reduced their capacity. The supply shortage remained serious, and General Walker decided to postpone his attack. On 14 November, General MacArthur's headquarters so notified the Department of the Army. When pressed for reasons, the Far East Command staff officers told their counterparts in Washington that the logistical estimate on which General Walker's decision was based was not available to GHQ. Meanwhile, Walker's forces took a few steps forward along the Ch'ongch'on River to positions they would use as a line of departure when they did reopen their general advance.[6]

The Eighth Army would need about 4,000 tons of supply per day in order to sustain the offensive northward. By 20 November, the efforts of all supply agencies began to pay off and achieved the

[3] Rad, CM-IN 8483, CINCFE (Bolté) to DA, 6 Nov 50.

[4] (1) Rad, AFOOP-OD56864, CG USAF to CG FEAF, 10 Nov 50. (2) Rad, AX 3359 B VCAP, CG FEAF to CG USAF, 12 Nov 50.

[5] Telecon, TT 3992, DA to GHQ, 9 Nov 50.

[6] Telecons, TT 4011, DA to GHQ, 14 Nov 50, and TT 4016, 15 Nov 50.

required figure.[7] General Walker, on 22 November, notified General MacArthur that the logistics problems in the forward area of the Eighth Army had been solved and that he could now support a renewed offensive.[8]

The X Corps Plan

Logistically, the X Corps on the east coast enjoyed a somewhat better status than Eighth Army. Supplies came in at Wonsan, Hungnam, and Iwon. Almond's combat units sat relatively close to those ports in early November. ROK units moving up the east coast were actually supplied by LST's operating over the beaches. The situation of the 7th Division and the 1st Marine Division, however, became progressively more difficult logistically as they moved inland away from the ports.[9]

The reverses suffered by the Eighth Army and the appearance of Chinese troops in front of his own troops sobered General Almond considerably. Whereas earlier he had pressed the Marines to push forward as rapidly as possible to the border, the brief Chinese intervention caused him to grow more cautious for a few days. But the virtual disappearance of the Chinese from the field had the same effect on him as it had had on General MacArthur. In any case, Almond was under orders to resume the advance. On 11 November, he again directed the Marines to advance to the north.[10]

The presence of Chinese forces at the front of the X Corps caused General MacArthur's staff to re-examine the scheduled operations of Almond's corps. The staff now assumed that the coming attacks would not be routine marches to the border. General Willoughby's intelligence report to the Department of the Army on 10 November showed that the enemy's offensive potential had been materially strengthened. Particularly significant was a Chinese build-up in the Changjin-Pujon Reservoir area. Willoughby told Department of the Army that this build-up posed a serious threat to Almond's forces not only in the immediate area but also in the coastal area along the northeast shoreline of Korea. "It is believed," Willoughby stated, "that this enemy concentration even now may be capable of seizing the initiative and launching offensive operations." He speculated that such operations might take the form of a concerted drive to the south in an effort to cut off U.N. forces then located to the north and east of Hungnam. Willoughby estimated that as of that date there might well be as many as 64,200 regular Chinese troops in Korea. By the next day, on 11 November, he had raised this figure to 76,800.[11]

---

[7] (1) Interv, Col Appleman with Col Albert K. Stebbins, EUSAK G–4, 4 Dec 53. (2) War Diary, EUSAK, G–4 Journal, Rad 7, 241015 Nov 50. (3) Interv, Col Appleman with Gen Allen, 15 Dec 53.

[8] Rad, GX 50025 KGIX, CG Army Eight to CINCFE, 22 Nov 50.

[9] Telecon, TT 3992, DA and GHQ, 9 Nov 50.

[10] X Corps Opns Order No. 6, 11 Nov 50.

[11] (1) Telecon, TT 3996, DA and GHQ, 10 Nov 50. (2) Telecon, TT 4000, DA and GHQ, 11 Nov 50. (3) Later analyses of Chinese troop movements and order of battle during this part of November show that, in fact, the Chinese had, as of this date, moved 300,000 men, organized into 30 divisions, into Korea. In front of the Eighth Army stood 180,000 men and 120,000 were concentrated in front of Almond's X Corps. See Appleman, *South to the Naktong, North to the Yalu*, pp. 768–69.

In the face of the enemy strength evident in General Almond's area and in the Eighth Army's zone, General Wright's JSPOG staff closely examined the original plan developed for X Corps operations. The Eighth Army attack was to be the main U.N. effort. Wright's staff looked at the X Corps' plans from the standpoint of "how can X Corps best assist Eighth Army?" The JSPOG planners had either not consulted or did not believe intelligence estimates forwarded to Washington by Willoughby, since their planning assumptions credited the Chinese with less strength than shown in Willoughby's reports of the same date. According to JSPOG assumptions on 12 November, the Eighth Army faced 18,000 Chinese troops, and the X Corps, 7,500. These troops were in addition to 50,000 North Koreans fronting the X Corps' path of advance. The Chinese were credited with the ability to reinforce at the rate of 24,000 men per day.[12]

As JSPOG officers saw it, Almond's plan called for an advance to the Korean border and destruction of enemy forces; keeping contact with the Eighth Army; protection of the Eighth Army's right flank against enemy forces sideslipping into it from the north; elimination of guerrillas; administration of his area. These projected operations would not provide direct assistance to the Eighth Army in its attack, but JSPOG officers noted that successful completion of X Corps plans would be of considerable incidental aid to the Eighth Army.[13]

Advantages seen in the X Corps' planned operation were that momentum of forces moving along an established direction would be retained; North Korean forces would not have time to dig in and resist; logistics difficulties within the X Corps would be minimized; and the route of the 1st Marine Division's advance (to Changjin thence north to the Yalu) would pose a threat of envelopment to the enemy. On the other hand, certain disadvantages would result if the X Corps carried out the operation as then planned. The X Corps' drive would not immediately affect enemy forces facing the Eighth Army. The direction of movement of the center of mass in the X Corps would be *away* from the main strength of the enemy. The most significant observation was the statement that by continuing to advance to the north, "X Corps incurs the danger of becoming seriously over-extended," and that if progress by the right flank of the Eighth Army was appreciably slower than X Corps', the left flank of the X Corps would be exposed.[14] For all practical purposes, that flank was already exposed.

A prophetic warning was contained in the portion of the staff study discussing the advance of the Marine elements of the X Corps:

> As the 1st Marines move toward Changjin they will tend to be extended. The left flank of the Marines will be on the mountainous ridge that divides the watersheds of the peninsula. It is generally impassable for heavy military traffic. However, prisoner reports show that the *124th CCF Division* entered North Korea at Manpojin and is now in the Choshin [Changjin] Reservoir area. If the 1st Marine Division attacks north beyond this route well ahead

[12] Staff Study, X Corps Assistance to Eighth Army, 12 Nov 50, JSPOG files.

[13] *Ibid.*

[14] *Ibid.*

of the Eighth Army it will be vulnerable to attacks on its flank and rear.[15]

This line of reasoning was quite in harmony with the views of General Smith, the Marine division commander. General Smith on 15 November received a visit from the chief of staff to COMNAVFE, Rear Adm. Albert K. Morehouse, whom Admiral Joy had sent to Korea on a liaison and inspection tour. At that time Smith, feeling that he was, as he expressed it, "talking in the family," expressed frank concern over what he considered General Almond's unrealistic planning and his tendency to ignore enemy capabilities when he wanted a rapid advance.[16]

Smith's views on the combat scene are further illustrated by a personal letter to General Cates, the Marine Corps Commandant, on the same day, 15 November. He frankly admitted that he felt Almond's orders were wrong and that he, as Marine commander in Korea, was not going to press his own troops forward rashly to possible destruction. "Our orders still require us to advance to the Manchurian border," Smith said. "However, we are the left flank division of the Corps and our left flank is wide open." Smith pointed out that there was no Eighth Army unit closer to his flank than eighty miles southwest. While the X Corps, according to Smith, could assure him "when it is convenient" that there were no Chinese on his flank, he observed, "if this were true, there could be nothing to prevent the Eighth Army from coming abreast of us. This they are not doing." [17]

Smith deliberately stalled on the advance because he did not like the prospect of stringing out his division along "a single mountain road close to 200 miles long." Smith's principles, which he followed all the way and which probably accounted for a good number of saved lives a month or so later, called for concentrating his entire division into a reasonable sector and developing as completely as possible his main supply route. He built under adverse conditions an airfield at Hagaru-ri, and, through a slower advance, took care of his flank security. He outposted the high ground along both sides of his main supply route at all times.[18]

JSPOG officers believed that if X Corps operations were to be effective in assisting the Eighth Army, only one general course of action lay open. Almond should attack to the northwest, thus threatening the rear of the Chinese formation facing the Eighth Army and forcing their withdrawal to avoid envelopment. If Almond called off his advance north, two divisions could be made available for this attack. Since the attack would probably develop on a narrow front as a struggle for control of the route of advance, concentration of forces for a co-ordinated attack would not be necessary. The attack could be launched at once using forces already in position.[19]

The JSPOG staff concluded that the X Corps must eliminate enemy forces in the reservoir area before any operations were feasible, and that once Changjin

[15] *Ibid.*

[16] *Aide-Memoire*, Gen Smith, p. 600.

[17] *Ibid.*, p. 609.

[18] See Lynn Montross and Capt. Nicholas A. Canzona, *U.S. Marine Operations in Korea, 1950–1953*, vol. III, *The Chosin Reservoir Campaign*, ch. VII.

[19] Staff Study, X Corps Assistance to Eighth Army, 12 Nov 50, JSPOG files.

was cleared it might be feasible to revise the X Corps-Eighth Army boundary and direct X Corps to attack to the northwest to cut the Manp'ojin-Kanggye road, which enemy forces, in all likelihood, were using as a main supply route. They recommended that no change in the projected operations of the X Corps be made immediately, but that the X Corps be directed to begin planning for an attack to the northwest to cut the enemy main supply route.[20]

It is apparent that the joint planning staff did not like the look of the situation in northeast Korea and did not completely indorse Almond's plan for operating there. But the planners hedged. Two factors may have caused them not to speak out against the plan. First, they must have known that MacArthur was set on attacking to end the fighting. Also, they had a no more acceptable solution to the problem than that under consideration.[21]

The advantages which the staff read into Almond's plans were so innocuous as to seem fabricated. On the other hand, the disadvantages, or more exactly, the dangers of Almond's intended advances, were plainly and honestly stated. An objective appraisal would have weighed the advantages against the disadvantages and found the scale tipped completely on the side of disadvantages and danger. Had this been done, it is entirely likely that MacArthur's advisers would have urged immediate changes in Almond's planned operations to include more limited objectives, more co-ordinated advances, and, possibly, even preparations for defensive action.

[20] *Ibid.*

[21] *Ibid.*

By personal letter to General Almond on 10 November, General Wright outlined the general plan to be carried out by the Eighth Army and relayed General MacArthur's desire that the X Corps do everything possible to assist Eighth Army. Then, on 15 November, and accepting the recommendation of the JSPOG staff, MacArthur directed Almond to develop, as an alternate feature of his operation, plans for reorienting the attack to the west upon reaching the vicinity of Changjin town, north of the Changjin Reservoir. This alternate operation would be executed upon order from General MacArthur.[22]

Meanwhile, General Almond had been doing some planning of his own and on 14 November sent a letter to General Wright which, in effect, was quite in line with the order to plan for a westward move after clearing the town of Changjin. General Almond told General Wright:

> I have your letter of 10 November relaying the CinC's directive that the X Corps be made fully familiar with Eighth Army's plan in order to be prepared for any possible change for a strong effort in coordinating with Eighth Army's attack. Two members of my planning staff have just returned from Eighth Army with a draft copy of General Walker's Operation Plan No. 15, yet to be published in final form. They discussed Eighth Army's plan at some length with General Walker and certain members of his staff.
>
> As you may already know, the Eighth Army plan is for a very deliberate and thorough advance to objectives distant only an average of some 20 miles North of present front line positions.

[22] (1) X Corps Comd Rpt, 27 Nov 50, p. 9. (2) Rad, CX 69009, CINCFE to CG X Corps, 15 Nov 50.

You will recall that during your recent visit with us at WONSAN we presented X Corps capabilities of making an all-out effort, with not less than two US divisions, to the west in the event of an enemy break-through or envelopment of Eighth Army's right flank. We have devoted continuing efforts in planning possible operations not only to further the CinC's overall objective of securing all of North Korea within our assigned zone as expeditiously as possible but also to assist Eighth Army's effort.

With the containment by Eighth Army of the Communist offensive in that area, coupled with the unchanged overall mission, it now appears to me to be inadvisable, at this stage of Eighth Army and X Corps operations, for X Corps forces to operate in any strength to the west. The principal reason for this conclusion is that the only two feasible vehicular routes to the westward in X Corps zone, short of CHOSIN Reservoir,[23] are the YONG-HUNG-TAEPYONG-NI and the WON-SAN-YANDOK roads. Since both of these routes enter the Eighth Army zone in rear of General Walker's present front lines, any advance in strength to the westward over them would appear to be a fruitless operation. Even contacting the Eighth Army right flank in the vicinity of ONYANG-NI with more than foot troops would require a major engineer road-building effort in the mountains to the eastward thereof.

In view of the foregoing, I am convinced that X Corps can best support Eighth Army's effort by continuing its advance to the north, prepared to move westward if desirable when X Corps elements are well north of CHOSIN Reservoir, and they will be prepared to trap and destroy any enemy forces engaging Eighth Army which depend upon a line of communication through MANPOJIN. North of CHOSIN Reservoir suitable lateral routes to the west appear to exist but these routes would have to be verified when that area is reached.

Thus, X Corps Operation Order No. 6, 11 November 1950, directing advance in zone to the north border of Korea is in accordance with Part II, CX67291, and is I believe, at present the most important contribution we can make to the overall operation in Korea. The success of this advance will result in the destruction of Chinese and North Korean forces in the reservoir area, which might otherwise be employed on the Eighth Army front, and will place X Corps units in a position to threaten or to cut enemy lines of communication in the Eighth Army zone. As a corollary, X Corps will secure the important hydroelectric power installations in its zone and will be well along toward completing its ultimate mission prior to the advent of severe winter conditions.

I fully appreciate the CinC's desire for us to assist the Eighth Army in every possible way. I trust that my analysis of present X Corps capabilities explains our views here and hope that energetic execution of my Operation No. 6 will place assistance to the Eighth Army before the cold weather now upon us is much more severe.[24]

General Willoughby continued to report a crucial build-up of forces in the Changjin-Pujon Reservoir area north of Hamhung-Hungnam. Even as Almond and Wright exchanged views on the best course of action for the X Corps, Willoughby informed Washington that his study revealed a great vulnerability of the open west flank of the X Corps and of the main supply route leading from Hungnam to the Changjin Reservoir. Almost 10,000 enemy troops had been spotted immediately west of this vital line. In addition, the enemy had the equivalent of four divisions in the Changjin-Pujon area. With this strength, the Chinese could counterattack to the south-

[23] Chosin Reservoir is the Japanese name for the Changjin Reservoir, and is the name by which U.N. forces best knew this body of water. The Pujon Reservoir, east of the Changjin, was also best known by its Japanese name, Fusen.

[24] Ltr, Gen Almond to Gen Wright, 14 Nov 50.

U.S. MEN AND TANKS *advance through the rubble-strewn streets of Hyesanjin, near the Manchurian border.*

east with troops from the Changjin area in an effort to isolate X Corps forces northeast of Hungnam, could conduct active guerrilla operations against corps lines of communications, could throw a combined offensive against X Corps using guerrillas and other forces, or could launch an offensive against the gap between the Eighth Army and X Corps by sideslipping to the southwest from the Changjin Reservoir area.[25]

On 20 November General Almond, acting on instructions from General MacArthur, warned his command that minimum forces only were to advance to the immediate vicinity of Korea's northern border. No troops or vehicles were to go beyond the boundary into Manchuria or the USSR, and no fire was to be exchanged with, or air strikes brought down on, forces north of the northern boundary. Damage, destruction, and

[25] Telecon, TT 4028, DA and GHQ, 17 Nov 50.

THANKSGIVING DAY SERVICES *being held for members of the U.N. forces on the bank of the Yalu river.*

disruption of power plants were to be avoided.[26]

Troops of the 17th Infantry, 7th Division, reached the Yalu River at Hyesanjin on 21 November. General MacArthur immediately congratulated General Almond, who, in turn, commended the 7th Division for "an outstanding military achievement." Almond's message ended on an optimistic note when he told Maj. Gen. David G. Barr, "The 7th Division has reached its objective and I am confident that you will hold it."[27]

Almond meanwhile ordered a plan made for a westward advance along the Hagaru-ri–Mup'yong-ni axis. He directed that the road to the Changjin Reservoir be developed as a corps supply road and that an RCT of the 7th Division be assigned to seize Changjin town and to protect the east flank of the 1st Marine Division. The two objectives,

[26] Rad, X 12811, CG X Corps to All Comdrs, 20 Nov 50.

[27] Rads, X 2867 and X 2859, CG X Corps to CG 7th Division, 22 Nov 50.

Changjin and Mup'yong-ni, were too widely separated to be assigned to a single division. General Almond also directed that the planners take into consideration that extreme winter temperatures of 30 to 40 degrees below zero Fahrenheit would severely restrict both friendly and enemy operations.

On 23 November, Colonel Chiles, X Corps operations officer, took this plan to Tokyo where he discussed it with General MacArthur's staff. On 24 November, General MacArthur directed that the plan be carried out with one modification, a shift of the proposed boundary between the X Corps and Eighth Army farther west and south in the zone of the 1st Marine Division. General Almond was told to designate his own D-day.[28] The details of the corps plan were passed on to General Walker and the Eighth Army staff by visiting GHQ officers on 24 November.[29]

General Almond ordered his troops to advance at 0800, 27 November. The final assignment of tasks directed the 1st Marine Division to seize Mup'yong-ni and advance to the Yalu, the 7th Division to attack from the Changjin Reservoir and advance to the Yalu, and the ROK I Corps to advance from Hapsu and Ch'ongjin areas to destroy the enemy.[30]

The Cautious View

Other friendly nations meanwhile were concentrating on the threatening situation developing along the Yalu border and took a consistently darker view of it than did the United States. On 13 November, the Australian Prime Minister informed the United States through diplomatic channels that the Australian Government now believed that Chinese intervention had created a new situation in Korea which called for careful examination. He recommended "military caution" and forecast that the consequences of Manchurian border incidents could be so grave that it might be best "temporarily to ignore Chinese Communist provocation to the extent possible."[31] From embassies located in Peiping came other warnings. The Swedish Ambassador to Communist China reported in mid-November that Chinese Communist movements toward Korea were on a large scale. The Burmese Embassy in Peiping at the same time expressed the view that the Chinese Communists were ready to go to any length to aid the North Koreans and that they were fostering mass hysteria based upon an alleged United States intention to invade Manchuria. The Netherlands on 17 November passed along to the United States Government information from Peiping that Chinese intervention in Korea was motivated by fear of aggression against Manchuria. If U.N. forces halted fifty miles south of the Yalu, the Netherlands believed, there would be no further intervention.

General MacArthur was prescient in his apprehension that steps might be taken to prevent his advance to the northern border of Korea. Strong sentiment was developing among other members of the United Nations and within the Department of State for a solution

[28] Rad, CX 69661, CINCFE to CG X Corps, 24 Nov 50.

[29] Rad, CX 69661, CINCFE to CG X Corps and CG Eighth Army, 23 Nov 50.

[30] X Corps Opns Order No. 7, 25 Nov 50.

[31] Intelligence Rpt, 13 Nov 50, in G-2, DA files.

to the problem of Chinese intervention through means other than those currently planned. General Bolté pointed out to General Collins on 20 November that the State Department was seriously considering a "buffer state" or neutralized zone as a means of stopping the military action in Korea and decreasing the possibility of world conflict. This idea, according to General Bolté, was being pushed within the Department of State with considerable vigor, to the extent that specific proposals by which the policy would be presented were being drawn up. The British Government had reportedly suggested that such an approach should be considered by the United Nations.[32]

Bolté left no doubt as to where he stood. He told Collins that he was as unalterably opposed to a buffer zone concept as was General MacArthur. Bolté felt that any buffer zone offer by the United States could seriously restrict the United Nations (and the United States) militarily without any resulting gain. He recommended strongly that General MacArthur's missions and directives not be changed.[33]

The move to halt the United Nations Command short of the international boundary took more definite form on 21 November with scheduling of a meeting between the representatives of the Department of Defense and the Department of State to discuss the possibility of negotiating with the Chinese Communists to end the fighting in Korea by establishing a demilitarized zone on one or both sides of the Korean-Manchurian frontier. General Bolté again addressed the Chief of Staff on the matter and again expressed strong feelings against any such method of curtailing military operations in Korea. "In light not only of the United Nations objectives in Korea," Bolté said, "but also of our national objectives world-wide, and until such time as CINCUNC indicates that he is unable to continue the action against the Chinese Communists, his directives . . . should not be changed, and a decision to halt the action in Korea short of the Korean frontier should not be made on military grounds." The Army's top planning officer felt that the only grounds on which MacArthur should be ordered to halt his advance would be that further offensive action would cause too great a risk of global war and conversely that cessation of the offensive would tend to minimize that risk. In General Bolté's opinion, a continuation of the action would not, of itself, engender risk of general war nor would a cessation of the action lessen such a risk. He held a rather optimistic view of the United Nations Command's combat potential, saying, "It is not envisaged that the Chinese Communists can succeed in driving presently committed United Nations forces from Korea, unless materially assisted by Soviet ground and air power." He believed that MacArthur had sufficient strength to hold any line in North Korea "in light of circumstances now prevailing." Bolté admitted that the drive to the border would no doubt increase the tenseness of the situation to some extent. But he emphasized that the decision to cross the 38th Parallel was based on the consideration that all

---

32 Memo, Bolté for CofS USA (Gruenther), 20 Nov 50, sub: Buffer State in North Korea, in G–3, DA file 091 Korea, Case 120.

33 *Ibid.*

of Korea should be cleared of Communist forces, and that attack from Manchuria would be recognized as an open act of military aggression. Further, the United Nations would actually have a better chance of localizing the conflict by driving all Communist forces from North Korea. A show of strength might well discourage further aggression where weakness would encourage it.[34]

General Bolté urged that if the Secretary of State suggested ". . . a new United Nations Security Council resolution calling for a demilitarized zone in North Korea to be administered by a United Nations body with Chinese Communist representation," the Defense Department oppose it. He concluded, prophetically, that ". . . history has proved that negotiating with Communists is as fruitless as it is repulsive. The present case is no exception." [35]

As a result of the conference and of further moves by other members of the United Nations a compromise solution was worked out. Assistant Secretary of State Rusk prepared a message for General MacArthur along general lines agreed to by the Department of Defense. He forwarded this message to the Joint Chiefs of Staff, noting that "we fully recognize that the Department of State does not have drafting responsibility with respect to this message, but we thought that a revised draft might provide the most convenient means for setting forth our views for the consideration of the Department of Defense and the Joint Chiefs of Staff." General Collins made a few alterations in the State Department draft and on 24 November the Joint Chiefs sent the revised message to General MacArthur.[36]

This message could in no way be considered a directive to MacArthur. At best, it was a tentative proposal for a course of action that left him with both the initiative and the responsibility for deciding which way the war should go. General Collins frankly told MacArthur that the question of halting short of the border had been raised because of the growing concern among other members of the United Nations. The United States was faced, in this as in other instances, with loss of support in the United Nations if it did not carefully consider the views of its allies in Korea. Collins warned that the United States anticipated proposals within the United Nations for resolutions which would place unwelcome restrictions on MacArthur's advance. Considerable sentiment existed among other nations in favor of establishing a demilitarized zone between the United Nations forces and the frontier in the hope of reducing Chinese Communist fear of military action against Manchuria and a corresponding sensitivity on the part of Russia with respect to Vladivostok.[37]

The consensus among American political and military leaders in Washington, crystallized at the meeting of Department of Defense and State officials, had been that no change should be made in MacArthur's immediate mission; but that the highest officials in the American Government should at once draft a

[34] Memo, G-3 DA for CofS USA, 21 Nov 50, sub: State-Defense High-Level Mtg on Korea, with Annex A.

[35] *Ibid.*

[36] Rad, WAR 97287, CofS USA (Collins) to CINCUNC, 24 Nov 50.

[37] *Ibid.*

course of action to permit the establishment of a unified Korea and, at the same time, reduce the risk of more general involvement. The State-Defense group had worked out in exploratory discussions certain military measures which, it seemed to them, might reduce the tension with Communist China and the Soviet Union, thus avoiding a rift between the United States and its allies. These measures, which if adopted would change MacArthur's mission, were transmitted to him.[38]

The measures assumed that MacArthur could push to the Yalu. General Collins suggested that after advancing to or near the Yalu MacArthur pull his forces back. Using a holding force of ROK troops, he would secure the terrain dominating the approaches leading from the mouth of the Yalu to the area held by the 7th Division near Hyesanjin. Other United Nations troops would fall back into reserve positions to support the South Koreans if necessary. This plan would be used only if effective enemy resistance ceased. The line held would be extended eastward through Ch'ongjin on the Sea of Japan with no advances being made by Almond's forces beyond this line. "It was thought," General Collins explained, "that the above would not seriously affect the accomplishment of your military mission." [39]

Only if it were militarily necessary would the United Nations troops destroy the hydroelectric installations in North Korea. The United Nations Committee for the Unification and Rehabilitation of Korea would deal at the appropriate time with appropriate representatives, presumably the Chinese Communists, to insure an equitable distribution of power from these installations. And in the event that the Chinese did not again attack in force, orderly elections could be held in North Korea and the country unified in line with United Nations plans. No decision had yet been made on procedures for handling the matter of entering northeastern Korea, which was extremely sensitive since dealings there would be with the USSR and not China.[40]

"While it is recognized," General Collins went on, "that from the point of view of the commander in the field this course of action may leave much to be desired, it is felt that there may be other considerations which must be accepted. . . ." Apparently, American authorities still felt that the Chinese were interfering reluctantly in Korea. General Collins postulated that this course "might well provide an out for the Chinese Communists to withdraw into Manchuria without loss of face. . . ." The Russians, too, might be reassured; and it was felt that Russian concern was at the root of their pressure on the Chinese to interfere in Korea.[41]

General Collins asked for MacArthur's comments on the proposals, to include timing and method of announcement if he agreed. He wanted to be sure that the measures did not impede the military operation, yet felt it important that the Chinese and Russians not misinterpret MacArthur's intention as aggression against their borders. General Collins concluded:

[38] *Ibid.*

[39] *Ibid.*

[40] *Ibid.*

[41] *Ibid.*

Since there are many political and military implications involved in these ideas and since other nations would be involved, no action along these lines is contemplated until full opportunity has been given for further consideration of your views, final decision by the President and, possibly discussion with certain other countries.[42]

General MacArthur turned thumbs down on the proposals. But his reply, in contrast to his earlier blast against any form of restriction upon his advance, was temperate. The tone of his reply approached, in some respects, patient forbearance. The anticipated move to halt his advance had come, not as an order, but as a suggestion which could become an order only after time-consuming negotiation. Walker's forces had already jumped off toward the border and well might reach it before further political action could be taken.

"The concern underlying the search for the means to confine the spread of the Korean conflict is fully understood and shared here, but it is believed that the suggested approach would not only fail to achieve the desired result but would be provocative of the very consequences we seek to avert," General MacArthur stated. He had just returned to Tokyo from a tour of the battlefront where he had flown over the Yalu River area in a "personal reconnaissance." This flight had convinced him beyond all doubt that it would be utterly impossible to stop upon commanding terrain south of the Yalu if he were to keep the lines of approach to North Korea from Manchuria under effective control. The terrain, ranging from the lowlands in the west to the rugged central and eastern sectors, could not easily be defended. Only along the river line itself, a line which he was not proposing to sacrifice once achieved, were there natural defense features to be found such as in no other defense line in all of Korea. "Nor would it be either militarily or politically defensive," he asserted, "to yield this natural protective barrier safeguarding the territorial integrity of Korea." [43]

General MacArthur feared, aside from the military foolishness of such a move, that the political results would be "fraught with most disastrous consequences." Any failure on the part of the United Nations Command to keep going until it had achieved its "public and oft-repeated" objective of destroying all enemy forces south of Korea's northern boundary would be viewed by the Korean people as betrayal. The Chinese and all other Asians would, he maintained, view it as weakness and appeasement of the Communist Chinese and Russians.[44]

He presented a novel secondary argument against establishment of any sort of buffer zone by pointing out that political tension between Manchuria and Korea required that the international boundary be closed to minimize bandit raids and smuggling. His study of Russian and Chinese propaganda caused him to doubt that either nation was actually concerned over the fate of the Yalu power installations. The ROK unit which had reached the Yalu at Ch'osan in October had found that the power plants there had

42 *Ibid.*

43 Rad, C 69808, CINCUNC (MacArthur) to DA for JCS, 25 Nov 50.

44 *Ibid.*

been shut down for a full month with much of the machinery and equipment removed, and nothing had been said by the Russians or the Chinese about the loss of power. "In view of these factual considerations," he said, "one is brought to the conclusion that the issue of hydroelectric power rests upon the most tenuous of grounds." [45]

General MacArthur continued his argument by emphasizing that the entry of Chinese Communist forces into the Korean conflict was a risk which the United States had taken with its eyes wide open when it sent troops into Korea. "Had they entered at the time we were beleaguered behind our Pusan Perimeter beachhead," MacArthur surmised, "the hazard would have been far more grave than it is now that we hold the initiative. . . ." United Nations forces were committed to seize the entire border area, and had already, in General Almond's sector, occupied a sector of the Yalu River. Yet, in his opinion, there had been no noticeable political or military reaction by the Chinese or Russians.[46]

He then outlined his plans for the future in Korea, telling the Joint Chiefs of Staff that as soon as his men consolidated positions along the Yalu River he would replace American troops with ROK forces. He would then order, through public announcement so the Chinese could not fail to hear, the return of American forces to Japan, and the parole of all prisoners of war to their homes, and would leave the unification of Korea and the restoration of the civil procedures of government to the people, with the advice and assistance of United Nations authorities.[47]

If this plan did not effectively appeal to reason in the Chinese mind, MacArthur maintained, ". . . the resulting situation is not one which might be influenced by bringing to a halt our military measures short of present commitments." But by resolutely meeting those commitments and accomplishing the publicly proclaimed military mission of destroying enemy forces in Korea, the United States could find its only hope of checking Soviet and Chinese aggressive designs before those countries were committed to a course from which ". . . for political reasons . . . they cannot withdraw." [48]

General Bolté urged the Chief of Staff to subscribe to these views and recommended that the Joint Chiefs of Staff reiterate their approval of the idea of a full force advance to the border. But events were to overtake any such action by the Joint Chiefs.[49]

MacArthur Attacks

Across most of the battlefront during mid-November the enemy seemed to be withdrawing. Cautious probings by U.N. units occasionally brought strong local reaction, but American commanders noted a definitely defensive trend. On 20 November, as the Eighth Army moved into position for the coming drive northward, the United Nations Command reported to Washington that the

[45] *Ibid.*

[46] *Ibid.*

[47] *Ibid.*

[48] *Ibid.*

[49] Memo, Gen Bolté for CofS USA, 27 Nov 50, sub: U.S. Courses of Action in Korea, in G–3, DA file 091 Korea, Case 121.

enemy had broken contact and was apparently withdrawing to positions farther north in Eighth Army's zone. But in the eastern sector, Almond's troops were still meeting resistance. "Recent reports seem to indicate," MacArthur's staff informed Washington on 20 November, "that the enemy is organizing the ground to take advantage of rough terrain, but it is still not clear as to just what this general limited withdrawal activity may portend." These officers noted that similar withdrawals by the enemy in the past had preceded offensive actions. "On the other hand," they pointed out to Army officials in Washington, "the sudden reversal coupled with limited withdrawals and considerable activity in the vicinity of strong defensive points may indicate a high level decision to defend from previously selected and prepared positions." [50]

Another enemy, North Korea's winter weather, had made its unwelcome appearance. On 14 November, the temperature across the entire front plummeted to readings ranging from ten degrees above zero in the west to twenty degrees below zero in northeastern Korea.

General Walker's orders to his commanders reflected a considerable degree of caution and some respect for the enemy forces facing Eighth Army. He directed a closely co-ordinated attack by phase line in order to have the army under control at all times for any sudden tactical change required by enemy action. The days of the reckless pursuit had apparently ended.[51] Most units reached their line of departure by 17 November. Since the logistical picture had improved and promised to improve more in the near future, Walker announced to his commanders that the attack northward would start on 24 November.[52]

The Joint Chiefs of Staff had been kept informed of the situation as it developed throughout the month, and on 18 November General MacArthur notified them that the Eighth Army would launch its attack as scheduled on 24 November. He emphasized that the delay in mounting the offensive had been caused by logistical difficulties, not enemy action. Rather euphemistically perhaps, in light of later events, he assured the Joint Chiefs that intensified air attacks by his air forces during the preceding 10-day period had been very successful in isolating the battle area, stopping troop reinforcement by the enemy, and greatly reducing his flow of supplies.[53]

There was an almost complete absence of enemy contact on the entire Eighth Army front as Walker's men assumed their starting positions on 22–23 November. General MacArthur, suspicious of this unusual quiet and somewhat worried over the gap between the X Corps and Eighth Army, ordered General Stratemeyer to patrol this gap with great care. But American pilots flying from twelve to sixteen sorties in daylight hours and a half-dozen sorties at night located no enemy forces in the gap.[54]

General Willoughby reported to the

[50] Telecon, TT 4036, DA and GHQ, 20 Nov 50.

[51] (1) EUSAK Opn Plan No. 15, 14 Nov 50. (2) War Diary, EUSAK, G–3 Sec, 11 and 14 Nov 50.

[52] RAD, 172100, CG EUSAK to CG IX Corps, CG X Corps, and CG ROKA, 17 Nov 50.

[53] Rad, C 69211, CINCUNC to DA, 18 Nov 50.

[54] (1) Rad, CX 69453, CINCFE to CG FEAF, 21 Nov 50. (2) Rad, CG FEAF to CINCFE, 22 Nov 50.

Department of the Army on the day before the attack that he felt the Chinese Communist Army was having supply problems of its own and intimated that, if the Chinese did try to stop Eighth Army, they would be at a disadvantage. He told Washington military authorities that the Chinese had "embarked on their Korean venture in some cases with only three days rations" and that constant contact with U.N. ground forces and the pounding from American air had undoubtedly depleted the enemy's ammunition reserves. "Constant United Nations pressure along the entire line during the past few weeks," Willoughby stated, "should make it perfectly clear to the Reds that this drain on fire power is certainly not apt to be decreased but increased." He did not consider it likely that the Chinese high command would make any appreciable effort to alleviate the supply shortages of their forces, ". . . as the Chinese have always been, by western standards, notoriously poor providers for their soldiers." On the day of the jump-off, 24 November, Willoughby's intelligence staff predicted that the U.N. forces were opposed, in Korea, by 82,799 North Korean soldiers and a Chinese Communist military force of between 40,000 and 70,935.[55]

In a communiqué issued only hours before Walker's divisions started northward, the United Nations commander sketched an optimistic picture of what he referred to as his "massive compression envelopment." He felt that the Air Force had sharply curtailed enemy reinforcement and resupply. General Almond's forces had "reached a commanding enveloping position cutting in two the northern reaches of the enemy's geographical potential," and Walker's forces were now to move forward to "complete the compression and close the vise." "If successful," General MacArthur declared, "this should for all practical purposes end the war." [56]

---

[55] Telecons, TT 4058 and TT 4063, DA and GHQ, 24–25 Nov 50.

[56] Communiqué No. 12, GHQ UNC, 24 Nov 50.

CHAPTER XV

Facing New Dilemmas

Intervention

The Eighth Army moved forward as scheduled on 24 November, and against light to moderate resistance registered gains of as much as twelve miles during the first thirty-six hours. (*Map IV*) MacArthur's G–2, General Willoughby, forecast confidently on 26 November that "Should the enemy persist in his present non-aggressive attitude and withdraw, he may find sanctuary behind the Yalu River." But in terms more prophetic than he knew, he added that "Should the enemy elect to fight in the interior valleys, a slowing down of the United Nations offensive may result." [1]

Beginning shortly after dark on 25 November, strong Chinese forces struck suddenly and hard at General Walker's central and eastern units. (*Map V*) The ROK II Corps, at Walker's right, scattered before the vicious onslaught. The IX Corps, in the center, reeled, held briefly, then gave ground. On Walker's left, the I Corps, under no pressure except at its east flank, withdrew in coordination with the IX Corps' rearward moves.[2]

Walker notified Tokyo at noon on 27 November that the Chinese were attacking in strength, but that it was too early to tell if the Chinese meant to sustain their attacks. On the following day, he reported that the enemy attack force numbered some 200,000, all of them apparently Chinese, and that he was no longer in doubt that the Chinese had opened a general offensive.[3] The Chinese broadened their offensive on 27 November with attacks against the X Corps. General Almond's Marine troops had scarcely begun their advance toward Mup'yong-ni on the 27th before they met strong resistance; and on the 28th Chinese units slipped southeastward past the Marines and cut their supply route.

This wide display of Chinese strength also swept away General MacArthur's doubts. "No pretext of minor support under the guise of volunteerism or other

---

1 Telecon, TT 4063, DA and GHQ, FEC, UNC, 26 Nov 50.

2 Details of the Chinese offensive and subsequent actions may be found in B. C. Mossman, Ebb and Flow, presently in preparation as part of the UNITED STATES ARMY IN THE KOREAN WAR series.

3 (1) Telecon, Gen Hickey and Col Landrum, 1225, 27 Nov 50, in GHQ, UNC files. (2) Rad, G 30065 KGOO, CG Eighth Army to CINCFE, 28 Nov 50.

U.N. CD POSITION, EVENING 26 NOV (X CORPS)
ELEVATIONS IN METERS
200
500
1000 AND ABOVE
10
20
30
40 MILES
10
20
30
40 KILOMETERS
Lin-chiang
Yalu R
CHINA
KOREA
Manp'ojin
Ch'osan
SUIHO RESERVOIR
SUIHO DAM
Yalu R
Sakchu
Mup'yong-ni
Changjin R
Pujon R
1 MAR XX
Toksil-li
16 NOV
PUJON (FUSEN) RESERVOIR
CHANGJIN (CHOSIN) RESERVOIR
Yudam-ni
Hagaru-ri
Kot'o-ri
Chinhung-ni
Samak-ko
Sudong
Majon-dong
Huich'on
Onjong
Unsan
ROK II XXX U.S. X
Chunghung-ni
Ch'ongch'on R
Taech'on
Sinup
Hamhung
Yongwon
U.S. IX XXX ROK II
I XXX IX
Tokch'on
Ch'ongju
Kunu-ri
Kwangch'on
14 NOV
Sinanju
U.S. I XXX ROK II
Taedong R
Pukchang-ni
Yonghung
EIGHTH XXXX U.S. X
Sukch'on
Sunch'on
Paeksong-ni
Sinch'ang-ni

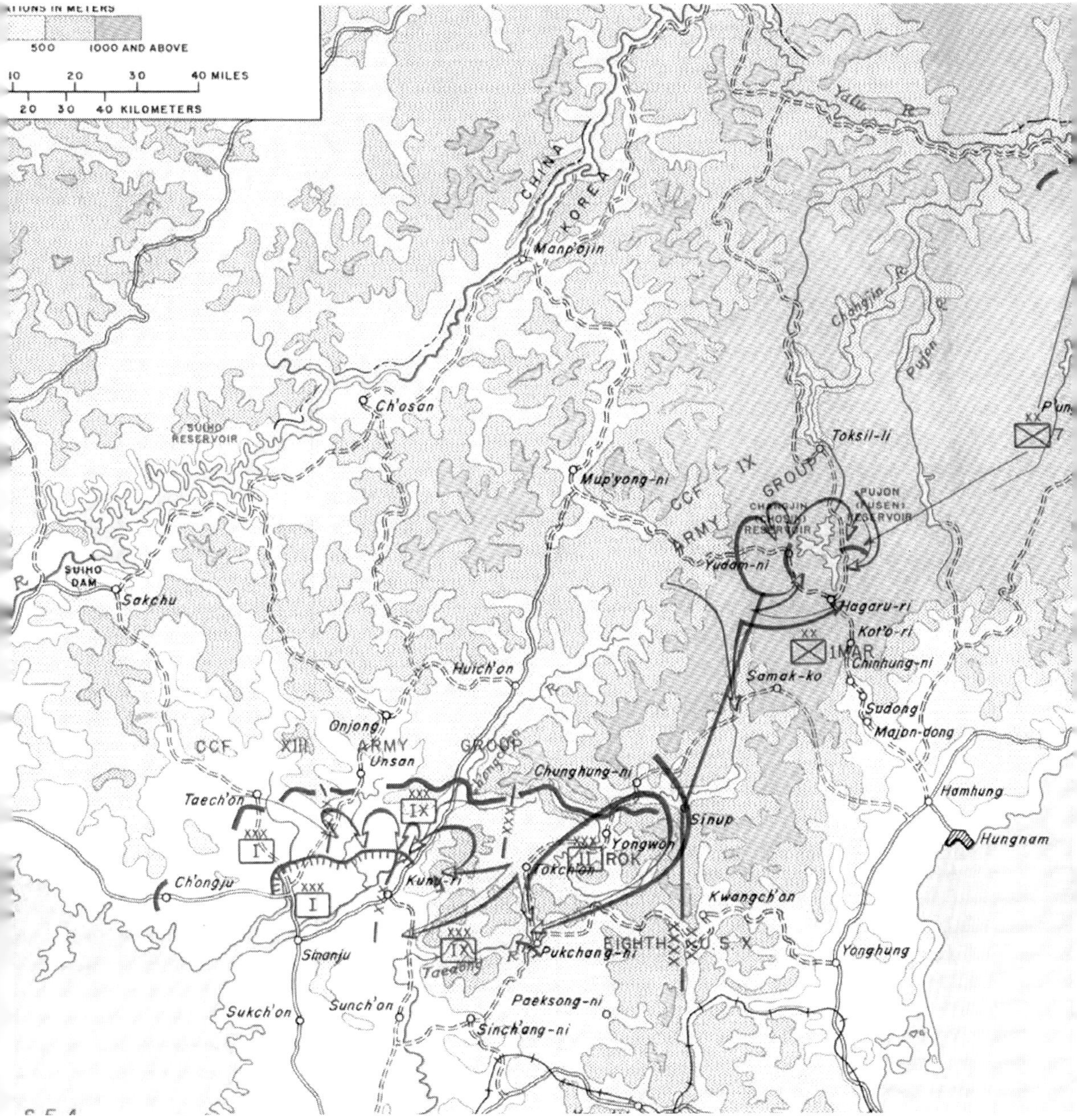

ATIONS IN METERS
500
1000 AND ABOVE
10 20 30 40 MILES
20 30 40 KILOMETERS
CHINA
KOREA
Yalu R.
Manp'ojin
Changjin R.
Pujon R.
Ch'osan
SUIHO RESERVOIR
Mup'yong-ni
Toksil-li
CCF IX ARMY GROUP
PUJON (FUSEN) RESERVOIR
CHANGJIN (CHOSIN) RESERVOIR
Yudam-ni
Hagaru-ri
Kot'o-ri
1MAR
Chinhung-ni
Samak-ko
Sudong
Majon-dong
SUIHO DAM
Sakchu
Huich'on
Onjong
CCF XIII ARMY GROUP
Unsan
Chunghung-ni
Taech'on
IX
Sinup
Yongwon
II ROK
Tokch'on
Kunu-ri
Ch'ongju
I
Kwangch'an
Hamhung
Hungnam
Sinanju
IX
Taedong R.
Pukchang-ni
EIGHTH
U. S. X
Yonghung
Sukch'on
Sunch'on
Paeksong-ni
Sinch'ang-ni

subterfuge now has the slightest validity. We face an entirely new war." [4] Instead of fighting fragments of the North Korean Army reinforced by token Chinese forces, Walker and Almond apparently now faced a total Chinese force of about 300,000.[5]

MacArthur claimed that Walker's 24 November advance had forced the Chinese to attack prematurely, theorizing that the Chinese originally planned to launch their offensive in the spring of 1951 when better weather and greater supply and troop resources would be at hand. But even if his claim were correct, the Chinese attack gave MacArthur no real advantage. For he was finding it increasingly difficult, and so admitted to Washington, to interdict enemy routes of reinforcement and resupply from Manchuria because the Yalu River was now freezing hard enough to permit the Chinese to cross without using bridges. Furthermore:

> It is quite evident that our present strength of force is not sufficient to meet this undeclared war by the Chinese with the inherent advantages which accrue thereby to them. The resulting situation presents an entire new picture which broadens the potentialities to world-embracing considerations beyond the sphere of decision by the Theater Commander.

Having thus shifted responsibility for the next decision to Washington, MacArthur announced that for the time being he intended to pass from the offensive to the defensive, making local adjustments as the ground situation required.[6]

As reflected in MacArthur's abrupt change in tactics, the opening episode of the Chinese offensive had reversed the course of the war. The Chinese opening success was due largely to the skillful execution of well-laid plans, in particular to the achievement of complete surprise. That surprise was not wholly the result of superior Chinese camouflage and march discipline. Intelligence received by MacArthur and his senior commanders had been incompatible and inconclusive. But this intelligence did provide clear warnings that Chinese forces were poised between United Nations troops and the northern border of Korea. Much of the surprise achieved by the Chinese stemmed from the tendency of U.N. leaders to discount these warnings.

As a defense of his own judgment and the efforts of the theater intelligence officers MacArthur insisted ". . . that the intelligence that a nation is going to launch war, is not an intelligence that is available to a commander, limited to a small area of combat. That intelligence should have been given to me." [7]

But at the national level, authorities declared that Chinese intentions had not been sufficiently clear to permit a definite judgment. General Bradley, in fact, maintained that the Department of Defense had had no intelligence that the Chinese would enter the war.[8]

The daily estimates given MacArthur by his own intelligence staff were supplemented by General Willoughby in private presentations. Whether Willoughby gave MacArthur different information from that contained in the daily estimates is not known, but beyond doubt, Willoughby's presentations amplified the

[4] Rad, C 69953, CINCFE to JCS, 28 Nov 50.
[5] *Ibid.*
[6] *Ibid.*

[7] *MacArthur Hearings*, p. 18.
[8] *MacArthur Hearings*, p. 759.

routine staff reports. On the assumption that the G–2's published estimates and personal briefings were similar, MacArthur must have learned of the enemy's capabilities and the order in which those capabilities might be employed. But he possibly found the reports also puzzling and contradictory. On 15 November, Willoughby's staff forecast that the most likely sequence of enemy moves would be (1) Conduct of offensive operations. (2) Reinforce with communist forces from outside Korea. (3) Conduct guerrilla operations. (4) Defend. Then, in amplification of this forecast, Willoughby's officers reported:

> Information received from Chinese Nationalist military sources, during the past few days gives strong support to an assumption that the Chinese Communists intend to "throw the book" at United Nations forces in Korea. . . . It is fast becoming apparent that an excessive number of troops are entering Northeast China. . . . Such a marshalling of troops cannot be explained in terms of redeployment . . . or demobilization. It seems doubtful that the Chinese Communists, if intending to intervene in Korea would wait until this late in the war. On the other hand it seems incredible that the Chinese Communists have deluded themselves with their own propaganda and fear a United States attack on Manchuria.[9]

Such contradictions could scarcely have been of much help to MacArthur in deciding for himself what the enemy most probably would do.

On 16 November, General MacArthur was told that the Chinese Communists had probably deployed twelve divisions of trained soldiers in Korea. Three days later he was told that "it would appear logical to conclude that Chinese Communist leaders are preparing their people psychologically for war." On the same day that the Eighth Army struck northward in its general offensive, Willoughby reported that "Even though Chinese Communist strategy may not favor an immediate full-scale war, preparations for such an eventuality appear to be in progress." [10]

On the second day of the attack, before the Chinese had fully committed themselves, Far East Command intelligence officers changed the predicted order of enemy courses of action, placing reinforcement from outside Korea at the top of the list, and the conduct of offensive operations in second place. But again MacArthur received contradictory estimates of Chinese intentions when he was told:

> Although too early for concrete evaluation, there are some indications of a withdrawal of Chinese Communist forces to the Yalu or across the border into Manchuria. . . On the other hand, there are many reports of Chinese Communist plans to strengthen their intervention forces now in Korea and all indications point to a heavy troop build-up in Northeast China and Manchuria. . . . Also there are many indications that the Chinese Communists will stubbornly defend reservoir and power installations along the Yalu. . . .[11]

These vacillatory daily reports contained too many qualifying clauses to permit a positive forecast. But in the sense that they indicated a continuing

[9] (1) DIS, GHQ, FEC No. 2989, 15 Nov 50. (2) General Willoughby's later explanation of intelligence failure sheds no light on the mystery but is of interest to readers wishing greater detail. See Willoughby, *MacArthur, 1941–1951*, pp. 378–90.

[10] (1) DIS, GHQ, FEC, No. 2990, 16 Nov 50. (2) DIS GHQ, FEC, No. 2993, 19 Nov 50. (3) DIS GHQ, FEC, No. 2998, 24 Nov 50. On 16 November, about thirty CCF divisions were in Korea.

[11] DIS, GHQ, FEC, No. 2999, 25 Nov 50.

Chinese build-up in Manchuria and Korea and that they did point out psychological and other preparations for an offensive against United Nations forces, these reports possessed some validity.

The reasons that prompted General MacArthur to persist in his drive to the border in the face of the very obvious Chinese potential to meet his advance with considerable military force, must remain conjectural. He was aware of the presence in Korea of substantial numbers of Chinese soldiers; and his own staff had warned him of the great Chinese potential for immediate reinforcement. He had never been told, however, that the enemy had as many divisions in Korea as actually were present.

MacArthur's determination to pursue his mission to the bitter end appears to have had its basis in three concepts. First, MacArthur apparently thought that the Chinese build-up and threatening posture were part of a gigantic bluff and that the Chinese, since they could not afford to go to war with the United States, would not attack his forces. The tenor of his message to the Joint Chiefs of Staff on 24 November, turning down their suggestion that he establish a holding line some distance short of the Yalu, clearly shows that he minimized the Chinese menace.[12] He felt, it is also clear, that it was pretty late in the day for the Chinese to be entering Korea, and that if they had been serious in their intentions they would have intervened when United Nations forces were still in the vicinity of the 38th Parallel. Willoughby, too, doubted ". . . that the Chinese Communists, if intending to intervene in Korea, would wait until this late in the war." [13]

MacArthur did not fear the Chinese and felt that in the event he was mistaken and the Chinese were not bluffing, his forces were capable of taking care of both the Chinese and the North Koreans. For, as noted earlier, he had pointed out to the Joint Chiefs of Staff that Chinese Communist entry was a risk taken in full knowledge of the situation. Chinese entry during the Pusan days would have been extremely dangerous. But now that the UNC held the initiative and had less area to defend the risk was much smaller.[14] Only a few days before, he had told the Joint Chiefs that his air forces had succeeded in isolating the battle area and in cutting off enemy reinforcement and choking off enemy supply.[15] MacArthur seems to have overestimated the power of his own command vis-à-vis the Chinese. Both the Eighth Army and the X Corps, he reasoned, were victorious, battle-tested military forces. His naval and air forces gave him complete control of the sea and air. Furthermore, Chinese troops, during World War II, had proven inferior to Japanese troops and thus, by inference, to American troops.

A significant factor was MacArthur's belief that his air power could isolate the battlefield. MacArthur still persisted in this view on the eve of the attack to the Yalu. He announced on 24 November, "My air force for the past three weeks, in a sustained attack of model coordination and effectiveness, successfully interdicted enemy lines of support from the

[12] Rad, C 69808, CINCUNC to DA, 25 Nov 50.

[13] DIS, GHQ, FEC, No. 2998, 15 Nov 50.

[14] Rad, C 69808, CINCUNC to DA, 25 Nov 50.

[15] Rad, C 69211, CINCUNC to DA, 18 Nov 50.

north so that further reinforcement therefrom has been sharply curtailed and essential supplies markedly limited." General Wright contends that this belief in the effectiveness of air power was one of General MacArthur's greatest weaknesses in dealing with the Chinese.[16]

But, from all indications, the overriding consideration in MacArthur's decision to push on to the Yalu was his firm conviction that his mission, "the destruction of the North Korean Armed Forces," dictated his line of action, and could be accomplished only by an advance to the border. This mission, in spite of noticeable tendencies on the part of Washington toward its modification, was not altered, largely because of MacArthur's vehement protests during November.[17] When the Joint Chiefs of Staff had told MacArthur on 8 November that ". . . this new situation indicates that your objective . . . the destruction of the North Korean armed forces may have to be re-examined,"[18] MacArthur retorted in extremely strong terms that any course short of complete destruction of the enemy would be tantamount to abject surrender and a breaking of faith with the peoples of Asia.[19]

There is little doubt that MacArthur ardently believed in his mission and that he was more than willing to call what he regarded as a Chinese bluff in order to carry out that mission. He may well have recalled those tenets of American military doctrine which hold that "the mission is the basic factor in the commander's estimate," and that "to delay action in an emergency because of incomplete information shows a lack of energetic leadership, and may result in lost opportunities. The situation, at times, may require the taking of calculated risks." This is borne out by his explanation later of his northward advance as a "reconnaissance in force." He stated the alternatives which faced him on 24 November. ". . . One," he testified, "was to ascertain the truth of the strength of what he [the enemy] had; the other was to sit where we were. Had we done that he would have built up his forces, and undoubtedly destroyed us. The third was to go in precipitate retreat, which would not have been countenanced, I am quite sure."[20]

MacArthur also vindicated his advance by insisting that ". . . the disposition of those troops [Eighth Army and X Corps], in my opinion, could not have been improved upon, had I known the Chinese were going to attack."[21] Actually, the Eighth Army, when hit by the Chinese, was deployed on a broad front with its right flank open and was supported by few reserves. Almond's corps was strung out in widely separated columns advancing through extremely rugged terrain. Not only was the X Corps' left flank unprotected, but Chinese forces of considerable strength had been reported on that flank.

Emergency Meeting—Tokyo

General MacArthur called an emergency council of war in Tokyo on the night of 28 November. Generals Walker

---

[16] (1) GHQ, UNC Communiqué No. 12, 24 Nov 50. (2) Interv, Gen Wright with Col Appleman, copy in OCMH.

[17] Rad, JCS 69808, CINCUNC to JCS, 25 Nov 50.

[18] Rad, JCS 96060, JCS to CINCFE, 8 Nov 50.

[19] Rad, C 68572, CINCFE to JCS, 9 Nov 50.

[20] *MacArthur Hearings*, pp. 20–21.

[21] *Ibid.*, p. 19.

and Almond, hastily summoned from Korea, joined MacArthur, Hickey, Wright, Willoughby, and Whitney at MacArthur's American Embassy residence. In a meeting which lasted from 2150, 28 November, until 0130, 29 November, the seven officers studied the possible countermoves in meeting the entry of the Chinese. MacArthur, feeling that above all he must save his forces, finally ordered Walker to make withdrawals as necessary to keep the Chinese from outflanking him and directed Almond to maintain contact with the Chinese but to contract the X Corps into the Hamhung-Hungnam area.

Since the Eighth Army seemed in greater danger than Almond's corps, the main theme of the conference appears to have been "What can X Corps do to help Eighth Army?" When General Almond held that his first mission was to extricate the Marine and Army forces cut off in the Changjin Reservoir area, MacArthur agreed but asked Almond what he could do to relieve the Chinese pressure on Walker's right flank. General Wright suggested that Almond might send the U.S. 3d Division west across the Taebaek mountain range to join Eighth Army and to attack Chinese forces moving in on Walker's right flank. Pointing out that the road across the Taebaek Range appeared on the map but was actually nonexistent, Almond objected that the bitter winter weather and the possibility of strong Chinese forces in the gap between the two commands would make any such relief expedition an extremely hazardous venture in which the whole 3d Division might be lost. But he agreed to the scheme if Eighth Army would supply the division after it crossed to the west side of the Taebaek Mountains. Walker made no such promise, and MacArthur made no immediate decision on the attack. He later ordered, then canceled, a drive by a task force from the 3d Division to link up with the Eighth Army right flank.[22]

In Washington, meanwhile, the Joint Chiefs of Staff accepted General MacArthur's appraisal of the situation and approved his plans for passing from the offensive to the defensive. For some time, Admiral Sherman, Chief of Naval Operations, had been expressing concern over MacArthur's operations in northeastern Korea. In his opinion, the employment of U.S. naval vessels in support of the X Corps so close to Vladivostok offered the Russians a tempting pretext for intervening if they were so inclined. He insisted that the X Corps should be withdrawn to a general consolidated defense line.

On 29 November, the Joint Chiefs of Staff told MacArthur to put aside any previous directives in conflict with his current plan to defend. After calling attention to the need for co-ordinating Eighth Army and X Corps operations, the Joint Chiefs suggested that MacArthur should close the gap, more than thirty airline miles in width in the beginning and now widening, between Walker and Almond and form a continuous defense line across the peninsula.[23] But MacArthur differed with the Joint Chiefs. According to him, X Corps

---

[22] (1) Rad, C 50106, CINCFE to CG X Corps, 30 Nov 50. (2) Interv, Col Appleman with Almond, copy in OCMH.

[23] (1) Rad, JCS 97592, JCS to CINCFE, 29 Nov 50. (2) Memo, CNO for JCS, 29 Nov 50, sub: Sit in Northeast Korea.

units "geographically threatened" the main supply lines of enemy forces bearing down upon the right flank of the Eighth Army. He maintained that the Chinese had been forced to commit an estimated eight divisions to ward off X Corps thrusts against their supply lines, thus depriving them of eight divisions to throw against the Eighth Army. So long as the X Corps stayed in this position, MacArthur insisted, the Chinese could not, with any degree of safety or assurance of success, penetrate to the south through the existing corridor. He pointed out also the great difficulties of closing the gap.

> Any concept of actual physical combination of the forces of the Eighth Army and X Corps in a practically continuous line across the narrow neck of Korea is quite impracticable due to the length of this line, the numerical weakness of our forces, and the logistical problems created by the mountainous divide which splits such a front from north to south.[24]

As to the immediate situation within the X Corps, General MacArthur informed his superiors that he had ordered Almond to pull his forces into the Hamhung-Hungnam sector. Almond had been specifically warned against allowing any piecemeal isolation and trapping of his forces. These forces were already fighting their way out of isolation and entrapment. MacArthur believed that, while the X Corps might seem overextended, the terrain conditions would make it extremely difficult for the Chinese Army to take any advantage of this fact.[25]

In a second message a few hours later, MacArthur gloomily predicted that the Eighth Army would not be able to make a stand in the foreseeable future and would ". . . successively have to replace to the rear." He had now concluded that the Chinese intended to destroy the U.N. forces completely and to secure all of Korea.[26]

General MacArthur's disclosure of plans for pulling the X Corps back into the Hamhung-Hungnam sector and his forecast of more withdrawals by the Eighth Army only increased the Joint Chiefs' concern. MacArthur's citation of the formidable mountainous terrain as a deterrent to enemy advances in strength was nullified, in the opinion of the Joint Chiefs of Staff, by the Chinese demonstrated ability to negotiate rugged land barriers. The compression of the X Corps, with the accompanying development of a progressively widening gap between Almond's and Walker's forces would, the Joint Chiefs told MacArthur on 1 December, afford the Chinese additional opportunity to move strong forces southward between the Eighth Army and X Corps. The first task, in their opinion, was to extricate the Marines and 7th Division troops from the Changjin Reservoir. But once that was done, they wanted the Eighth Army and X Corps "sufficiently coordinated to prevent large enemy forces from passing between them or outflanking either of them." Following their custom of not directing MacArthur's tactical disposition by specific orders, the Joint Chiefs of Staff discreetly but pointedly suggested once again that he join his forces in a defensive line across the peninsula.[27]

[24] Rad, C 50095, CINCUNC to DA for JCS, 30 Nov 50.
[25] *Ibid.*
[26] Rad, C 50105, CINCFE to DA, 30 Nov 50.
[27] Rad, JCS 97772, JCS to CINCFE, 1 Dec 50.

But MacArthur remained solidly against any junction of the Eighth Army and X Corps at this time. Joining the two forces, he explained to his Washington superiors, would produce no significant added strength. It would, on the other hand, endanger the freedom of maneuver deriving from their separate lines of supply by sea.[28]

General MacArthur again offered other objections. He called the development of a defense line across the waist of the peninsula infeasible because of the numerical weakness of his forces and the distances involved. He called to the attention of the Joint Chiefs of Staff the necessity of supplying his two major forces from ports within their respective areas; and he reminded them of the geographical division of the peninsula into two compartments by the Taebaek mountain range which ran north and south between the Eighth Army and X Corps. A continuous line across the narrow waist would be approximately 120 airline miles, or 150 road miles, in length. MacArthur explained:

> If the entire United States force of seven divisions at my disposal were placed along this defensive line it would mean that a division would be forced to protect a front of approximately twenty miles against greatly superior numbers of an enemy whose greatest strength is a potential for night infiltration through rugged terrain. Such a line with no depth, would have little strength, and as a defensive concept would invite penetration with resultant envelopment and piecemeal destruction.

MacArthur apparently had changed his mind about the ability of the Chinese to operate over the rough terrain in the gap between the X Corps and Eighth Army.[29]

MacArthur doubted that the Joint Chiefs fully realized the great changes wrought by the Chinese entry. He tabulated for them the latest of his intelligence agencies' estimates of enemy strength. Twenty-six Chinese divisions had been identified in combat and an additional 200,000 men were either in reserve or being committed. This formidable array of enemy strength was further augmented by the remnants of the North Korean Army which were being reorganized in rear areas. ". . . There stands, of course, behind all this, the entire military potential of Communist China." [30]

The terrain on which the fighting was taking place was having a twofold effect on the course of battle. It diminished the effectiveness of MacArthur's air arm in trying to channelize and interrupt the Chinese system of supply. Secondly, the rough ground aided the enemy in his dispersion tactics. These drawbacks, MacArthur maintained, greatly reduced the normal benefits which would be expected from complete control of the air. His naval potential, too, was greatly minimized by the concentration of enemy forces in areas inaccessible to naval gunfire. Under these circumstances, MacArthur held, the potential destructive force of the United Nations combined arms was greatly reduced and the question was becoming more and more one of the relative combat effectiveness of ground forces.[31]

Commenting on the condition of his

---

28 Rad, C 50332, CINCUNC to DA for JCS, 3 Dec 50.

29 *Ibid.*

30 *Ibid.*

31 *Ibid.*

own forces, MacArthur pointed out that while they so far had exhibited good morale and marked efficiency, they had been in almost incessant combat for five months and were mentally fatigued and physically battered. Moreover, with the exception of the 1st Marine Division, each American division then in Korea was at least 5,000 men understrength. The Chinese troops, on the other hand, appeared to be fresh, very well organized, splendidly trained and equipped, and apparently in peak condition for actual operations.[32]

MacArthur concluded that unless he promptly got ground reinforcements of the greatest magnitude, his command would be forced either into successive withdrawals with diminishing powers of resistance after each such move, or into taking up beachhead positions which, while insuring a degree of prolonged resistance, would afford little hope of anything beyond defense. He charged that his directives were now completely outmoded. The strategic concepts which had been evolved for operations against the North Korean Army were not suitable for continued application against the full power of the Chinese. Without being specific, MacArthur then called for sterner measures than he was then authorized to employ. "This calls for political decisions and strategic plans in implementation thereof, adequate fully to meet the realities involved," he declared. "In this, time is of the essence, as every hour sees the enemy power increase and ours decline."[33] In clear terms, the United Nations commander issued a prognosis which expressed his pessimism unmistakably. He told the authorities in Washington:

> This small command, actually under present conditions, is facing the entire Chinese nation in an undeclared war, and, unless some positive and immediate action is taken, hope for success cannot be justified and steady attrition leading to final destruction can reasonably be contemplated.[34]

Faced with General MacArthur's strong objection to a defensive line across the peninsula, the Joint Chiefs of Staff yielded to his judgment. On 4 December, after obtaining President Truman's approval, they told MacArthur that they now regarded the preservation of his forces as the primary consideration and agreed to the consolidation of forces into beachheads.[35]

Eight thousand miles away from the fighting, however, the Joint Chiefs of Staff could not fully sense conditions in Korea. Consequently, General Collins flew to the Far East for conferences with MacArthur, Walker, and Almond and for a firsthand view of the battle. Collins intended to find out from MacArthur what chance he had to defend successfully, what general line or area he could hold, and for how long. Secondly, Collins wanted to obtain MacArthur's opinion of a cease-fire.[36]

Collins Visits Korea

Collins arrived in Tokyo on the morning of 4 December, conferred briefly with General MacArthur, then flew to

[32] *Ibid.*

[33] *Ibid.*

[34] *Ibid.*

[35] (1) Rad, JCS 97917, JCS to CINCFE, 4 Dec 50. (2) Truman, *Memoirs*, II, 393.

[36] Draft msg for JCS Representative, 1 Dec 50, in G-3, DA file 091 Korea, Case 127/8.

Korea for talks with General Walker and to inspect the Eighth Army's lines. Walker's troops had been withdrawing southward as agreed at the 28 November Tokyo conference, and although enemy pressure had lessened, were, at the time of Collins' inspection, dropping back below Sukch'on and Sunch'on to positions not far north of P'yongyang. Walker had already told MacArthur that he could not hold P'yongyang and estimated that the enemy would unquestionably force him to pull south of the 38th Parallel to the vicinity of Seoul.

Walker told Collins that he could continue the withdrawal without serious losses unless he were ordered to defend the Seoul-Inch'on area. If this happened the Chinese could encircle him. Walker felt, and General Collins concurred, that an evacuation from Inch'on would be very costly. If evacuation became necessary, Walker wanted to withdraw from Pusan, not Inch'on. He was confident he could get his forces safely into the Pusan area, and even considered it possible that he could hold there indefinitely if the X Corps reinforced him.[37]

Walker's troops passed below P'yongyang on 5 December, destroying many supplies there and falling back to new positions to the south. On the next day, General Collins flew to Hamhung to see General Almond. He found Almond confident that he could hold the Hamhung-Hungnam area for a considerable time without serious losses, and that he could withdraw successfully and cheaply when so ordered. Collins agreed with Almond's estimate.

Returning to Tokyo for a final conference on 6 December, Collins met with General MacArthur, Admiral Joy, and General Stratemeyer, and with key staff officers, Hickey, Willoughby, and Wright, for a full discussion of what moves to take against the Chinese. As a framework for their talks, they projected three hypothetical situations covering the next few weeks or months.

In the first, they posited that the Chinese would continue their all-out attack, but with MacArthur forbidden to mount air attacks against China; that no blockade of China would be set up; that no reinforcements would be sent to Korea by Chiang Kai-shek; that there would be no substantial increase in MacArthur's U.S. forces until April 1951 when four National Guard divisions might be sent MacArthur; and that the atomic bomb might be used in North Korea. General MacArthur spoke strongly, charging that placement of such limitations on his command while it remained under strong Chinese attack would represent essentially a surrender. Under these conditions the question of an armistice would be a political matter, helpful perhaps, but certainly not requisite from a military standpoint. His forces would have to be withdrawn from Korea in any case, and the United States should therefore not be hasty in seeking an armistice under these conditions. He agreed with Walker and Almond, as did General Collins, that the United Nations forces could be safely withdrawn from Pusan and Hungnam respectively, with or without an armistice.

Under the second set of conditions, the conferees assumed a situation in which the Chinese attack would continue, but with an effective naval

[37] Memo, Gen Collins for JCS, 8 Dec 50, sub: Rpt on Visit to FECOM and Korea, 4–7 Dec 50.

blockade of China put in effect, air reconnaissance and bombing of the Chinese mainland allowed, Chinese Nationalist forces exploited to the maximum, and the atomic bomb to be used if tactically appropriate. Given these conditions, General MacArthur said he should be directed to hold positions in Korea as far north as possible. He would, in this case, move Almond's X Corps to Pusan to join the Eighth Army in an overland movement.

Under the third postulate, that the Chinese would agree not to cross south of the 38th Parallel, MacArthur felt the United Nations should accept an armistice. The conditions of the armistice should preclude movement of North Korean forces, as well as of Chinese, below the parallel; North Korean guerrillas should withdraw into their own territory; the Eighth Army should remain in positions covering the Seoul-Inch'on area, while X Corps pulled back to Pusan; and a United Nations commission should supervise the implementation of armistice terms. He viewed these as the best arrangements that could be made, unless the United Nations should decide to act under the second postulate. He reiterated a firm belief that the Chinese Nationalists should send troops to Korea without delay and that other powers in the United Nations should increase their contingents to a total U.N. strength of at least 75,000. He concluded by telling General Collins that unless substantial reinforcements were sent quickly to his command, the United Nations Command should pull out of Korea.[38]

MacArthur Is Warned

General MacArthur indeed felt that he was being forced to fight the Chinese with his hands tied. His resentment displayed during talks with General Collins had already welled over into public channels, giving rise to official concern and laying the groundwork for later controversy. Shortly before his 6 December meeting with Collins, MacArthur in an interview with the editors of *U.S. News and World Report* had severely criticized the restrictions placed upon his command. He called the continuing prohibition against hot pursuit and bombing Chinese bases in Manchuria "an enormous handicap, without precedent in Military History."[39]

He also had sent a message to Hugh Baillie, president of United Press, in which he again criticized the national policies under which he was operating. Disturbed by MacArthur's actions, President Truman on 5 December ordered the Joint Chiefs of Staff to inform all unified commanders that any public statement concerning foreign policy should be cleared with the Department of Defense before issuance. He aimed this action directly at MacArthur.[40] The

[38] (1) *Ibid.* (2) Memo, CofS FEC for Gen Collins, 4 Dec 50, GHQ, UNC Comd Rpt, Dec 50, Annex 4, Part III, "I" 19433.

[39] *U.S. News and World Report* (December 8, 1950), pp. 16–22.

[40] (1) New York *Times*, December 2, 1950. (2) Rad, W 98310, DA to CINCFE, 8 Dec 50. (1) and (2) are reproduced in *MacArthur Hearings*, pp. 3532–35. (3) The President later charged that MacArthur's repeated statements had led many into the impression that the United States had changed its policy. This the President would not allow. Therefore on 5 December he issued an order to all government agencies that until they received further written notice from him all speeches, press releases, or other public statements concerning foreign policy would be cleared by the Department of State before issuance. In a second such notice he admonished all

effect of these instructions became apparent almost immediately when MacArthur on 9 December submitted for approval by the Department of Defense a communiqué which he proposed to release to the press. MacArthur stated that his forces had successfully completed tactical withdrawals and were now waiting for "political decisions and policies demanded by the entry of Communist China into the war." He continued, "The suggestion widely broadcast that the command has suffered a rout or debacle is pure nonsense." He charged that "Advance notice of the Chinese decision to attack was a matter for political intelligence which failed. . . . Field intelligence was so handicapped that once the decision to commit was made, this new enemy could move forward . . . without fear of detection. . . ." [41]

The Joint Chiefs of Staff took exception and replied in a chiding vein that the proposed release did not conform to the President's instructions and that in the future he should confine such communiqués to completed phases of military operations. "Discussion of foreign and military policy referenced to press comments and comments relative to political or domestic matters should not be included in military communiqués issued in the field," they pointed out.[42]

officials overseas, including military commanders, to use caution in their public statements, to clear all but routine statements with their departments, and to engage in no direct communication on military or foreign policy matters with newspapers, magazines, or other publicity media in the United States. See Truman, *Memoirs*, II, 383.

[41] Rad, C 50736, CINCFE to DA, 8 Dec 50.

[42] Rad, JCS 98410, JCS to CINCFE, 9 Dec 50.

High Level Conferences

Collins, upon his return to Washington, told the Joint Chiefs of Staff, "If the United Nations decision is not to continue an all out attack in Korea and if the Chinese Communists continue to attack, MacArthur should be directed to take the necessary steps to prevent the destruction of his forces pending final evacuation from Korea." [43]

No one was more aware of the need for wise decisions than was the President, who now had to make them. The military collapse in Korea had effectively erased any illusion that the current national policies would continue to serve national interests. The crises evolving from the military failure in Korea and the even more ominous catastrophe of open and unrestrained military action by the Chinese had shocked Washington as well as Tokyo. But this was not the time for hasty judgments. In the tinder-dry international atmosphere the wrong decision could fan the brush fire in Korea into a flaming worldwide holocaust. Since a bad decision would be worse than no decision at all, the President and his advisers moved slowly and with great thoroughness and caution in their search for the right answers.

President Truman first learned of the Chinese assault early on 28 November when General Bradley gave him MacArthur's first reports of the crushing blows being dealt his forces. The Joint Chiefs of Staff, already alerted by the increasing resistance to the Eighth Army's advance, had been watching the situation closely all during the preceding day. President Truman called immediately

[43] Memo, Collins for JCS, 8 Dec 50.

for a meeting of the National Security Council.[44]

General Bradley told the council that the Joint Chiefs of Staff considered the new turn in Korea to be very serious but not so devastating as newspaper reports indicated. He stressed the dangers of MacArthur's command being attacked from Manchurian airfields, but advised against authorizing MacArthur to bomb those airfields.

Secretary of Defense Marshall recommended in strongest terms that neither the United States nor the United Nations become involved in a general war with China. He was joined in this view by all the service secretaries and by the Joint Chiefs of Staff. Reflecting the emphasis which U.S. military planners were placing on the defense of western Europe, General Bradley warned that if the United States allowed itself to become embroiled in an all-out fight with China it would not be able to continue the build-up of forces in Europe. General Marshall added that it was essential for the United States, in dealing with this new and very serious aggression, to keep strictly within the framework of the United Nations, regardless of the difficulties which might arise.

The council was in general agreement that the Chinese intervention made it extremely urgent for the United States to build up its military forces and to enlarge its efforts to procure both men and materials. The President also agreed with this view and with the necessity for sending to Congress a supplementary budget to take care of the increased costs of greater military readiness.

The Secretary of State told the council that the Chinese attack on U.N. forces had moved the United States much closer to general war. Always to be kept in mind in approaching the Korean problem, he stressed, was that the real antagonist, the power behind the scene in Korea and elsewhere, was Russia. Therefore, any action that President Truman might eventually take must be taken with full knowledge that a war with Russia could be the result. If, for example, the United States successfully bombed Chinese airfields in Manchuria, the President believed that Russia would have cheerfully entered the fight.[45]

The foreign policy of the United States was, and had been for the past three years, predicated on containing the USSR within its 1947 limits. Now, if the United States took action in or against Chinese territory and entered the USSR's perimeter of special influence and interest, it would risk a war it might not win. Acheson believed that there were a number of ways in which the United States could damage the Chinese without going to war with them, although he did not enumerate these means. But he had concluded that it would be best for the United States to find some way to end the fighting in Korea.

Over the next several days, President Truman held more meetings with his top advisers and with Congressional leaders but made no decisions on courses of action. MacArthur, in his call for political decisions, had not explained what he meant by the term. But certainly new

[44] (1) Truman, *Memoirs*, II, 385–87. (2) The following section is based, in addition to specific citation, on an interview by the author with Mr. Truman at Independence, Missouri, in June 1961.

[45] Truman, *Memoirs*, II, 387–88.

national policies must now be established with regard to such questions as whether to retaliate against China and whether to attempt to negotiate a cease-fire. Vital in answering these questions was the determination of USSR intentions, of the price which the United States could afford to pay for a cease-fire, of the effects of voluntary withdrawal from Korea, and of the effect of all these matters upon America's allies, particularly those of western Europe.[46]

Although General MacArthur had not yet expressed strong public views on the subject, his obvious opinion that the United States should strike back at China, particularly manifest in his readiness to attack the Yalu installations in early November, had aroused the nations allied with the United States in NATO. NATO had existed more than a year and a half, but its military forces were still in the planning stages. The forces of the member nations in western Europe were at the time few in number, poorly equipped, badly trained, and inadequately supported. They would be virtually defenseless against the Soviet Union, or even some of its satellite nations. Without American aid, in event of an attack from the east, the nations of western Europe could anticipate much the same fate that had befallen South Korea in the first days of the Korean War. These nations, particularly England, France, and the Benelux countries, had therefore welcomed the Presidential decision to oppose the North Korean venture with armed force. Until then there had been real doubt among these nations that the United States would actually fulfill its NATO obligation should one of them be attacked.[47] These doubts had been largely removed by the President's action in June.

But the extension of the Korean War through Chinese intervention and the very possibility of increased and more drastic action by the United States, brought another fear that if the United States became involved in a war with Communist China, American commitments to NATO would, through sheer necessity, go by the board. China then might have little difficulty in persuading Russia to move into western Europe; and without U.S. resistance to this aggression, Russia could take all of Europe at little cost.[48]

---

[46] Like so many other decisions of the Korean War, the decision to evacuate Korea defied classification as either political or military. Ostensibly military, the decision nevertheless held profound political implications such as its effect on the South Koreans, the Japanese, the Formosans, and U.S. allies in other areas of the world.

[47] Article 5 of the Charter of the North Atlantic Treaty stated, "The Parties agree that an armed attack against one or more of them in Europe or North America shall be considered an attack against them all; and consequently they agree that, if such an armed attack occurs, each of them, in exercise of the right of individual or collective self defense, recognized by Article 51 of the Charter of the United Nations, will assist the Party or Parties so attacked by taking forthwith, individually and in concert with the other Parties, such action as it deems necessary, including the use of armed force, to restore and maintain the security of the North Atlantic area."

[48] Sentiment in the NATO nations was influenced measurably by exaggerated speculation in the foreign press. President Truman was aware that foreign newspapers were speculating openly about the American reaction, quoting some of the more hawkish senators and talking about General MacArthur's ill-concealed disapproval of American policy. Some of these papers actually predicted that the United States would ignore the United Nations and plunge into a war with China. Most Europeans had heard only that there was opposition to NATO and to sending American troops to Europe. See Truman, *Memoirs*, II, 394–95.

The NATO nations, people and leaders alike, distrusted General MacArthur's strategic judgment. They feared that his stature and influence might enable him to appeal so forcibly to the American people for more drastic military action as to override the more temperate approaches to the Chinese which seemed to be favored by Washington.

At times, even Washington officials set NATO nerves on edge with public statements. On 30 November, at a press conference, President Truman remarked, no doubt extemporaneously, that the use of the atomic bomb was under active consideration, unintentionally implying to some oversensitive observers that its use would be left to the discretion of General MacArthur. Even though subsequently he attempted to subdue the storm of protest and consternation which followed by pointing out that only he could authorize use of the atomic bomb and that he had not given such authorization, he could not avoid the real issue that any decision to use the bomb would be a United States, not a United Nations, decision.[49]

The United Kingdom was predominant among the anxious advocates of the NATO viewpoint. The most respected leaders of that nation, including Winston Churchill and Anthony Eden, and of the Commonwealth of Nations, were seriously disturbed by rumors that MacArthur wanted stern measures against China. As a result of the mounting tension which, conceivably, could have shattered NATO and the western bloc of the United Nations as well, Clement Attlee, Prime Minister of the United Kingdom, proposed conferences between himself and President Truman. These were quickly arranged and scheduled to begin on 4 December.[50]

In the interim, Attlee conferred with leaders of the other Commonwealth nations and with the French Premier and Foreign Minister. Apparently, he was also to represent the viewpoints of these nations in his talks with President Truman.

On the day before Attlee's arrival in Washington, and primarily in preparation for his visit, the Secretary of Defense, the Secretary of State, Presidential Adviser Averell Harriman, and the Joint Chiefs of Staff met to discuss the possible courses of action open to the United States. One matter of paramount concern was how to solve the current crisis, and at the same time preserve solidarity in the United Nations, especially with the British Commonwealth nations. At this meeting, the nation's top authorities reached general agreement that the military posture of the United States should be strengthened without delay. The Army staff was already making studies to determine what increase in production schedules and in Army forces and personnel should be made.[51]

With particular regard to Korea, one suggestion was that the United States should press the United Nations for a resolution calling for a cease-fire on the condition that the Eighth Army leave

---

[49] (1) New York *Times*, December 1, 1950. (2) Truman, *Memoirs*, II, 395–96.

[50] The President received a message from the British Prime Minister, Clement Attlee, on 30 November, asking to visit Washington and to discuss, face to face, the meaning of events in Korea and the possible courses of future action. See Truman, *Memoirs*, II, 393.

[51] Rad, WAR 97929, DA to CINCFE (Personal), for Collins from Haislip, 3 Dec 50.

North Korea and pull back across the 38th Parallel. Any such resolution would obviously have to be passed before the Chinese pushed the Eighth Army across the parallel by force. There was also much concern over the price that the Chinese would demand for agreeing to a cease-fire.

Another point discussed was whether the United Nations forces, in the absence of a cease-fire, should evacuate as soon as they had withdrawn into beachheads or wait until the enemy forced them out. The conferees also examined the no less important question of whether the United States should attack Communist China by air and sea after the United Nations were forced out of Korea. No definite recommendations for the President as to courses of action evolved from this meeting. It was agreed that such recommendations must await the return of General Collins, then still in Korea, as well as the results of the conferences which were to begin next day between President Truman and Prime Minister Attlee.

The British Prime Minister was thought to be particularly disturbed by President Truman's remarks on possible use of the atomic bomb. The British position was that the atomic bomb should certainly not be used without consultation—and probably not without agreement—with them and perhaps other members of the United Nations; and they were strongly opposed to its use in China.[52]

Just before the arrival of Attlee, Department of State officials examined the reasons behind his unusual request for a conference with President Truman and Attlee's attitude on the Korean situation. The occasion for Mr. Attlee's visit, they concluded, was "the sudden change in the situation in Korea." [53]

The growing British concern over the U.S. foreign and defense policies stemmed in part from what Department of State officials described as "the deterioration in the position of the West vis-à-vis the Soviet Union." The British did not entirely trust the discretion of the United States. Their concern was heightened by uncertainty as to the consequences of some United States policies and actions. This concern was not peculiar to the British but was known to be shared by other western powers.[54]

The Department of State forecast that Attlee would express to Truman the genuine fear shared by all British peoples that the United States was drifting toward a third world war and that even though an open war with Russia might be avoided, the United States would become more completely embroiled in an exhausting war with Communist China. The two particularly sensitive points in this connection were the immediate situ-

[52] (1) Rad, DA (Haislip) to CINCFE (Collins via Larsen), unnumbered, 3 Dec 50. (2) The reaction of the Chinese, as described by the Indian Ambassador, to the statement of President Truman seems to have been opportunistic. "It was the next morning (the 1st of December)," Pannikar recalls, "that Truman announced that he was thinking of using the atom bomb in Korea. But the Chinese seemed totally unmoved by this threat. . . . The propaganda against American aggression was stepped up. The 'Aid Korea to resist America' campaign was made the slogan for increased production, greater national integration, and more rigid control over anti-national activities. One could not help feeling that Truman's threat came in very useful to the leaders of the revolution to enable them to keep up the tempo of their activities." See Pannikar, *In Two Chinas: Memoirs of a Diplomat*, pp. 116–17.

[53] JCS 2176/1, 3 Dec 50, Incl B.

[54] *Ibid.*

ation facing U.N. forces in Korea and the U.S. policy on Formosa.[55]

The Department of State officials told the Joint Chiefs of Staff:

> With respect to Korea there is profound concern that actions have been and may continue to be taken which unnecessarily aggravate the situation and bring us closer to war with China. Germane to this is the rather widespread British distrust of General MacArthur and the fear of political decisions he may make based on military necessity. Bearing on this is the British belief in the buffer area and their stand against attacks across the Yalu. Also involved is the fear of the effect on Asiatics of use of the Atomic Bomb or even open consideration of its use.[56]

The Joint Chiefs of Staff were in complete agreement with that portion of the Department of State's conclusion which stated:

> We believe that the British are very sincere in their concern over the above matters and that they should be handled with full understanding and appreciation of that fact. Although we approach them with understanding and sympathy and meet them wherever reasonably possible, we should not give them cause to think that we are fully satisfied with British actions and policies. In particular, the occasion should be taken to emphasize to them the importance and urgency of getting along with the defense effort. They are inclined sometimes to regard the world situation as primarily a United States-Soviet problem and therefore to keep the sights for their own efforts too low.[57]

Insofar as the specifics of action in Korea were concerned, the Department of State recommended to the President, and the Joint Chiefs of Staff concurred, that he should make no commitment to Attlee restricting the freedom of action of the United States on use of the atomic bomb. The President should tell Attlee that the United States did not desire to use the atomic bomb and stress that the United States fully realized the dire consequences of using the bomb. He should also tell the British Prime Minister that the United States desired and expected to move in step with the British in meeting the current crisis.[58]

The Department of State proposed that President Truman should discuss with the British Prime Minister two possible courses of action in Korea. The first of these involved a withdrawal of U.N. forces to a line on the 38th Parallel in conjunction with a possible cease-fire agreement. The second course was the evacuation of all of Korea. In the event of military necessity, the Department of State held, the X Corps should withdraw from Korea to Japan and an attempt should be made to stabilize the situation by a political cease-fire agreement, with the line of demarcation between forces along the 38th Parallel. The Department of State wanted President Truman to tell Attlee that MacArthur intended to assemble his forces into three beachheads: in the Seoul-Inch'on area, at Hamhung, and at Pusan. The X Corps could be evacuated to Japan in any way that proved militarily practicable.[59]

The proposal held:

> Before the Chinese Communists have reached the 38th Parallel in strength we should try to establish a cease-fire on the basis of the 38th Parallel with the armies

[55] *Ibid.*
[56] *Ibid.*
[57] *Ibid.*

[58] (1) *Ibid.* (2) Memo for Secy Defense, 4 Dec 50, sub: Use of Atomic Bomb.
[59] JCS 2176/1, 3 Dec 50, Incl B.

separated by a demilitarized zone. The principal purpose of this effort would be to deny success to aggression and to consolidate an overwhelming majority of the United Nations members behind the cease-fire effort. Arrangements for a cease-fire on the basis of the 38th Parallel would not, however, be conditioned on agreement to other issues, such as Formosa and the seating of Communist China in the United Nations.

During the cease-fire effort (apparently before a cease-fire had been agreed to by both parties), the United Nations would retire on the Seoul-Inch'on area, but would not begin any evacuation until the results of the cease-fire were determined.[60]

While the Joint Chiefs of Staff and their planning staff agreed that a cease-fire might be militarily advantageous for the United Nations Command under conditions then obtaining, they wanted to be sure of two things. First, the considerations offered the Chinese in exchange for a cease-fire agreement must not be too great, and, secondly, the United Nations commander must not be operationally restricted. Such a plan as the Department of State proposed, dictating not only the area into which the Eighth Army would retire but also restricting the conditions under which MacArthur might evacuate his troops, was unacceptable. The Joint Chiefs of Staff, in revising the Department of State proposals, cut out any reference to the evacuation of the X Corps. "Arrangements for this cease-fire," the Joint Chiefs of Staff maintained, "must not impose conditions which would jeopardize the safety of United Nations forces." In other words, MacArthur must be free to withdraw at any time. They also objected to the Department of State provision that would have compelled the Eighth Army to withdraw on the Seoul-Inch'on area.[61]

In the Department of State's recommendation to President Truman regarding the possible necessity of evacuating Korea was the explanation that the Department's position did ". . . not exclude the possibility of some military action which would harass the Chinese pending their acceptance of a United Nations settlement for Korea and would not exclude any efforts which could be made to stimulate anti-Communist resistance within China itself, including the exploitation of Nationalist capabilities." [62]

It is significant, in view of the early December date, that the Joint Chiefs of Staff seized upon this discreetly worded hint of retaliatory measures and reworded it, not only in stronger terms, but by adding several possible retaliatory measures later proposed by General MacArthur, to include a naval blockade of China and bombing of Chinese lines of communication outside of Korea.[63]

President Truman and Prime Minister Attlee met at the White House on 4 December and on each day thereafter for five days. Also present at these meetings were Secretary of Defense Marshall and Secretary of State Acheson as well as the British Ambassador to the United States. Discussion was frank, open, and occa-

---

60 *Ibid.*

61 (1) JCS 1776/167, with Incls, 3–4 Dec 50. (2) Memo, ACofS G–3 for CofS USA, 3 Dec 50, sub: (JCS 1776/7) Korea. Both in G–3, DA file 091 Korea, Case 129.

62 *Ibid.*

63 *Ibid.*

sionally heated, but ended in agreement on most major issues.[64]

It was very clear that the British, and by inference the other NATO nations, while they had no intention of deserting the United States, could not reconcile themselves to what they believed to be its unrealistic and extremely dangerous policy in relation to Red China. Attlee took the position at first that there was no choice under current conditions but to negotiate with the Chinese Communists, with such negotiations most certainly extending beyond Korea and amounting to the surrender of Formosa to the Communists, a grant of United Nations status to the Chinese Communists, and the recognition of their government by the United States as the price for a cease-fire and the withdrawal of Chinese troops from the Korean peninsula.[65]

President Truman emphasized that it was not the American policy to desert its friends when the going got rough. He pointed out that the United States did not make distinctions between little aggressions and big aggressions. President Truman's position coincided with that of the Joint Chiefs of Staff when he told Attlee that if a cease-fire were proposed, the United States would accept it, but the United States would pay nothing for it. If a cease-fire were not accepted, or if it were accepted and the Chinese later started fighting again, the United States would fight as hard as it could.

Acheson agreed that an immediate cease-fire in Korea would be of advantage to the United Nations. But to buy that cease-fire in the fashion suggested would be unacceptable to the United States. The American leaders were opposed to Attlee's suggestions on the ground that they would actually reward China for her acts of aggression and would seriously weaken the position of the United States in the Far East, politically as well as militarily. As for U.S. retaliation against China, no promises were made by American authorities that they would not take more active measures such as blockade or bombardment of the mainland; but Attlee was assured that few of the President's key advisers were urging this course and that, as a basic principle of its policy, the United States was determined to avoid any enlargement of the conflict if at all possible.

Following their meeting on 7 December, Attlee and President Truman agreed that there would be no general voluntary evacuation of Korea at that time. General Collins, having returned from Korea, on 9 December briefed the two heads of government on the military situation. After the briefing he told reporters that MacArthur's forces would be able to take care of themselves without further serious losses.[66]

After more discussions between the two heads of state, certain agreements were reached. Among these was agreement that neither the United States nor the United Kingdom would object to

[64] (1) This resumé of the discussions is based on messages sent to General MacArthur by the Department of State informing him of the progress: Rad, CM-IN 18584, State to SCAP, 8 Dec 50; Rad, CM-IN 19784, State to SCAP, 12 Dec 50. (2) Truman, in his memoirs, gives a good deal of information on the trend of these talks and of his private conversations with Attlee. See Truman, *Memoirs*, II, 396–413, and Acheson, *Present at the Creation*, pp. 480–85.

[65] Rad, CM-IN 18584, State to SCAP, 8 Dec 50.

[66] New York *Times*, December 7, 1950 and December 9, 1950.

any appeal by Asiatic nations to the Chinese Communists for a cease-fire. It was agreed that the objective of both nations was to achieve a free and united Korea. A cease-fire and peaceful solution of the current conflict with the Chinese Communists was desirable in the immediate future if it could be secured on honorable terms. There was no disagreement on the matter of keeping Communist China out of the United Nations, and there was agreement that the Chinese Communists would not be granted any payment for a peaceful solution in Korea, such as Formosa or Indochina. If no solution could be obtained, American and British troops would fight in Korea until they were forced out.[67]

[67] Rad, CM-IN 19784, Secy State to SCAP, 12 Dec 50.

CHAPTER XVI

The Brink of Disaster

Reinforcement Prospects

When General MacArthur had warned General Collins on 6 December that unless material reinforcements were sent within a reasonable time his forces in Korea should be evacuated as soon as possible, Collins had replied that MacArthur could expect no significant numbers of additional troops in the near future. Collins had not spoken idly, as was borne out by Department of the Army officers in Washington who were then examining the Army's currently available means of strengthening MacArthur's command.

General Bolté had proposed on 3 December that the 82d Airborne Division, the only combat-ready division in the United States, be sent to Japan immediately where it would be available to protect the American base in Japan, or to fight in Korea if needed. He felt that the division could move without its drop equipment since the troops probably would be used in conventional ground operations only. Bolté maintained that the division could reach Japan within thirty-four days after being alerted. Since the division was approximately 2,000 men overstrength, a cadre of that size could be left in the United States to activate another airborne division. The Army's G–4, Maj. Gen. William O. Reeder, admitted that, from the logistic standpoint, the division could be sent, but he did not concur in sending it. General Ridgway recommended to the Acting Chief of Staff, General Haislip, that any decision on sending the unit be postponed until General Collins returned from Korea. On 8 December, after Collins' return, the shipment was disapproved.[1]

Aside from the 82d Airborne Division, the Army's means for reinforcing MacArthur with combat-ready divisions were slight indeed. Two National Guard divisions, the 28th and the 43d, could be readied for shipment to Korea by June 1951; two more, the 40th and 45th, could be sent by July 1951; by August, the 4th Infantry Division could join the others; and by September 1951 the 2d Armored and 11th Airborne Divisions could be in Korea. But the current U.S. ground force capabilities were so limited that the situation in Korea was likely

[1] Memo, ACofS G–3 for CofS USA, 3 Dec 50, sub: Movement of 82d Abn Div to FEC, with handwritten comments on original, in G–3, DA file 320.2 Pac, Case 59.

to run its course before significant deployments (other than the 82d Airborne Division) could be made. Furthermore, "The greatly increased possibility that global war will eventuate from the current crisis makes it mandatory that the United States make no further deployments that are not in consonance with the strategic concept of the current emergency war plan." [2]

Replacements

Furnishing individual replacements was an equally insoluble problem. When the Chinese struck, each of MacArthur's divisions had lacked about 30 percent of its men and officers. This weakness, acceptable during operations against an enemy of inferior strength, became intolerable after the Chinese intervened.[3] MacArthur made this very clear on 28 November when he appealed for more than twice the number of replacements then scheduled for his command. He had been notified that 33,000 replacements would arrive in December. He claimed that he now needed 74,000 replacements to compensate for losses suffered in the Chinese attack and to bring his units up to strength. This figure did not include losses anticipated for January.[4]

The Department of the Army recognized MacArthur's need but could increase neither the number nor the rate of replacement shipments. In fact, the promised 33,000 could not even be provided. The best estimate of shipments during December was placed at 23,000.[5]

Chinese Nationalist Forces

In search of additional troops, MacArthur had reminded the Joint Chiefs of Staff of Chiang Kai-shek's July offer to send 33,000 troops to serve under him in Korea. This offer had been turned down on his own advice and on the advice of the Joint Chiefs of Staff. But MacArthur felt that the Chinese intervention put an entirely different light on the offer, and on 28 November applied to the Joint Chiefs of Staff for the services of Chiang's troops. He felt that the original refusal had been prompted or at least influenced by the belief that the use of Chinese Nationalists in Korea might give the Chinese Communists an excuse for coming into the war. Another reason had been the need for the Chinese Nationalists to conserve their strength to meet threatened attacks against Formosa by the Chinese Communists. Neither reason remained valid, MacArthur informed the Joint Chiefs of Staff. He maintained that the Chinese force on Formosa was the only source of trained manpower available to him for early commitment against his new enemy. He estimated that these troops could reach Korea within two weeks and in far greater strength than the 33,000 originally offered. "I strongly recommend," MacArthur urged the Joint Chiefs, "that the theater commander be authorized to negotiate directly with the Chinese Government authorities on Formosa for the movement north and

---

2 Memo, ACofS G–3 for CofS USA, 3 Dec 50, sub: Further Reinforcements for Korea, in G–3, DA file 091 Korea, Case 125.

3 Briefing, Gen Gaither to Army Comdrs' Conference, 4 Dec 50, in G–3, DA file 337, Case 12/2.

4 (1) Rad, CX 69983, CINCFE to DA, 28 Nov 50. (2) Rad, WAR 97786, DA to CINCFE, 1 Dec 50.

5 Rad, WAR 97786, DA to CINCFE, 1 Dec 50.

incorporation in the United Nations Command of such Chinese units as may be available and desirable for reinforcing our position in Korea." [6]

Washington authorities did not share these views. They felt that the introduction of Chinese Nationalist forces into the Korean conflict would precipitate a full-scale war with Communist China and might trigger a global war for which the United States was unprepared. Furthermore, the use of Chiang Kai-shek's men would likely be unacceptable to some, if not all, of the United Nations members with troops in Korea. The Commonwealth nations, for instance, would very probably refuse to have their forces employed alongside Chinese Nationalist troops. In case of a general war with China, moreover, it would be better to use Nationalist forces on the mainland rather than in Korea. Also, in view of an increasingly critical supply situation, complicated by recent substantial losses in Korea, Washington was reluctant to equip Chinese Nationalist troops for Korea. Politically, the move would commit the United States to the Chinese Nationalist regime to an unacceptable extent. In any case, Washington doubted that the employment of 33,000 Chinese Nationalist troops, which represented the only firm offer made, would decisively influence the situation in Korea.[7]

The Joint Chiefs of Staff gave no immediate, definite answer to MacArthur, merely replying that they were considering the proposal. But they warned that the matter could have a worldwide impact which might disrupt the unity of the nations associated with the United States in the United Nations and even isolate the United States from its allies.[8]

On 18 December, General MacArthur made another attempt to procure major reinforcements, although not for Korea, when he asked that the four National Guard divisions called to active duty in September be sent to Japan at once. He pointed out that a recent build-up of USSR propaganda interest in Japan and the increasing tempo of international Communist pressure upon the remaining free segments of Asia were alarming the Japanese. In order to provide reasonable safeguards against any USSR thrust at Japan, he urged that these four divisions be moved to Japan to complete their training. The Joint Chiefs of Staff told MacArthur that it did not appear probable that the National Guard divisions could be sent him, although General Collins, then away on a trip to Europe, would have to give the final decision. Every effort would continue to bring MacArthur's units to full strength and to keep them there. "Meanwhile," the Joint Chiefs of Staff suggested, "you may wish to consider moving a portion of X Corps to Japan without prejudice to future disposition." After General Collins' return, the Joint Chiefs of Staff agreed that since no decision had been made at the governmental level as to the future United States course of action in Korea, no additional divisions would be deployed to the Far East for the time being.[9]

---

6 Rad, C 50021, CINCFE to JCS, 28 Nov 50.

7 Draft Memo for JCS Representative, 1 Dec 50, with Annex 1, in G-3, DA file 091 Korea, Case 127/8.

8 Rad, JCS 97594, JCS to CINCFE, 29 Nov 50.

9 (1) Rad, C 51599, CINCFE to DA, 18 Dec 50. (2) Rad, CM-OUT 99274, JCS to CINCFE, 19 Dec 50. (3) Rad, JCS 99616, JCS to CINCFE, 23 Dec

Operation Pink

Not only had the Eighth Army and X Corps suffered numerous casualties in the Chinese onslaught of late November and early December—they also had lost considerable amounts of supplies and equipment through enemy action, abandonment, and voluntary destruction. The widely separated units of X Corps, particularly those in the Changjin Reservoir area, had left behind or destroyed bedding, tentage, ordnance equipment, signal equipment, and engineer equipment as they made their way back to Hamhung. General Almond's supply staff estimated that the X Corps needed equipment for refitting one-half a division as a result of the Chinese attacks against his forces in the Changjin Reservoir area. General Walker's supply chief told the GHQ G–4 on 1 December that the Turkish brigade had lost a large part of its equipment and that at least one RCT of the 2d Division would need considerable re-equipping.

From his depot stocks in Japan, General MacArthur ordered the immediate shipment to Korea of major items of equipment for two RCT's. As one of the few concrete steps which it could take to remedy the situation, the Department of the Army started immediate action to furnish MacArthur's command with full equipment for an infantry division. On 4 December, the Department's G–4, General Larkin, ordered the immediate preparation, loading, and shipping on a priority basis to the Far East Command a complete division set of equipment less aircraft, general-purpose vehicles, ammunition, and certain other items either not readily available for shipment or not essential to combat operations. This shipping operation was code-named Operation PINK. Upon being informed of the action being taken, MacArthur asked that equipment for two divisions, rather than one, be sent. But the Department of the Army told him that, because of the pending augmentation of the Army and the requirements of units ordered to his command, further emergency shipments of equipment could not be made.

Operation PINK took place in an atmosphere of the greatest urgency. Four ships loaded at San Francisco, California, and four at Seattle, Washington, commencing on 5 December. On 9 December, these eight ships, partially combat-loaded, sailed for the Far East. Much of the equipment they carried came from Mutual Defense Assistance Pact (MDAP) stocks, from the Special Reserve, and from troops located near San Francisco and Seattle. Among the items rushed to MacArthur were 140 medium tanks.[10]

50. (4) Truman says of this transaction, "The Joint Chiefs of Staff and General Marshall held a series of meetings with State Department officials, trying to find some way to meet the problem. Reinforcements were simply not available. . . . The military chiefs thought that we might consider ways to withdraw from Korea 'with honor' in order to protect Japan. The State Department took the position, however, that we could not retreat from Korea unless we were forced out." See Truman, *Memoirs*, II, 432. The author has found no record of these meetings in files available to him, nor does any indication of this type of thinking by the Joint Chiefs of Staff appear elsewhere in their discussions.

10 (1) Telecon, TT 4088, DA and FEC, 2 Dec 50. (2) Rad, WAR 97969, DA to CINCFE, 4 Dec 50. (3) DF, ACofS G–4 to ChTechSvcs, 4 Dec 50, sub: Emergency Shipment of One Div Set of TO&E Equipment to FEC, in FEC, G–4 file G–4/D5 WAR 98480, 9 Dec 50. (4) Rad, WAR 98907, DA to CINCFE, 14 Dec 50.

Possibility of Evacuation

During General Collins' absence from Washington in early December, other leading officers of the Army staff, including Generals Haislip, Ridgway, Gruenther, Bolté, and Reeder, studied the situation resulting from Chinese intervention and its impact on the United States Army. These officers were uncertain of the extent of the Korean crisis, but judging from the situation map and MacArthur's gloomy reports, concluded that United Nations forces stood in some danger of being overrun and destroyed.

General Bolté, mindful of MacArthur's warning, that unless large numbers of ground troops were sent him at once he would be forced to withdraw his divisions into beachheads, declared that MacArthur should be directed to pull out of Korea. He pointed out that the additional forces needed by MacArthur simply did not exist in the United States or in other member nations of the United Nations. Hence, if MacArthur continued to fight in Korea, his command might be destroyed.[11]

Bolté was convinced that the Chinese intervention had considerably enhanced Russian capabilities in any global war, and that the United States must take immediate countermeasures against this Russian advantage. Even in the most optimistic circumstances MacArthur's forces in Korea, including seven U.S. divisions, would be neutralized and useless in any effort to counter a USSR attack on the United States or its allies. Consequently, Bolté recommended that MacArthur's mission be modified at once to permit him to evacuate Korea as soon as possible.[12]

The greatest concern and one shared by American military and political leaders lay in the possibility that the Chinese intervention in Korea was only the first step in a USSR move to conquer the world. Throughout December, these authorities did what they could and considered what they could further do to place the United States in the best possible position to meet the global war that seemed so imminent. General Bolté set the stage for preparation at the military level early in December when he urged that American unified commands be alerted and authorized to put their current emergency war plans into effect in case of attack.[13]

On 6 December, the Joint Chiefs of Staff took the action recommended by Bolté. They told their unified commanders worldwide that Chinese intervention had greatly increased the possibility of general war. "Take such action," the unified commanders were directed, "as feasible to increase readiness without creating an atmosphere of alarm." [14]

At the Department of the Army level, Army officials in early December considered the initiation of full mobilization without delay.[15] Full mobilization was judged premature, but intermediate steps

---

[11] (1) MFR, 3 Dec 50, sub: Korean Sit. (2) Memo, Gen Bolté for CofS USA, 3 Dec 50, sub: Course of Action To Be Taken as a Result of Developments in Korea, both in G–3, DA file 091 Korea, Case 134.

[12] *Ibid.*

[13] Memo, Gen Bolté for CofS USA, 3 Dec 50, sub: Courses of Action To Be Taken as a Result of Developments in Korea, in G–3, DA file 091 Korea, Case 134.

[14] Rad, JCS 98172, JCS to All Comdrs, 6 Dec 50.

[15] MFR, 3 Dec 50, sub: Korean Sit, G–3 091 Korea, Case 134.

were taken to increase the strength of the Army and greatly broaden the mobilization and production bases. The National Security Council, acting on recommendations of the Joint Chiefs of Staff, had approved on 22 November a military program for fiscal 1951 providing an army of sixteen combat divisions within a total Army strength of 1,263,000. By 30 June 1954, the Army would reach a strength, through gradual expansion, of 1,353,000, with eighteen combat divisions.[16]

Secretary of the Army Pace expressed the views of many Army officials when he told the Army Policy Council on 6 December that as a result of the Chinese intervention Americans were now living in a world essentially different from the kind of a world they had been living in a week before. The Army's requirements of 6 December were quite different from those of 30 November. He emphasized that the Army's program of an orderly build-up was not good enough nor fast enough to meet the emergency situation. General Ridgway, speaking for General Collins in the latter's absence, told Pace that the Army staff had prepared a plan for quick expansion to a 21-division army of 1,530,000 strength. Pace approved this concept for planning purposes only.[17]

Still, the immediate threat posed by Chinese action and the larger threat of possible global war gave some impetus to Army expansion. On 5 December, the Joint Chiefs of Staff recommended an accelerated rate of Army expansion with the 18-division force originally scheduled for June 1954 to be created by June 1952. The National Security Council agreed to this action on 14 December. As a result of this acceleration policy, the Department of the Army in December called two more National Guard divisions, the 31st and 47th, to active Federal service, beginning in January 1951.[18]

Various military authorities, including General Collins and General Ridgway, had expressed the opinion that the President should proclaim a national emergency. Such a proclamation would place in force the statutory provisions and authorizations normally granted the President in time of war and facilitate the expansion of the nation's armed forces and industrial facilities in support of these forces.[19] On 15 December, in a radio address to the nation, President Truman declared that a state of national emergency existed. On the following day, he affixed his signature to a proclamation which said, in part:

> NOW, THEREFORE, I, HARRY S. TRUMAN, President of the United States of America, do proclaim the existence of a national emergency, which requires that the military, naval, air, and civilian defenses of this country be strengthened as speedily as possible to the end that we may be able

[16] JCS 2101/25, 22 Nov 50.

[17] Min, 51st mtg Army Policy Council, 6 Dec 50, in CofS, DA file 334 (APC).

[18] (1) Memo, Gen Ridgway for Gen Haislip, 7 Dec 50, sub: Briefing for the Chief of Staff, in CofS, DA file 337, Case 14. (2) MFR, 11 Dec 50, sub: Conference, in G-3, DA file 320.2, Case 45. (3) History of DA Activities Relating to the Korean Conflict, 25 June 1950-8 September 1951, ACofS G-3, Army War Plans Br, p. 4. (4) MFR, 11 Dec 50, sub: Augmentation of the Army to Eighteen Divs in G-3, DA file 320.2, Case 45.

[19] (1) Min, 51st mtg Army Policy Council, 6 Dec 50, in CofS, DA file 334 (APC). (2) MFR, 11 Dec 50, sub: Conference, in G-3, DA file 320.2, Case 45. (3) PL 450, 82d Congress, 2d session, H. J. Resolution 477 as amended by PL 12, 83d Congress.

to repel any and all threats against our national security. . . .[20]

Washington officials asked MacArthur to sound out opinion within his command on the President's proclamation. MacArthur replied that because of limitations of time and the "far flung distribution" of his various subcommands no real sampling of reactions to the speech could be obtained. But, as far as he could judge, the reaction was favorable. "There can be little doubt, however," MacArthur concluded, "but that most would concur in my own personal opinion that the crucial realities of the nation's present circumstances impel the immediate and complete mobilization of our full military potential." [21]

X Corps Is Ordered Out of North Korea

The possibility that evacuation might be forced by enemy pressure was being considered in Tokyo at this same time. On 6 December, General Wright gave General MacArthur a detailed study of the problem of quitting Korea should it become necessary. Wright pointed out that an evacuation through Inch'on would be slow and dangerous. Pusan, on the other hand, offered every advantage for speedy and efficient outloading of men and equipment. At Pusan, twenty-eight ships could be berthed around the clock while Inch'on could handle only LST's and similar assault craft and then only for two 4-hour periods each day. Pusan had pier-crane facilities for all types of heavy lift, while Inch'on had none. The turnaround time from Japan to Pusan, moreover, was only one-fourth that from Japan to Inch'on. By conservative estimate, General Wright believed that all U.N. units and equipment could be taken out of Pusan five times as fast as from Inch'on. Wright realized, too, that air operations against the Chinese would be more effective as the enemy moved deeper into Korea. If the evacuation took place from Pusan, it could be inferred that MacArthur's forces had delayed successively and that rear airfields would be maintained and protected. Once these forces reached the old Pusan Perimeter, Japanese airfields could be used to continue effective support of the evacuation.[22]

Following this line of reasoning, General Wright then recommended that Almond's corps be sealifted from the northeastern portion of Korea at the earliest practicable date and relanded at Pusan or P'ohang-dong. Wright further recommended that X Corps be absorbed by the Eighth Army. Thus strengthened, the Eighth Army would make withdrawals in successive positions—if necessary, to the Pusan Perimeter.[23]

General MacArthur was most reluctant to place Almond under Walker's command, but yielded to what appeared to him to be the overriding wisdom of consolidating his strength in Korea. On 7 December, he approved General Wright's recommendations and notified

[20] (1) Presidential Proclamation 2914, 16 Dec 50. (2) *MacArthur Hearings*, p. 3520.

[21] (1) Rad, DA 99090, DA to CINCFE, 17 Dec 50. (2) Rad, C 51515, CINCFE to DA for Haislip, 18 Dec 50.

[22] (1) Memo, Gen Wright for CofS, GHQ, UNC, 6 Dec 50. (2) Comd Rpt, GHQ, UNC, Dec 50, Annex 4, Part III, "A" 741.

[23] *Ibid.*

both Walker and Almond of his decision.[24] He told them:

Current planning provides for a withdrawal in successive positions, if necessary, to the Pusan area. Eighth Army will hold the Seoul area for the maximum time possible short of entailing such envelopment as would prevent its withdrawal to the south. Planning further envisions the early withdrawal of X Corps from the Hungnam area and junction with Eighth Army as practicable. At such time, X Corps will pass to command of the Eighth Army.[25]

The Joint Chiefs of Staff told MacArthur at once that they approved his action and that they felt Almond's corps should be withdrawn from Hungnam as early as practicable.[26]

At the 28 November meeting in Tokyo, Almond had told MacArthur that he could hold at Hungnam forever if he were so ordered. Considered without reference to the plight of the Eighth Army in the west, the presence of the X Corps on the Chinese flank could have proven of considerable military value. But military considerations in northeast Korea had become secondary. The Joint Chiefs of Staff shortly instructed MacArthur to ignore that region.

Upon returning from the Tokyo consultations, Almond had directed his forces to retire upon Hungnam. The first order of business had been to extricate the Army and Marine units cut off by the Chinese around the Changjin Reservoir. Fighting as they withdrew and aided by concentrated close air support, these Marine and Army troops reached Hungnam on 11 December. By the same date, most of Almond's remaining forces had reached the same area.

MacArthur converted his 7 December withdrawal plans into orders on the 8th, and on the 11th, flew into Yonp'o Airfield near Hungnam to hear in person Almond's plan for taking the X Corps out of northeastern Korea.[27] Almond told MacArthur that his corps could clear Hungnam by 25 December and close in Pusan by 27 December. The total tonnage to be outloaded from Hungnam would reach 400,000 tons. To move this amount of tonnage by water, 75 cargo vessels, 15 troopships, and 40 LST's would be required. About 500 tons of men and equipment would have to be airlifted each day from 14 to 18 December. Almond believed that the withdrawal would be orderly, and that enemy forces in the area were too limited to interfere with the movement. No supplies or organic equipment would be destroyed or left behind.[28]

On 12 December, MacArthur notified the Department of the Army that the X Corps had started withdrawing, with the ROK 3d Division already en route by water to Pusan. The plan for the remainder of the evacuation provided for the contraction of the corps defense perimeter around the Hamhung-Hungnam area as the corps units departed in phases.[29]

There were no manuals to rely on in

[24] (1) Comments, Gen Hickey on Appleman MS, 14 Feb 56. (2) Rad, CX 50635, CINCFE to All Comdrs, 7 Dec 50.

[25] *Ibid.*

[26] Rad, JCS 94800, JCS to CINCFE, 8 Dec 50.

[27] (1) Rad, CX 50801, CINCUNC to All Comdrs, 8 Dec 50. (2) Comd Rpt, U.S. X Corps, Dec 50. (3) Rad, CX 50801, CINCUNC to CG X Corps, 9 Dec 50.

[28] Memo, by Gen Almond, 11 Dec 50, sub: Movement of X Corps to the Pusan Area, GHQ, UNC Comd Rpt, Dec 50, Annex 4, Part III, "I" 1475.

[29] Rad, CX 51102, CINCFE to DA, 12 Dec 50.

TRACKED AMPHIBIOUS VEHICLES *of 3d Infantry Division coming off the beach at Hungnam, 24 December 1950.*

the planning and carrying out of the evacuation of such great numbers of troops and such great quantities of equipment from an area under constant enemy pressure. There was no time, either, for research or experimentation. Unlike Dunkerque, the evacuation plan called for the removal of all equipment and supplies.[30]

Department of the Army officials were apprehensive lest Almond's force leave behind supplies and equipment which would be of value to the enemy. On 19 December, in a teleconference with the Far East Command's representatives, they asked what plans had been made for the evacuation or destruction of X Corps' supplies and whether or not it appeared that these plans could be carried out successfully. These officials were reassured by General MacArthur's staff, who told them that the evacuation plan called for the transfer of X Corps' supplies to the Eighth Army area. Such items as were excess or damaged but repairable would be sent to Japan. Any

[30] X Corps Special Rpt, Hungnam Evacuation, December 1950.

USS *BEGOR* LIES AT ANCHOR *ready to load the last U.N. landing craft as a huge explosion rips harbor installations at Hungnam.*

supplies which could not be loaded out in time would be destroyed. Most of the stocks of food and ammunition would have been used up by troops covering the evacuation. "The prospect of successfully implementing the evacuation," Department of the Army officials were reassured, "are excellent." [31]

Fortunately, and for reasons best known to themselves, the Chinese made no concerted effort to overrun the beachhead, although light scattered thrusts suggesting reconnaissance in preparation for larger operations were made by them throughout the evacuation operation. As the corps perimeter contracted, naval gunfire, artillery, and air support were intensified against the possibility of enemy attempts to build up forces for major assaults. The 3d Division provided the last defensive force for the perimeter, and, on 24 December, strongly supported by naval gunfire and carrier-based marine and naval aircraft, successfully executed the final withdrawal from the Hungnam beachhead. Extensive demo-

[31] (1) Telecon, TT 4147, DA and GHQ, 19 Dec 50. (2) Rad, C 51686, CINCFE to DA, 20 Dec 50.

litions of bridges and installations of military value were carried out at the last minute. According to X Corps officers, no serviceable equipment or supplies were abandoned, and all personnel were evacuated.[32] All together, 193 shiploads of men and matériel were moved out of Hungnam Harbor aboard Navy transports. Approximately 105,000 fighting men, 98,000 Korean civilians, 17,500 vehicles, and 350,000 tons of bulk cargo were removed from the beachhead.[33]

As Almond's troops evacuated their holdings in northeast Korea, the Eighth Army continued to withdraw to the south. Chinese pressure did not force these displacements. They were part of Walker's general withdrawal plan to form a continuous defensive line across Korea at the most advantageous position. The Eighth Army still had no contact with the Chinese, but intelligence reports showed that the enemy was moving into P'yongyang. By mid-December, the Eighth Army occupied a vague line extending along the south bank of the Imjin River, through Yongp'yong, Hwach'on, and Inje, to Yangyang on the east coast.[34]

The Chinese did not follow up Walker's withdrawal as closely as had been expected. In mid-December, ROK units in east central Korea had been attacked by North Korean troops. These North Korean forces appeared to be engaged in a screening and reconnaissance mission for the Chinese. Since the North Koreans were operating on their home grounds it was natural they should be used for this purpose. Throughout the next several days, more and more North Koreans appeared in front of the Eighth Army, apparently probing the line on behalf of the Chinese. American authorities were frankly puzzled by the actions of the Chinese. General Willoughby had several theories which he passed along to Washington. "Due to the depth of the withdrawal executed by Eighth Army," he said, "it is evident that the enemy, lacking any great degree of mobility has been unable to regain contact." He surmised that the Chinese had expected Walker to make a strong stand north of P'yongyang and that when he failed to do so the Chinese had been thrown off-balance. "There is little doubt but that he is now regrouping his forces under the screen of North Korean units," the UNC intelligence chief asserted, "preparatory to renewing the offensive at a time of his own choosing." It had been reliably reported that the Chinese had entered P'yongyang soon after it was deserted by the Eighth Army, but the whereabouts of the main body of Chinese forces in late December remained a mystery.[35]

On 19 December, Willoughby again expressed his puzzlement, telling Washington, "The whereabouts of the Chinese Communist forces and the reasons why these units have remained so long out of contact continue in the speculative realm." He felt that an offensive was not immediately forthcoming since the lack of contact with the Chinese and the

[32] X Corps Special Rpt, Dec 50, sub: Hungnam Evacuation, 9–24 Dec 50.

[33] Comd Rpt, GHQ, UNC, Dec 50.

[34] Comd Rpt, Hq, EUSAK, Dec 50, p. 62, plate 6, line B.

[35] (1) Telecon, TT 4135, DA and GHQ, 14 Dec 50. (2) Telecon, TT 4142, DA and GHQ, 18 Dec 50.

VEHICLES TO BE LOADED ABOARD SHIPS *at Hungnam during withdrawal from that town.*

relatively light pressure being exerted by North Korean forces against the Eighth Army did not point in that direction. In order to find out just what was going on, General MacArthur directed Walker to conduct aggressive ground reconnaissance to a considerable depth through the North Korean screen with particular attention to finding probable routes of enemy advance, locations, strengths, and to capture Chinese prisoners for interrogation.[36]

General Walker was killed in a vehicle accident near Uijongbu, Korea, on the morning of 23 December 1950. General Milburn, the I Corps commander, became acting commanding general of the Eighth Army. The possibility that Walker might be killed had been discussed earlier by General MacArthur and General Collins during the latter's visits to the Far East. General MacArthur had told Collins that if Walker were lost he wanted General Ridgway, Deputy Chief of Staff for Operations and Administration, Department of the Army, and one of General Collins' key assistants in Washington, as Eighth Army commander. MacArthur was familiar with Ridgway's fine combat record in World War II and realized that Ridgway, in his position at the Department of the Army, was in extremely close touch with the Korean situation and capable of stepping in at once. Upon Walker's death, an

[36] (1) Telecon, TT 4147, DA and GHQ, 19 Dec 50. (2) Telecon, TT 4156, DA and GHQ, 21 Dec 50.

immediate call was made from Tokyo to General Collins, who obtained clearance from Secretary Marshall and President Truman on the designation of Ridgway as the new Eighth Army commander.[37] General Ridgway left Washington almost immediately and reached Tokyo at midnight on Christmas Day.

No American outside the Far East knew more about the Korean situation than General Ridgway. In his position as deputy chief of staff for administration all reports, studies, and recommendations on Korea at the national level had passed through his hands. It was he who had taken much of the action to speed the shipments of units and replacements to MacArthur in July and August. He had gone with the President's special representative, Mr. Harriman, to Tokyo in August during the touch-and-go battles around Pusan. He had conferred with MacArthur and seen the Eighth Army's plight at first hand.

Ridgway's whole career had prepared him to command the Eighth Army. As a young officer he had served in China and in the Philippines. During World War II he had commanded an airborne division, later, a corps. He had led his troops brilliantly through Sicily and Normandy, through the Battle of the Bulge, and to the Baltic at the war's end. After the war, in a variety of staff and command assignments, Ridgway had gained valuable knowledge of Communist methods, purposes, and strategies. He was convinced they had to be stopped in Korea.

Ridgway took command of forces in Korea that had suffered a month of reversals. No major decisions as to future courses of action had yet been reached although numerous exploratory steps had been taken. As in November, once the enemy relaxed pressure, the nation's planners seemed to slacken their efforts to find a solution. The problems facing the United States and the United Nations Command were more political than military. Such decisions as whether or not to take action against the Chinese aggressors outside Korea had been raised but not answered. The problem of whether or not to evacuate had also been raised, but had been put aside as a result of lessening enemy pressure. The same fact had befallen the closely related cease-fire problem. The question of whether or not to reinforce the Far East Command was half answered by the nation's inability to do so, and the rest of the answer was obscured by the fog of indecision surrounding the core of the problem, "What is the best course of action now."

Ridgway Takes Over

Christmas found the Eighth Army halted uneasily near the 38th Parallel, awaiting its new commander and the new enemy. Signs were increasing that the Chinese were closing the gap and were advancing down the peninsula in a co-ordinated effort to feel out the Eighth Army's defenses before launching another major attack. A tense calm hung over the battle area. In a telephone report from Korea on 26 December, General Allen, Chief of Staff, Eighth Army, told GHQ officers, "We got another army. Pick up another one about every day. They are just build-

[37] *MacArthur Hearings*, pp. 588, 1201–02.

ing up. Don't know when they will hit. That is all we have. Otherwise dead quiet." [38]

MacArthur told Ridgway that the best he could hope for was a tactical success, possibly holding and defending South Korea. He remarked, "We are now operating in a mission vacuum while diplomacy attempts to feel its way. . . ." Any substantial military success by Ridgway's Eighth Army would greatly strengthen the hands of the diplomats. Tactical air power had proven disappointing to MacArthur, who now charged that it could not isolate the battlefield or stop the flow of enemy reinforcements into the battle. MacArthur reiterated that the Chinese were dangerous opponents and that the entire Chinese military establishment was coming into Korea to win. Touching on his recommendations to Washington, MacArthur remarked that the Chinese mainland was wide open in the south for attack by forces on Formosa. He had recommended that such an attack be made since it would relieve the pressure in Korea.[39]

Ridgway found MacArthur discouraged by the swing of events in Korea and ready to turn over to him a great deal of authority and latitude in directing combat operations. MacArthur indicated to Ridgway that he was to be both empowered and expected to plan and carry out all military operations of the United Nations forces in Korea. He told Ridgway that he was to act as he thought best. "You will make mistakes in Korea," MacArthur said, "we all do. But I will take full responsibility." He also told Ridgway that the X Corps would pass to his control as soon as it arrived in South Korea.[40]

When Ridgway questioned MacArthur more specifically as to his authority in directing operations in Korea, including a possible attack, MacArthur simply said, "Matt, the Eighth Army is yours." "No field commander could have asked for more," Ridgway says of this full grant of authority.[41]

General Ridgway thus went into Korea carrying a carte blanche to employ the Eighth Army as he found best and without reference to Tokyo for instructions. Ridgway could attack, defend, or withdraw; the decision was left to him. But while he was not required and never did ask confirmation of his actions, he did notify MacArthur in detail of his intentions. But MacArthur never questioned him. Whereas Walker had been kept under close supervision and control, Ridgway was not.[42]

---

[38] General Matthew B. Ridgway, The Korean War, Issues and Policies, June 1950–June 1951, pp. 350–53, MS, copy in OCMH.

[39] (1) *Ibid.* (2) See also Matthew B. Ridgway, *The Korean War* (New York: Doubleday and Co., 1967), pp. 82–83.

[40] *Ibid.*

[41] *Ibid.*

[42] General Ridgway recalls that between the time he assumed command and March 1951, after the Eighth Army had begun its northward march, General MacArthur stayed aloof from tactical decisions and visited Korea only twice. Thereafter he visited Korea and Ridgway weekly. Ridgway also expressed the belief that, had he thought it necessary, he could have led the Eighth Army out of Korea without recrimination from his superiors. General Hickey, who was acting chief of staff under MacArthur and, later, chief of staff under Ridgway, was of the opinion that Ridgway was under no restrictions on withdrawing, and that it was Ridgway's decisions and actions that eventually tipped the balance and kept the Eighth Army fighting in Korea. See Intervs, Appleman with Ridgway, Oct 51, and Appleman with Hickey, 10 Oct 51.

Fear that the Eighth Army might evacuate Korea obsessed the South Koreans. Admittedly, the specter had substance. The final decision awaited the outcome on the battlefield but the prospects were not bright at the moment. One of Ridgway's first acts on reaching Korea was to call on President Rhee in Seoul and to assure him, "I am glad to be here and I aim to stay." To the men of his new command, Ridgway announced bluntly, "You will have my utmost. I shall expect yours." [43]

During his conference with General MacArthur, Ridgway had asked for and received permission to attack in order to regain lost ground, and when he went to Korea he fully intended to attack as soon as possible. But he found not only his major commanders but also his Eighth Army staff extremely skeptical of such an attack. They were not, in Ridgway's words, "offensive minded." Under the circumstances and in view of these attitudes, Ridgway decided against an attack in the immediate future, at least during the remainder of December. But he ordered plans made at once for offensive operations and he set about instilling an "attack" spirit into his staff. "I skinned Eighth Army staff officers individually and collectively many times to have them do what I wanted," Ridgway later recalled. "I told them heads would roll if my orders were not carried out"; and he warned his staff, "I am going to attack to find out where the enemy is since G–2 cannot give me clear evidence." [44]

With characteristic directness, Ridgway began forcing the army to turn its eyes to the front. Step by step, in deliberate and carefully conceived actions and orders, he bore down on his new command. By example and by exhortation, he began shaking his staff, commanders, and men out of the defeatist mood. Where toughness was required, he was tough; where persuasion was indicated, he persuaded; and where personal example was needed, he set the example.

A Decision Hangs Fire

The enemy, meanwhile, had completed his concentrations and other preparations for attacking the Eighth Army, and on the night of 31 December introduced the New Year with a general offensive south of the 38th parallel. (*Map VI*) The Chinese attacked on a 44-mile front stretching east from Kaesong on Ridgway's left flank to a point northwest of Ch'unch'on on the east central front. The main effort came down the Yongch'on-Uijongbu-Seoul axis, obviously aimed at the seizure of Seoul and Inch'on. General Ridgway, in reporting the attack to General MacArthur, predicted that the Chinese invasion of South Korea was a prelude to an attempt by the Chinese to drive his command from the Korean peninsula by sheer manpower. "The Army Eight," Ridgway told MacArthur, "will continue

[43] Ridgway's activities and reactions during the first several weeks after his assumption of command are well covered in his manuscript, The Korean War, Issues and Policies, June 1950–June 1951; in his books, *Soldier* and *The Korean War*; in various interviews and conversations with the author; and in interviews with Dr. John Miller, jr., Major Owen Carroll, and Mr. B. C. Mossman, 30 November 1956, copies in OCMH.

[44] Interv, Appleman with Ridgway, Oct 51.

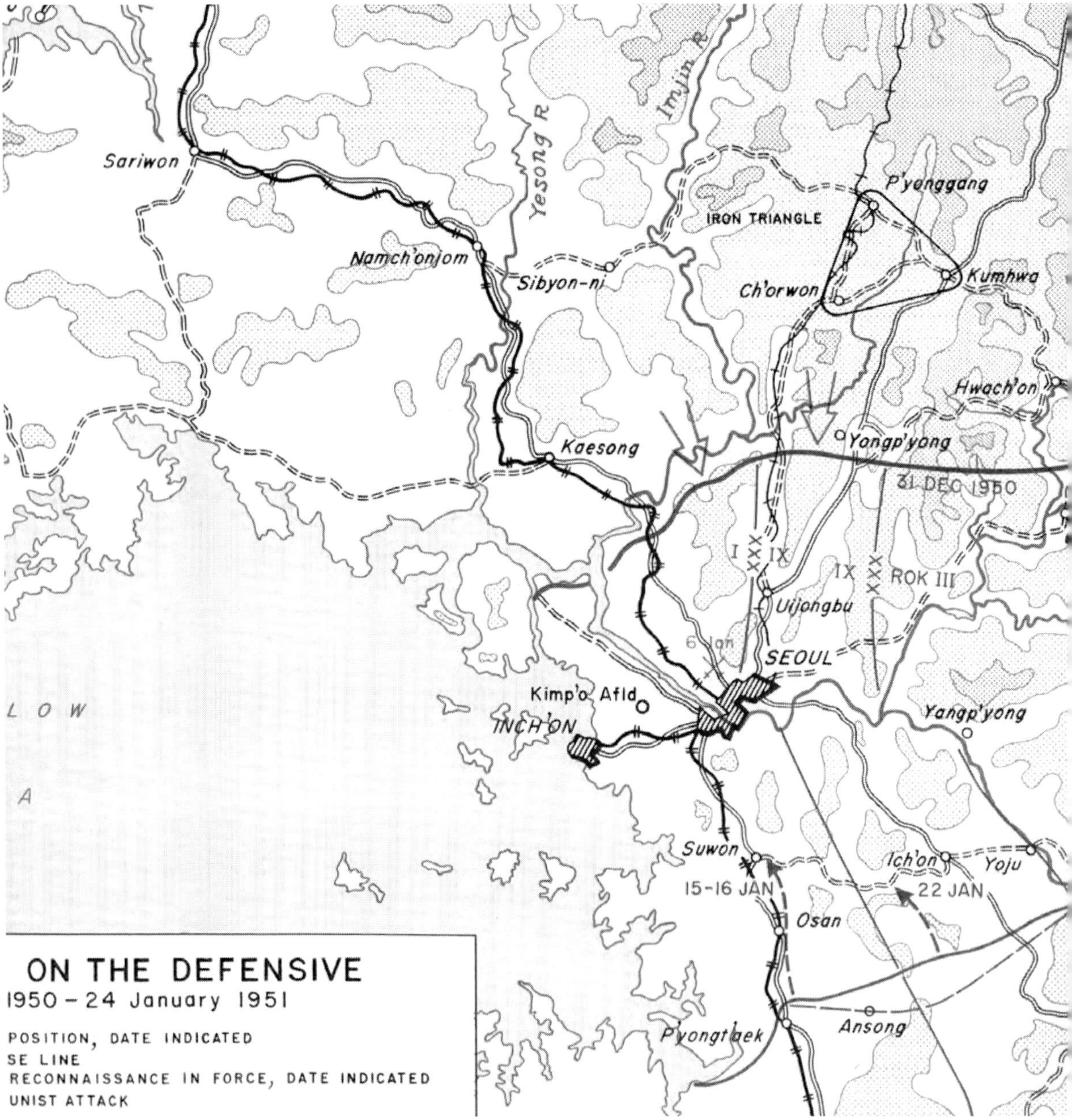
Sariwon
Namch'onjom
Sibyon-ni
Yesong R
Imjin R
IRON TRIANGLE
P'yonggang
Kumhwa
Ch'orwon
Hwach'on
Yongp'yong
Kaesong
31 DEC 1950
I
IX
ROK III
Uijongbu
6 Jan
SEOUL
Kimp'o Afld
INCH'ON
Yangp'yong
Suwon
Ich'on
Yoju
15-16 JAN
22 JAN
Osan
Ansong
P'yongt'aek
L O W
A
ON THE DEFENSIVE
1950 - 24 January 1951
POSITION, DATE INDICATED
SE LINE
RECONNAISSANCE IN FORCE, DATE INDICATED
UNIST ATTACK

KOREAN REFUGEES *slog through snow outside of Kangnung, northwest of Samchok, 8 January 1951.*

its present mission, inflicting the maximum punishment and delaying in successive positions while maintaining its major forces intact."[45]

The great strength of the Chinese assault in the west and the imminent danger of a breakthrough and envelopment down the east central corridors, defended largely by ROK units, forced General Ridgway reluctantly to direct certain withdrawals in early January. On 1 January, Ridgway ordered his western divisions to fall back from the Imjin River to a line slightly north of the Han River that formed a deep bridgehead around Seoul. But when the enemy swiftly followed up this withdrawal, Ridgway on 3 January decided to move south of the Han and to abandon Seoul. He was determined that this rearward move would be fought as a delaying action and so instructed his corps commanders. On 4 January, the Eighth Army started back to a line extending from P'yongt'aek on the west coast eastward to the coastal village of Samch'ok.[46]

Reports reached General Ridgway that

[45] Msg, C 52524, CINCFE to DA (quoting Ridgway), 2 Jan 51.

[46] Comd Rpt, EUSAK, Jan 51, Narrative, p. 56.

in withdrawing from the Han River some of his major units failed to damage the enemy materially or even delay him appreciably. Some units actually had broken contact with the enemy to fall back. Ridgway addressed his corps commanders sternly on this matter, emphasizing that he expected them to exploit fully every opportunity to damage the enemy.[47]

The Joint Chiefs of Staff Consider Options

Even before this January demonstration of Chinese power, the Joint Chiefs of Staff had concluded that the Chinese Communists had enough strength to drive MacArthur out of Korea. But they wanted MacArthur to stay if he could. A quick, massive build-up of the forces in Korea, much greater than that for Operation CHROMITE five months earlier, might keep the Eighth Army from being shoved into the Sea of Japan; but a major build-up, especially a quick one, was out of the question in view of shortages of combat divisions in the United States and the worsening world situation. On 30 December, the Joint Chiefs of Staff scotched any vestigial hopes which MacArthur might have held for additional ground forces by telling him that they would not send any more American divisions to fight in Korea at that time.[48]

Chinese successes in Korea had, concomitantly, increased the threat of a general war, encouraging further Communist military moves against other sensitive areas and heightening the tensions between the Soviet bloc and those nations allied with the United States. The Joint Chiefs of Staff pointed this out to MacArthur and told him bluntly, "We believe that Korea is not the place to fight a major war." If more American divisions were sent to Korea, American commitments throughout the world, including protection of Japan, would be seriously jeopardized.[49]

This news came as no surprise to MacArthur. Collins had told him substantially the same thing three weeks earlier when he had pressed for an increase in the strength of other United Nations contingents to a total of 75,000 men. The Joint Chiefs now told him that this could not be done either. "It is not practicable to obtain significant additional forces for Korea from other members of the United Nations," they said.[50]

In the minds of the Joint Chiefs of Staff, the best way for MacArthur to keep from being pushed off Korea was to fight and to fight hard. If the Eighth Army fought and killed enough enemy troops, Chinese and North Korean commanders might give up any attempt to drive the United Nations out of Korea as too costly. The Joint Chiefs of Staff expected no miracles. But if the Eighth Army could, without losing too many men and too much equipment, stop and hold the Chinese, not necessarily north of the 38th Parallel, MacArthur would have done his nation a great service. For the prestige, both military and polit-

[47] *Ibid.*, p. 12.
[48] Rad, JCS 99935, JCS (Personal) for MacArthur, 30 Dec 50.
[49] *Ibid.*
[50] *Ibid.*

ical, which the Chinese Communists had lately acquired by defeating the United Nations Command in North Korea, was exceedingly detrimental to the national interests of the United States; and MacArthur could deflate that prestige by staging a military comeback.

The events of the past month had shown clearly that General MacArthur's military mission assigned on 27 September stood in need of revision. The Joint Chiefs revised it in these words: "You are now directed to defend in successive positions. . . ." These positions were those which MacArthur had already described to his major commanders on 7 December. In addition to defending these positions, MacArthur was to damage the enemy as much as possible, "subject to the primary consideration of the safety of your troops." [51]

But the Washington authorities fully realized that mere words and military directives would not halt the Chinese and that enemy pressure might, in spite of MacArthur's best efforts, force him to evacuate Korea. They saw, too, that it was advisable to determine, in advance if possible, the last reasonable opportunity for MacArthur's command to evacuate in an orderly fashion. This was especially important since the enemy threatened not only Korea, but, in league with the Soviet Union, posed, by no great stretch of the imagination, a real threat to Japan. With Japan gone, MacArthur's command could only fall back on Okinawa, Formosa, or the Philippines.

The Joint Chiefs of Staff told General MacArthur:

> It seems to us that if you are forced back to positions in the vicinity of the Kum River and a line generally eastward therefrom, and if thereafter the Chinese Communists mass large forces against your positions with an evident capability of forcing us out of Korea, it then would be necessary, under those conditions, to direct you to commence a withdrawal to Japan.[52]

When the Joint Chiefs sought MacArthur's ideas on the timing of such a withdrawal, MacArthur assured them that there was no need to make a decision for evacuation until his forces were actually forced back to what he called the "beachhead line." Since the term *beachhead line* could be interpreted several ways, the Joint Chiefs asked MacArthur to be more specific. General Collins had brought back from the Far East a marked map showing nine possible defensive positions to be occupied by the Eighth Army in its withdrawal down the peninsula. One line marked positions held by the Eighth Army along the Naktong River in early September, and the Joint Chiefs asked MacArthur if this line was the beachhead line he had in mind. He stated that it was, but pointed out that exactly where the line would run should be regarded as completely flexible. "In an actual evacuation under pressure there would be progressive further contractions to a final inner arc," he told them. "The operation would probably be generally similar to that at Hungnam." General Collins still did not understand which line MacArthur meant. He reminded MacArthur that three lines on his marked map could be interpreted as a beachhead line, and asserted that when

---

[51] *Ibid.*

[52] *Ibid.*

the Eighth Army had been forced back to the northernmost of the three, the time for final decision would have arrived.[53]

Evacuation Plans

The rumor of a United Nations withdrawal from Korea spread quickly among men and officers of the ROK Army. General Ridgway pointed out to General MacArthur on 8 January that the apprehension among ROK soldiers as to their future was dangerous and could seriously affect his command. Ridgway suggested that MacArthur make a public statement which would serve to banish the fears of the ROK fighting forces. MacArthur passed this suggestion to the Joint Chiefs of Staff with the comment that "A reassuring statement by me such as General Ridgway suggests is impossible unless and until the basis for such a statement is established by policy determination at governmental level." [54]

MacArthur had already directed his staff to continue planning the evacuation procedures. Since an actual evacuation would be largely a Navy task, General Wright, the G–3, turned to the Navy for advice. Rear Adm. Arleigh A. Burke, Deputy Chief of Staff, COMNAVFE, on 7 January addressed Wright on the problems and factors to be considered. If the evacuation took place from Pusan, there was a strong likelihood it would be done under enemy pressure. There would be little similarity between a Pusan evacuation and the removal of forces from the Hungnam beachhead. Both in scale and difficulty, the Pusan operation would surpass that at Hungnam. Consequently, the length of time required to move troops and supplies would be much greater, and Burke therefore urged the early completion of advance plans. He advised Wright to designate at once the division which would hold the final perimeter at Pusan. Burke recommended the 1st Marine Division since it had special training in naval procedures, including the requirements for naval gunfire support, and had proven its combat effectiveness on more than one occasion. For employment on the intermediate perimeter, which would probably be manned by two divisions, Burke felt that any of the Army divisions would do.[55]

No divisions were ever designated for these duties, since by the middle of January the military situation gave General Wright some reason to believe that a forced withdrawal might not materialize. By 16 January, in fact, Wright was willing to speculate that, unless political considerations required or indicated withdrawal as the best course of action, it would be possible for the United Nations Command to remain in Korea as long as higher authority dictated. He hesitated to establish an evacuation target date even for planning purposes since, in his mind, the proper date would be dictated by enemy action and "political considerations." He did estimate

[53] (1) *Ibid.* (2) Rad, C 52391, MacArthur (Personal) for JCS, 30 Dec 50. (3) Rad, DA 80149, Collins (Personal) for MacArthur, 3 Jan 51. (4) Rad, C 52586, CINCUNC to DA, 3 Jan 51. (5) Rad, DA 80253, Collins (Personal) for MacArthur, 4 Jan 51. (6) Rad, C 52712, MacArthur (Personal) for Collins, 4 Jan 51. (7) Rad, JCS 80680, JCS (Personal) for MacArthur, 9 Jan 51.

[54] Msg, C 52964, CINCFE to JCS, 8 Jan 51.

[55] Memo, COMNAVFE for Gen Wright, 7 Jan 51, sub: Plans for Possible Re-Employment From Pusan Area, GHQ, UNC, G–3 files.

the time required for a complete withdrawal. On the basis of tonnage to be removed from the peninsula, estimated at 2,000,000 metric tons, the best possible out-loading time, using all possible ports of exit within the contracted defense perimeter, would be fifty days.[56]

As a further step in evacuation planning, General Collins, while on another visit to the theater, informed General MacArthur on 15 January that if a UNC evacuation became necessary, President Truman wanted all members of the ROK Government, ROK Army, and ROK police forces taken out. General MacArthur expressed satisfaction with this directive, stating that he thought it essential. Plans for the evacuation were immediately begun, and when Collins returned from Korea on 19 January the situation was laid before him. General Hickey pointed out that more than a million Koreans would have to be evacuated under the President's order. This figure included 36,000 ROK governmental officials and their dependents, 600,000 ROK police, and 260,000 ROK soldiers. These latter two groups had about 400,000 dependents. As to the place to which these people would be removed, Collins and MacArthur agreed that as many ROK soldiers as possible would be placed on the off-shore island of Cheju-do in order to maintain, after evacuation, a legal status for continuing to fight in Korea.

A possible complication in planning and achieving any evacuation of ROK personnel as directed by President Truman rested in a recent ROK petition to the United States for aid in strengthening ROK forces. In December, the ROK Minister of Foreign Affairs had asked Secretary of Defense Marshall to "release to us all the light arms which are available, in order that our young men may hurl themselves in the face of the advancing enemy." At the same time, the Korean Ambassador, Dr. John M. Chang, had urged the Department of State to arm the so-called Korean Youth Corps, which Chang claimed consisted of 500,000 young men, all eager to fight the Chinese. The Department of State recommended to the Department of Defense that this be done, but only after these Korean youths had been formed into organized units under the control and discipline of the military authorities in Korea.[57] General Marshall directed the Joint Chiefs of Staff to look into these requests and to obtain General MacArthur's views on their propriety.

In General MacArthur's mind, the whole problem of giving more arms to the ROK Government centered not on whether these units could be created and armed but on whether any advantage was to be gained by so doing. MacArthur emphasized that large numbers of small arms had already been given ROK police units, antiguerrilla security forces, and special ROK organizations for use in enemy-held territory. But friendly guerrilla forces lacked strong-willed leadership and were accomplishing little in enemy rear areas. Enemy guerilla

[56] Memo, Gen Wright for CofS GHQ, UNC (Gen Hickey), 16 Jan 51, sub: Disposition of U.N. Forces in Korea in Event of Withdrawal From Korea, G–3, GHQ, UNC files.

[57] (1) Ltr, B. C. Limb, ROK Minister of Foreign Affairs, to Gen George C. Marshall, Secy Defense, 12 Dec 50, in G–3, DA file 091 Korea, Case 149. (2) Ltr, H. Freeman Mathews, Depy Under Secy State, to Maj Gen James H. Burns, OSD, 12 Dec 50, in G–3, DA file 091 Korea, Case 137.

units, on the other hand, continued to operate effectively throughout South Korea.[58]

General MacArthur believed that checking the enemy would depend upon setting up a defense with U.S. divisions deployed in depth and in mutually supporting positions. This observation strongly indicates that MacArthur felt, in early January, that his forces would have to withdraw back to the Pusan Perimeter or even farther. He expressed the opinion that, because of the probably restricted area of the battlefield in which the United Nations forces might be fighting in the near future, and the greater value per rifle that might be gained by arming the Japanese National Police Reserve, training and arming of additional ROK forces appeared questionable. He recommended that the extra South Korean manpower be used to replace losses in existing ROK units, concluding:

> The long range requirement for or desirability of arming additional ROK personnel appears to be dependent primarily upon determination of the future United States military position with respect to both the Korean campaign and the generally critical situation in the Far East.[59]

MacArthur had once again taken the opportunity to point out to the Washington officials that he did not feel their policy was sufficiently clear. Further, if evacuation became necessary he did not want a bigger ROK Army to evacuate.

The big question in MacArthur's mind, now as before, was whether there was to be a change in national policy that would make evacuation unnecessary. If there was such a change, and the steps which MacArthur had proposed were taken, evacuation would not be necessary. But if the nation's leaders appeared unwilling to make this policy change, MacArthur felt that eventual evacuation was inevitable and that there was no reason why the Joint Chiefs of Staff should not issue their evacuation directive to him right away. If some slight chance existed that national policy might be changed, even if not immediately, General MacArthur felt that the Joint Chiefs of Staff could delay issuing the evacuation directive until the Eighth Army had been forced back to the ninth and final marked position he had drawn around Pusan in the hope that an extension of military activity against the Chinese might be allowed and evacuation staved off.[60]

---

[58] Rad, C 52879, CINCFE to DA for JCS, 6 Jan 51.

[59] Msg, C 52879, CINCFE to DA for JCS, 6 Jan 51.

[60] (1) Rad, JCS 99935, JCS (Personal) for MacArthur, 29 Dec 50. (2) Rad, C 52391, MacArthur (Personal) for JCS, 30 Dec 50. (3) Rad, DA 80149, Collins (Personal) for MacArthur, 3 Jan 51. (4) Rad, C 52586, CINCUNC to DA, 3 Jan 51. (5) Rad, DA 80253, Collins (Personal) for MacArthur, 4 Jan 51. (6) Rad, C 52712, MacArthur (Personal) for Collins, 4 Jan 51. (7) Rad, JCS 80680, JCS (Personal) for MacArthur, 9 Jan 51.

CHAPTER XVII

The Search For Policy

The changes in policy to which General MacArthur constantly referred involved decisions by the President to take, or not to take, certain specific military actions against the Chinese. General MacArthur first suggested such measures in conversations with General Collins early in December, and throughout the term of his command in Korea insisted with increasing emphasis that the actions he had sponsored be carried out.

MacArthur's view of the world situation, with particular emphasis on his own theater, was simple in its approach but exceedingly complex in its implications. He reasoned that the Chinese had, omitting only the formality of open declaration, gone to war against the United Nations Command. The Chinese were prosecuting this war, in MacArthur's view, with all the resources at their disposal and were being supported logistically by the Soviet Union.

On 30 December, he posed four retaliatory measures that he believed feasible and that would require a relatively small commitment of military forces. The first was to blockade the China coast; the second, to destroy Communist China's war industries through naval gunfire and air bombardment; the third, to reinforce the troops in Korea with part of the Chinese Nationalist garrison on Formosa; and the fourth, to allow diversionary operations by the Nationalist troops against vulnerable areas of the Chinese mainland. These measures, he was certain, could not only relieve the pressure on United Nations forces in Korea but could indeed severely cripple Communist China's war-making potential and thus save Asia from a Communist engulfment that otherwise faced it. While he realized that such actions previously had been rejected for fear of provoking Communist China into a major war effort, he now insisted that Communist China was already fully committed and that the retaliatory steps therefore could not prompt it to greater efforts. He also realized that there might be some danger of Russian interference if the courses he described were adopted. But he discounted this risk, reasoning that any Russian decision to start a general war would be reached solely on a basis of Russia's own estimate of relative strength and capabilities of the United States and itself.[1]

If Communist China was permitted to

[1] Rad, C 52391, MacArthur (Personal) for JCS, 30 Dec 50.

get away with what he called its "flagrant aggression," and if the United Nations Command evacuated Korea without attacking the Chinese mainland, MacArthur believed that the Asian peoples, including the Japanese, would be greatly dismayed. He implied that the United States would lose so much face with these peoples that a material reinforcement of the Far East Command would be necessary even to hold the littoral island defense chain, including Japan.[2]

MacArthur pointed out that the evacuation of his forces from Korea under any circumstances, forced or otherwise, would at once release the bulk of the Chinese Army then occupied in Korea and leave them free to attack other areas—quite probably areas of far greater importance than Korea itself. MacArthur claimed:

> On the other hand, the relatively small command we now have in Korea is capable of so draining the enemy's resources as to protect the areas to the south which would in itself be possibly a greater contribution to the general situation than could be made by such a force disposed in other areas for purely defense purposes, but not possessing the power to pin down and localize so massive a part of the enemy's potential as now committed in Korea.

The ROK Army, if a general evacuation took place, would disintegrate or become of negligible value. Japan itself would become extremely vulnerable following the loss of Korea.[3]

MacArthur again assailed the refusal of the Joint Chiefs of Staff to send reinforcements to his aid in Korea. This refusal had been based on the possibility of a greater need for these forces, in case of emergency, in areas more strategic than Korea. MacArthur explained to his superiors that sending additional forces to the Far East would foster rather than hinder the development of military resources in those strategic areas, particularly in western Europe. "I understand thoroughly the demand for European security and fully concur in doing everything possible in that sector," MacArthur continued, "but not to the point of accepting defeat anywhere else—an acceptance which I am sure could not fail to insure later defeat in Europe." He noted that the preparations for the defense of Europe were, by the most optimistic estimates, based upon a condition of readiness two years in the future, and he argued that sending him more American divisions could not possibly prejudice these preparations. Rather, it would insure thoroughly seasoned forces for later commitment to Europe synchronously with Europe's own build-up of military strength.[4]

Touching briefly upon the Joint Chiefs' tactical estimate of the situation in Korea and the danger of forced evacuation, General MacArthur agreed that their estimate was sound under the con-

[2] *Ibid.*

[3] *Ibid.*

[4] (1) *Ibid.* (2) On 19 December 1950, President Truman had, at the request of the NATO Council, appointed General of the Army Dwight D. Eisenhower Supreme Allied Commander, Europe. Eisenhower was to establish in western Europe an integrated allied command to which the member nations of NATO would contribute such forces as they were able. MacArthur knew of this. At this date, however, the plans for an allied defense of Europe, including the extent of United States participation had not been prepared. For the story of this planning and the later build-up of NATO forces in western Europe, see Sir Hastings L. Ismay, Secretary-General of the North Atlantic Treaty Organization, *NATO: The First Five Years, 1949–1954* (Utrecht: Bosch-Utrecht, 1955).

ditions then existing. These conditions were, as he enumerated them: no reinforcements, continued restrictions upon Chinese Nationalist action, no military measures against continental China, and the concentration of China's strength solely upon the Korean sector.[5]

The various actions MacArthur recommended against the Chinese outside of Korea were quite plainly acts of war. The United Nations was not committed, and all indications pointed toward a great reluctance on the part of its member nations to commit themselves to a war with China. The United States could endorse MacArthur's recommendations on its own behalf, but the President lacked authority to send any but American forces against China outside of Korea. If he ordered MacArthur to carry out the recommended actions, the United States, *ipso facto*, would be at war with China. Thus far, the Chinese Government had not declared war against the United States and had, in fact, disclaimed responsibility for the actions of Chinese armies in Korea. While this was purely a technicality it was an important one.

Confining the fighting in Asia to a limited arena in Korea and preserving the unity of the bloc of nations allied with the United States against Communist aggression were basic principles of established national policy. With these aims uppermost in their minds, the nation's top policy-makers weighed and analyzed each of the actions proposed by General MacArthur to determine whether or not the benefit to be derived from it would justify the great risk of causing the Korean fighting to mushroom and of alienating allied powers.[6]

The Retaliatory Measures: An Examination

American leaders studied the specific courses of possible retaliation against Communist China carefully, seeking in each case to determine how effective it would prove if applied; if it were practicable; the effect of its application on the unity of United States allies; and, looming larger than all the other points, if it would cause a general war. The existing national policy approved by President Truman stated, "The United States should not permit itself to become involved in a general war with Communist China." [7] Russia and Communist China had concluded a "peace and universal security" treaty which was made public on 15 February 1950. This treaty could be invoked by either party against "a state which indirectly or directly unites with Japan in acts of aggression," or where "important international questions touching on the mutual interests of the Soviet Union and China exist." [8] Hence, the treaty materially increased

---

[5] Rad, C 52391, MacArthur (Personal) for JCS, 30 Dec. 50.

[6] President Truman's thoughts on this controversial subject are pertinent. In his view, the United States was in Korea in the name of and on behalf of the United Nations. The unified command under General MacArthur was a United Nations Command and neither the President nor MacArthur would have been justified in exceeding the mission originally established by the United Nations General Assembly. Unrestricted military action against China, however attractive, had to be avoided if for no other reason than that it was a huge booby trap.

[7] JCS 1924/35.

[8] Rpt by the JSSC to the JCS on Possible Action in Event of Open Hostilities Between the United States and China, 6 Dec 50, in G-3, DA file 381 China, Case 1/3.

the possibility of Russian intervention in the event retaliatory measures were invoked against China.

Naval Blockade of China

The United States had imposed an economic embargo on Communist China on 3 December. All exports from the United States to Communist China had been banned with the exception of those authorized by validated export licenses; and no more such licenses were issued. On 14 December, the President authorized the Departments of State and Treasury to work out the application of controls over Chinese Communist assets in the United States. On 17 December, these assets were brought under control by a blocking order from the Department of the Treasury. On 16 December, the Department of Commerce had issued orders prohibiting United States ships and aircraft from visiting Chinese Communist ports.[9] But these economic sanctions were never fully effective because other nations, including members of the United Nations with forces in Korea, did not strictly observe them.

General MacArthur had recommended that the coast of Communist China be placed under a naval blockade, an entirely different matter from an economic embargo. For the imposition of a naval blockade implies the existence of a state of war between the blockader and the blockaded. It would have to be limited to ports and coastal lands belonging to or occupied by the enemy. It could not bar access to neutral ports or coasts. It would have to be applied equally to all ships of all nations. Hence, if the United States were to impose a naval blockade of the Chinese mainland unilaterally and without the full co-operation of the United Nations, numerous undesirable complications could arise.

Unless Port Arthur, Dairen, and Hong Kong were blockaded, the whole procedure would prove ineffective. Russia undoubtedly would demand unlimited access to the Dairen and Port Arthur areas, over which it exercised military rights, and other privileges under Sino-Soviet treaties. The British most certainly would refuse any American blockade of Hong Kong. From a political standpoint, unilateral action on a naval blockade probably would set the United States apart from its allies and promote the view that a war with China would be simply a United States war.

On the other hand, Communist China was extremely vulnerable to a properly enforced naval blockade since China depended to a great degree on imports for the materials of war as well as for other goods. The Joint Chiefs of Staff therefore were very much interested in a United Nations naval blockade as an instrument of pressure against China. They felt that if the United Nations should declare a naval blockade, the Russians would respect it just as they had the United Nations blockade of Korea. If the United States undertook its own blockade, however, the Russians might conceivably oppose it with military action.[10]

---

[9] Note by Secy Commerce, sub: Position of the United States Regarding a Blockade of Trade With China, JCS 2118/3, in G–3, DA file 091 Korea, Case 27/6.

[10] The Joint Chiefs of Staff later testified that they were opposed to any naval blockade of China which was not fully sanctioned and approved by the United Nations. Not only did they believe that

Employment of Chinese Nationalist Troops in Korea

General MacArthur had insisted on several occasions, most recently on 30 December, that troops from Chiang Kai-shek's forces on Formosa be sent to Korea to fight. The Joint Chiefs of Staff had consistently refused this request.

The strongest arguments against using these troops lay in the political ill effects of such action. As mentioned earlier, if Chinese Nationalist troops were employed in Korea by the United States, the United States would have to accept the re-establishment, to a large degree, of the World War II relationship between Nationalist China and the United States. Recognition of this acceptance would be widespread both nationally and internationally. The United States would have to reconsider, and possibly revise, its announced policy toward Formosa. This policy, in effect the neutralization of Formosa, was based on President Truman's announcement on 26 June 1950 in connection with the war in Korea. Because they believed that employment of Nationalist troops in Korea would cause the Chinese Communists to behave with even greater militancy, a majority of the United Nations members would probably reject a United States proposal to that effect.

There was also a strong possibility that any change in the American attitude toward Chiang involved in use of his troops would be interpreted by western European nations as reducing the defense of Europe to a lower priority. Further, the move would be certain to make it much harder to obtain a political solution to the Korean conflict through negotiation with the Chinese and North Koreans.

Chinese Nationalist Operations Against the Mainland

As one of his proposals for relieving the Communist Chinese pressure against his troops in Korea, General MacArthur had suggested that the United States "Release existing restrictions upon the Formosan garrison for diversionary action (possibly leading to counter-invasion) against vulnerable areas of the Chinese mainland." If MacArthur's proposal were taken at face value it would have meant that the order, sent to him on 29 June 1950 to insure by naval and air action that Formosa would not be used as a base of operations against the Chinese mainland by the Chinese Nationalists, need only be rescinded. From then on, it would be up to the Chinese Nationalists. But the problem was much more complex than that. In addition, there were a number of possible interpretations of the term "diversionary action," ranging from guerrilla action to full-scale invasion by Chiang's forces, supported by American naval and air power. Authorities in Washington studied this proposal from every conceivable angle to determine if it could be profitably carried out.[11]

such a blockade would "leak like a sieve" but that the dangers of alienating the British and of getting into a shooting war with the Russians were too great to accept. MacArthur, on the other hand, did not think that an American blockade, including Port Arthur and Dairen, would ". . . materially affect in any way the great decisions that would be involved in bringing the Soviets into a global war." See Rad, JCS 80680, JCS (Personal) for MacArthur, 9 Jan 51; and *MacArthur Hearings*, pp. 261, 355, 483, 1188, 1517.

[11] Rpt, JIC to JCS, sub: Estimate on the Effectiveness of Anti-Communist Guerillas Operating in China, JIG 318, in G-3, DA file 091 China, Case 41.

The military capabilities of the Chinese Nationalist forces on Formosa were extremely limited. General MacArthur was quite aware of this, having visited Formosa on 31 July 1950. In comparison to mainland China, with its population of 452,000,000 and an army of over 2,000,000 men, the Nationalists on Formosa had a population under their control of only 7,500,000 and an army of 428,000 men. The Nationalist Army had comparatively few arms; and these were a mixture of American, Japanese, Russian, and German weapons which were poorly maintained. The ratio was one individual weapon for every two and a half men. American leaders were under no illusion that the Nationalists could mount any sort of significant attack against the Chinese mainland unless the United States furnished the materials and transportation. Nor could this be done easily and quickly, even if the United States should decide to divert resources from other vital areas to support operations by Chiang Kai-shek.[12]

Considerations in Bombing China

Direct air and naval surface attacks on the Chinese mainland were probably the most immediate way of striking a hard blow against the Communists. These were also the actions most likely to precipitate a full-scale war.[13] All of the nations allied with the United States against Communist aggression in Korea were strongly opposed to direct attack on China. Since China had no great industrial centers, the most profitable targets would be military and air installations, railroads, and shipping facilities. But experience in World War II had shown that in spite of the best intentions and most accurate bombing, the civil population suffered along with such targets; and any heavy loss of civilian life undoubtedly would be sure to turn many Asiatic nations against the United States. There was little question, moreover, that China, if faced with this bombing, would call upon the USSR to come to its rescue. Most American leaders were therefore not willing to risk bombing China except as a last resort.

In all the discussions of "privileged sanctuary" enjoyed by the Chinese in Manchuria no mention had been made by MacArthur, or by the Joint Chiefs of Staff for that matter, of a similar privileged sanctuary enjoyed by the United Nations Command in Japan. Both naval and air operations against Korea were mounted from Japanese bases, and Japan was the main staging area from which thousands of U.N. troops were sent to fight in Korea. Consequently, if the United States bombed Manchuria to destroy enemy bases, the Chinese might bomb Japan. Whether the Chinese possessed such a capability was certainly a moot point; but it seemed reasonable to assume that with Russian help it would not take them long to acquire such a capability.

President Truman stated that he had never been able to believe that MacArthur, seasoned soldier that he was, did not realize that introducing Chinese Nationalist forces into mainland China

---

[12] JCS 2118/15, 29 Jan 51, in G–3, DA file 381 China, Case 6/3.

[13] Other than broadly hinting that the atomic bomb would be effective in Korea, MacArthur did not recommend officially or, as far as is known, unofficially, that the decision be taken to use the atomic bomb against either the North Koreans or the Chinese, in or out of Korea.

would be an act of war. Certainly, a commander who had been in the forefront of world events for thirty-five years must realize that the Chinese people would react to the bombing of their cities in much the same manner as the people of the United States would have done. The President did not believe, either, that MacArthur with his knowledge of the Orient could really think that he could cut off the vast flow of materials from Russia merely by bombing Chinese cities. The next step would have to be the bombing of Vladivostok and the Trans-Siberian railroad. Because he was sure that MacArthur could not possibly have overlooked these considerations President Truman was left with the simple conclusion that MacArthur was ready to risk general war. The President was not.[14]

Because they were not privy to MacArthur's intentions or to the instructions given him, British officials grew concerned that he might do something that would cause the conflict to spread beyond Korea. When these misgivings were brought to the attention of President Truman, he attempted to allay British fears by assuring Prime Minister Attlee:

> There has not been any change in the agreed United States-United Kingdom position that resistance to aggression in Korea should continue in Korea unless and until superior force required evacuation of our troops. Present tactical situation does not reflect any change in this position but rather essential adjustments to cover increased jeopardy to United Nations troops resulting from recent marked decrease in effectiveness of sorely tried South Korean divisions.[15]

The Joint Chiefs of Staff sent MacArthur an interim denial of his proposals on 9 January. They told him that his suggestions were being carefully considered but that, for the time being at least, little chance existed for a switch in the national policy. The blockade of the China coast, for instance, if imposed, would not take place until the United Nations Command had either stabilized the situation in Korea or had evacuated the peninsula. Nor would American authorities undertake such a blockade without British approval, in deference to the extensive British trade with China through Hong Kong. The Joint Chiefs felt also that any blockade required the concurrence of the United Nations Organization.[16]

The naval and air attacks which MacArthur wished to launch on the Chinese mainland would, in the opinion of the Joint Chiefs at this time, be authorized only if the Chinese attacked American forces outside of Korea, but no decision would be made on the matter until the eventuality arose. Nor did the Joint Chiefs, doubtful that Chiang Kai-shek's troops could have any decisive effect on the outcome of the Korean campaign, intend to approve their use in Korea. They noted that these troops might have a greater usefulness elsewhere in the future.[17]

Neither did they believe that MacArthur should or could count on action outside of Korea to ease the pressure on his forces. They directed him to defend in successive positions, inflicting the greatest possible damage on enemy forces,

---

[14] Truman, *Memoirs*, II, 415–16.

[15] Rad, State to SCAP (including quotation of Truman msg to Attlee), 12 Jan 51.

[16] Rad, JCS 80680, JCS (Personal) for MacArthur, 9 Jan 51.

[17] *Ibid.*

"subject to primary consideration of the safety of your troops and your basic mission of protecting Japan." At the same time, they granted him authority to withdraw from Korea to Japan if in his judgment evacuation was essential to avoid severe losses of men and matériel.[18]

The Joint Chiefs of Staff had given MacArthur two major interlocking courses of action to follow. Whereas he was to defend Korea, this defense was secondary to his mission of saving his troops from destruction and protecting Japan from invasion. The second course, withdrawal, must have been, in the minds of the Joint Chiefs, the natural sequel of the first. But MacArthur chose to interpret the directives strictly and found them, therefore, incompatible. Arguing that both directives could not be carried out simultaneously, MacArthur on 10 January asked for clarification of his orders. He tied to this request another hint that American political objectives needed looking into. He said:

> In view of the self-evident fact that my command as presently constituted is of insufficient strength to hold a position in Korea and simultaneously protect Japan against external assault, strategic disposition taken in the present situation must be based upon the over-riding political policy establishing the relativity of American interests.

It seemed that he was asking the Joint Chiefs to decide which of his missions they considered most important when, in fact, they already had told him.[19]

General MacArthur pointed out to the Joint Chiefs of Staff that his command originally had been sent to Korea to oppose the North Korean Army. There had been no intent that the United Nations Command should engage the armies of Communist China, MacArthur claimed; and he doubted very seriously that his troops would have been sent to Korea at all if it had been foreseen that they would have to fight the Chinese.[20]

His men were capable of holding a beachhead line in Korea for a limited time, the United Nations commander believed, but not without losses. Whether or not these losses could be termed "severe" depended, MacArthur said, "upon the connotation given the term." He angrily decried the unfavorable publicity given the withdrawals of the Eighth Army and X Corps. "The troops are tired from a long and difficult campaign," he complained heatedly,

> embittered by the shameful propaganda which has falsely condemned their courage and fighting qualities in the misunderstood retrograde maneuver, and their morale will become a serious threat to their battle efficiency unless the political basis upon which they are asked to trade life for time is clearly delineated, fully understood and so impelling that the hazards of battle are cheerfully accepted.

With these words, MacArthur seemed to be asking, in the name of his troops, that the measures he had recommended be put into effect or that an explanation be rendered to him and his men.[21]

Citing the limitations under which he was being required to carry on the campaign against the Chinese—namely, no reinforcements, continued restrictions upon Chinese Nationalist military action, no measures permitted against China's

[18] *Ibid.*

[19] Rad, C 53167, MacArthur (Personal) for JCS, 10 Jan 51.

[20] *Ibid.*

[21] *Ibid.*

continental military potential, and the concentration of China's military force in the Korea-Manchuria sector—MacArthur asserted that the military position of his forces in Korea would soon be untenable. He strongly recommended that, under these conditions and in the absence of any overriding political consideration, his troops should be withdrawn from the peninsula just as rapidly as it was tactically feasible to do so.[22]

The final factor in deciding what course to follow, in MacArthur's judgment, was just how far the United States was prepared to go in order to keep a position in Korea. If the primary interest of the United States in the Far East lay in holding a position in Korea and in pinning down a large segment of the Chinese military potential, "the military course is implicit in political policy and we should be prepared to accept whatever casualties result and any attendant hazard to Japan's security." The decision to remain in Korea or to withdraw was not a matter for him to determine, MacArthur contended.

> The issue really boils down to the question of whether or not the United States intends to evacuate Korea and involves a decision of the highest national and international importance, far above the competence of a theater commander, guided largely by incidents affecting the tactical situation developing upon a very limited field of action.[23]

Since the directives he had received from the Joint Chiefs of Staff left the initiative of the decision to evacuate in the hands of the enemy, MacArthur wanted to know if the present objective of United States political policy was to maintain a military position in Korea indefinitely, for a limited time, or to minimize losses by evacuating as soon as possible. "As I have pointed out before," he concluded, "under the extraordinary limitations and conditions imposed upon the command in Korea, its military position is untenable, but it can hold for any length of time, up to its complete destruction, if over-riding political considerations so dictate." [24]

The Joint Chiefs did not change their directives to General MacArthur despite his objection that he did not understand them. They did attempt to explain them to him. They made it quite clear that, after studying all the factors which he had recently presented, they were under no illusion that the United Nations Command could stave off a sustained major effort by the Chinese for any great length of time. But they wanted MacArthur to stay in Korea as long as possible and to kill as many Chinese as possible before pulling out for Japan. This would be in the national interest since it would gain further time for essential diplomatic and military consultations with other United Nations members. The Joint Chiefs told MacArthur:

> It is important also to United States prestige world-wide, to the future of the United Nations and the North Atlantic Treaty Organization and to efforts to organize anticommunist resistance in Asia that Korea not be evacuated unless actually forced by military considerations, and that maximum practicable punishment be inflicted on Communist aggressors.[25]

The Joint Chiefs could not judge the morale of MacArthur's troops from

[22] *Ibid.*
[23] *Ibid.*
[24] *Ibid.*
[25] Rad, JCS 80902, JCS (Personal) for MacArthur, 12 Jan 51.

Washington, they freely admitted. But they were quite concerned about the effect on his men, especially on ROK soldiers, if news of imminent evacuation should reach them. In JCS opinion, any instructions to evacuate would become known almost at once, despite security measures, and any resulting collapse of ROK resistance could seriously endanger the Eighth Army's ability to reach a secure beachhead about Pusan and hold it long enough for actual evacuation. "Your estimate is desired," they told MacArthur, "as to timing and conditions under which you will have to issue instructions to evacuate Korea." Meanwhile, their current directives remained in effect.[26]

The President was deeply disturbed by this. MacArthur was saying, in effect, that the course of action decided upon by the National Security Council and the Joint Chiefs of Staff and approved by the President was not feasible. He was saying that his forces would be driven off the peninsula or, at the very least, suffer heavy losses. MacArthur had always been kept informed but apparently few of the important papers had really found their way to his desk. President Truman therefore resolved to send a personal letter to General MacArthur setting forth the political aspects of the situation from the standpoint of the nation's leaders.[27]

"I want you to know," President Truman wrote MacArthur on 13 January, "that the situation in Korea is receiving the utmost attention here and that our efforts are concentrated upon finding the right decisions on this matter of the gravest importance to the future of America and to the survival of free peoples everywhere." Mr. Truman took special care to emphasize that what he said did not constitute a directive. He merely wanted to let MacArthur know what was being considered in Washington. Mr. Truman called upon MacArthur for assistance in solving some of the problems facing the United States. "We need your judgment as to the maximum effort which could reasonably be expected from the United Nations forces under your command to support the resistance to aggression which we are trying rapidly to organize on a world-wide basis," the President told MacArthur, and enumerated the political advantages which would come with a United Nations victory in Korea.[28]

President Truman cautioned MacArthur obliquely on the latter's proposals for more direct action against China. He warned:

> Pending the build-up of our national strength, we must act with great prudence in so far as extending the area of hostilities is concerned. Steps which might in themselves be fully justified and which might lend some assistance to the campaign in Korea would not be beneficial if they thereby involved Japan or Western Europe in large-scale hostilities.[29]

The President fully appreciated the seriousness of the United Nations Command's military position in Korea at that time and was in no way minimizing the danger. He recognized that continued resistance in Korea might not be militarily possible; but he suggested that, if MacArthur thought it practicable, resistance might still be continued, after

[26] *Ibid.*

[27] Truman, *Memoirs*, II, 434–37.

[28] Rad, JCS to CINCFE, JCS 81050, Truman (Personal) for MacArthur, 13 Jan 51.

[29] *Ibid.*

an evacuation, from offshore islands such as Cheju-do. In any event, Truman continued, ". . . it would be important that, if we must withdraw from Korea, it be clear to the world that that course is forced upon us by military necessity and that we shall not accept the result politically or militarily until the aggression has been rectified." Concluding, President Truman lauded MacArthur for his conduct of the campaign. "The entire nation is grateful for your splendid leadership in the difficult struggle in Korea and for the superb performance of your forces under the most difficult circumstances." [30]

The Joint Chiefs Visit Tokyo Again

General MacArthur's professed failure to understand their directives and his statement that troop morale was at a low ebb convinced the Joint Chiefs of Staff that it was time for another face-to-face talk with the Far East commander. General Collins and General Vandenberg, the designated representatives of the Joint Chiefs of Staff, landed in Tokyo on 15 January.

In the first conference, held on the 15th, MacArthur explained to Collins and Vandenberg that his confusion over his directives had arisen because it was not clear to him how long and under what conditions he was expected to keep his forces in Korea. Nor had he felt that his directives explained clearly enough his responsibility for defending Japan. But MacArthur then read to Collins and Vandenberg the President's letter, which, he maintained, had removed all doubts as to his responsibilities and missions. He interpreted the President's words as a directive to remain in Korea indefinitely.[31]

Collins, after hastening to point out to MacArthur that the President's message was not a directive, as had been clearly stated therein, declared that, at a conference with Mr. Truman just before their departure from Washington, it had been generally agreed that the decision to evacuate Korea should be delayed as long as possible without endangering the Eighth Army or the security of Japan. The United States objective was to permit the longest possible time for political action by the United Nations and the fullest opportunity to inflict the maximum punishment on the Chinese.[32]

Collins held that even if a decision were made to send reinforcements to Japan, it would take at least six weeks for them to arrive and that, in the interim, MacArthur's basic mission of defending Japan would remain unchanged. MacArthur countered by declaring, with some emotion according to General Collins, that his command should not be held responsible for the defense of Japan and still be required to hold a line in Korea. He maintained that Russian forces in Sakhalin and in the Vladivostok area had the capability of attacking Japan; and because this threat was always present, he urged that the four National Guard divisions he sent to Japan to help in its defense. MacArthur said that he understood that these divisions had been mobilized for this purpose; but Collins pointed out that MacArthur had

30 *Ibid.*

31 Memo, Gen Collins for the JCS, sub: Rpt on Visit to FECOM and Korea, Jan 51.

32 *Ibid.*

already been advised that these divisions had not been called up for that purpose and refused to make any commitment on sending the requested units.[33]

Encouraging Signs in Korea

The first real chance for a co-ordinated, though limited, attack since the abortive advance of 24 November, developed in mid-January and General Ridgway quickly took advantage of it. An enemy build-up was discovered north of the Eighth Army's defensive line between Osan and Suwon, and on 14 January General Ridgway ordered an armor-supported co-ordinated attack against this enemy concentration.

He decided on this attack against the advice of his staff. "To a man, the Eighth Army staff was against offensive action north and I alone had to make the decision," Ridgway stated.[34] Ridgway's purpose was to kill as many enemy soldiers as possible and then to withdraw to main positions, leaving a covering force in the area. The attack, known as Operation WOLFHOUND, jumped off on 15 January and inflicted some enemy casualties. The attack was most notable, however, as a sign that the Eighth Army was no longer entirely on the defensive and as a harbinger of the offensive spirit that General Ridgway was bent on developing in his new command.[35]

Generals Collins and Vandenberg arrived in Korea while Operation WOLFHOUND was in progress. General Collins spent two days with General Ridgway, touring the front lines and talking with corps and division commanders. Both the Army Chief of Staff and the Eighth Army commander made statements of great significance at a press briefing held on 16 January in Taegu. General Collins told the newsmen, "As of now, we are going to stay and fight," while General Ridgway seconded this by saying, "There is no shadow of doubt in my mind that the Eighth Army can take care of itself in the current situation." [36]

When Collins returned to Tokyo on 17 January, he sent a most encouraging report to his fellow members of the Joint Chiefs of Staff in Washington. He told them that the Eighth Army was in good shape and improving daily under General Ridgway's leadership. He had found morale very satisfactory, all things considered. The weakest link in the United Nations team was the ROK component. General Collins considered this force still capable of holding off North Korean units, but believed it lacked confidence and instinctively feared the Chinese. He had seen no signs of dissatisfaction or collapse in the ROK Army, but warned that such reactions could develop quickly in case of a serious reverse.[37]

---

[33] *Ibid.*

[34] Interv, Appleman with Ridgway, Oct 51.

[35] Rad, CX 101066 KGOO, CG Army Eight to Corps Comdrs, 14 Jan 51.

[36] Comd Rpt, EUSAK, Jan 51, Narrative, p. 75. While it is axiomatic that a field commander must keep a stiff upper lip and issue optimistic statements under the most adverse conditions in order to strengthen the will of his forces and to avoid giving comfort to the enemy, these statements, both by Collins and Ridgway, were not to that end. There is every indication that Ridgway believed exactly what he said at this point and that Collins, trusting Ridgway's judgment, was for the first time since late November satisfied that a successful stand could be made by Eighth Army troops.

[37] Rad, C 53613, Collins (Personal) for Bradley, 17 Jan 51.

What he had seen of the enemy made General Collins optimistic. The Chinese had made no major move to push south from the Han River, and when counter-attacked had usually fled. He had detected signs also of enemy supply difficulties and indications of a lowered morale among the Chinese. "On the whole," Collins reported, "Eighth Army is now in position and prepared to punish severely any mass attack." [38]

General Vandenberg, meanwhile, had inspected Air Force installations in Korea. In making both aerial and ground reconnaissance, a most remarkable procedure for a man of his high position, he flew by helicopter twelve miles in front of the main U.N. positions and joined a ground patrol.[39]

Both officers met with MacArthur in Tokyo once more before leaving for the United States. Collins read to MacArthur the message that he had sent to General Bradley forecasting a more favorable future for the United Nations Command. General MacArthur agreed that things did indeed look brighter and, after reviewing the military situation as he now saw it, stated that his forces could hold a beachhead in Korea indefinitely. He felt that with continued domination of the sea and air by the United Nations, and with the enemy's lengthening lines of communication, the Chinese would never be able to bring up enough supplies to enable them to drive his forces from Korea. But he reiterated strongly his belief that the decision to evacuate Korea was a purely political matter and should not be decided on military grounds.[40]

The effect of Collins' cheering report on the nation's leaders and on the national policy can hardly be exaggerated. For the first time since late November, authorities in Washington saw reasonable hope that catastrophe might be forestalled in Korea and that all was not as black as had been painted. Recommendations and plans for national policy that had been predicated on almost complete United States helplessness to continue the action in Korea faded in significance.

Clearly, the man most responsible for bringing about this radical change in the situation was General Ridgway. There is little question that when Ridgway took command of the Eighth Army in December, he was under no restrictions as to making further rearward movements. He could have continued falling back without serious recrimination, in view of the prevailing belief in Washington and GHQ that enemy strength was great enough to force the United Nations Command out of Korea. His leadership turned the tide, kept the Eighth Army fighting in Korea, and paved the way for advances that were soon to come.

---

[38] *Ibid.*

[39] *MacArthur Hearings*, p. 329. Vandenberg's excursion beyond the lines is one of the most remarkable sidelights of the entire Korean conflict. While speaking well for his courage, it reflects some doubt on his judgment. Probably no other single individual was in possession of a greater wealth of knowledge of the status, plans, and developments of the United States Air Force at this time. If the Communists were allowed to choose the individuals they would most like to have at their mercy, Vandenberg would doubtless have ranked well toward the top of their list; and that Vandenberg risked capture is certainly apparent.

[40] Memo for JCS, sub: Consultation with Gen MacArthur, 15–18 Jan 51, sgd by Collins and Vandenberg.

The 12 January Memo

The alternatives facing the United States in meeting the Chinese moves in Korea and elsewhere in Asia remained under constant review. In early January, a series of studies, prepared by the Joint Chiefs of Staff to develop recommended measures for the consideration of the National Security Council, enabled the nation's military leaders to crystallize their views on actions that should be taken. These views, at first glance, seemed strikingly similar to those held by General MacArthur. The essential difference lay in the timing of some of the recommended measures. MacArthur wanted all the military actions against the Chinese to take place at once in order to halt the Chinese drive in Korea. The Joint Chiefs of Staff, taking a longer view, attached to some of their recommendations conditions which would not have stopped the Chinese in Korea but would have held them to Korea. Nevertheless, there is in the recommendations made to the National Security Council much justification for General MacArthur's contention later that the Joint Chiefs of Staff supported him in his demands for direct action against China.

Admiral Sherman had sparked the movement in the Joint Chiefs of Staff for taking a stronger stand against Chinese aggression. Sherman told the other members that, so far as he was concerned, a state of open hostilities existed between the United States and China. He felt that since the Chinese, with Russian logistic support, had intervened so effectively in Korea, the time had come for the United States to re-examine its objectives and, particularly, the restrictions which had heretofore been accepted as necessary to prevent a spreading of the conflict. Sherman made certain specific recommendations as to actions which the United States should sponsor and which, after due consideration, the other members of the Joint Chiefs accepted and endorsed.

The Joint Chiefs forwarded their recommendations to the Secretary of Defense on 12 January. Among them was a JCS agreement that the United States should support the South Korean Government as much and as long as practicable, even an exile government, if the United Nations Command were forced to evacuate Korea. The preservation of U.N. combat forces was the most important consideration; but, if possible, the United States should stabilize the situation in Korea. If that were not possible, the United Nations Command would be evacuated to Japan. Major U.S. ground forces in the Far East should not be increased, but limited to those already engaged. If, however, the Chinese should prove unable to force the United Nations Command out of Korea, two of the recently mobilized National Guard divisions might be sent to Japan for defense of that nation. The economic blockade of China should be intensified at once. Further, preparation for an effective naval blockade of China should take place immediately. As soon as the United Nations Command's position in Korea was stabilized, or in the event of a forced evacuation, the naval blockade should be established. They did not specify, however, that it should be a U.N. blockade. The Joint Chiefs recommended further that all restrictions on air reconnaissance of Chinese coastal areas and the Manchurian base be removed at once. They also recommended that restrictions on operations of Chinese Nationalist forces be removed and that the United States furnish logistic support to those forces and to Nationalist guerrillas in China. The Joint Chiefs concluded their recommendations with the suggestion that damaging naval and air attacks be mounted against objectives in Communist China if and when the Chinese attacked U.S. forces outside of Korea. They did not recommend that Chiang Kai-shek's troops be used in Korea.[41]

During his January visit, General Collins had read the 12 January paper to MacArthur. MacArthur apparently did not note the stipulated conditions since he told Collins that he concurred with all the proposals contained in the Joint Chiefs' paper. Later, during the hearings on the relief of General MacArthur in May 1951, this study became rather notorious as the "January 12 Study," with General MacArthur claiming that the views expressed in this paper by the Joint Chiefs of Staff coincided with his own recommendations and that, therefore, he and the Joint Chiefs were in agreement on actions to be taken against the Chinese. But MacArthur's recommendations for immediate reprisal and the Joint Chiefs' recommendations of reprisals only under certain future conditions were hardly identical.

Consideration of the actions recommended against China by General MacArthur on 30 December, and largely supported by the Joint Chiefs of Staff in their 12 January memorandum, revealed differences of opinion among the senior members of the National Security Council staff. In each case, these differences stemmed from divergent attitudes with respect to the question of U.N. support. The representatives of the Secretary of State and the Secretary of Defense agreed with the Joint Chiefs of Staff on the desirability of blockading China and supporting attacks by Nationalist forces on the mainland, but they did not want to take such action unless it was done in co-operation with other friendly nations. Secretary of the Army Pace, in commenting on this position to the Secretary of Defense, said:

> It appears to me that the split views . . . generally revolve around the question of obtaining approval of cooperation of other friendly nations. While I agree that we should make every attempt to obtain their cooperation we should not permit the lack of their cooperation to deter us from a course of action that would contribute to a successful prosecution of the war.

Secretary Pace recommended to the Secretary of Defense that he support the Joint Chiefs of Staff position on these matters.[42]

The measures recommended by the Joint Chiefs of Staff encountered opposition in the National Security Council and were not approved, although discussion of the various courses continued.[43] On 24 January, the President met with the National Security Council, at which time the recommendations of the Joint

[41] (1) Memo, CNO for JCS, 3 Jan 51, sub: JCS 2118/5. (2) Memo, JCS for Secy Defense, 12 Jan 51, sub: JCS 2118/10. (3) *MacArthur Hearings*, p. 1532.

[42] Memo, CofS for Secy Army, 16 Jan 51, sub: U.S. Action to Counter Chinese Communist Aggression.

[43] Memo, JCS for Secy Defense, 23 Jan 51, sub: Recommended Policies and Actions in Light of the Grave World Sit, in G-3, DA file 381 Korea, Case 55.

Chiefs of Staff and the counterrecommendations of the National Security Council Senior Staff were reviewed. But no decision was reached. Mr. Truman then directed a continuation of the study by the Secretary of State and the Secretary of Defense in connection with a joint review of American politico-military strategy.

General Marshall has stated that as a result of the encouraging view of the military situation brought back by Collins and Vandenberg, the courses of action contained in the Joint Chiefs' January 12 study went into virtual discard. "As the result of this change in the military situation from that which prevailed during the early part of January," Marshall testified:

> It . . . [became] unnecessary to put into effect all of the courses of action outlined in the Joint Chiefs' memorandum of January 12. None of these proposed courses of action were vetoed or disapproved by me or by any higher authority. Action with respect to most of them was considered inadvisable in view of the radical change in the situation which originally had given rise to them.[44]

---

[44] *MacArthur Hearings*, p. 324.

CHAPTER XVIII

The United Nations Strike Back

By late January 1951, local successes by United Nations forces and a renewed offensive spirit within Ridgway's command had altered the combat scene and had improved the outlook. No longer was the threat of forced evacuation so real. Nor was the need for new decisions on national policy so pressing.

Attempts by United Nations leaders to arrange a cease-fire in Korea continued fruitlessly throughout the winter. A 14 December resolution by the U.N. General Assembly had established a Cease Fire Group, but otherwise had led to nothing. The Chinese rejected every overture to negotiate except on their own terms. In mid-January, the First Committee of the General Assembly established several principles as the basis for a cease-fire: withdrawal of all non-Korean forces from Korea; free elections under United Nations supervision and arrangements for interim administration; and, after a cease-fire, a conference including representatives of the United States, USSR, and Communist China on settlement of Far Eastern problems. The United States voted for this arrangement even though some American authorities were very skeptical about it.

A statement of these principles was sent to the Chinese Communist Government with an invitation to negotiate a cease-fire. The Chinese countered with a few principles of their own which included their acceptance into the United Nations Organization and the withdrawal of American forces from the Formosa area. The Chinese must have known the United Nations would not agree to these terms, and therefore were probably not surprised when the terms were rejected.[1]

Since the beginning of the Korean War, relations between the United States and its most important allies, Britain and France, had been strained to some degree, particularly after the October crossing of the 38th Parallel. The western nations had not yet reconciled their divergent points of view on the conduct of the campaign, relations with the Chinese Communists, and the disposition of Formosa. But as a result of the Chinese Communist rejection of the United Nations cease-fire proposal, there for the first time appeared to be some ground for a common allied approach to the

---

[1] Department of State, *U.S. Policy in the Korean Conflict, July 1950–February 1951* (Washington, 1951), pp. 27–37.

problems posed by the Chinese.[2] In following up this slight advantage, American leaders undertook to gain acceptance of the American viewpoint by René Pleven, French Prime Minister, when he visited President Truman in Washington on 29–30 January 1951.[3]

In anticipation of Pleven's visit, Department of State planners prepared for the President a statement of American objectives in Korea and the probable paths toward those objectives. This compilation of views amply illustrated that while the American Government had a broad pattern for Korea, no specific means to work out this pattern had yet been developed.

The Korean venture was of necessity a partnership arrangement. Most of the partners who had to be consulted by the United States, the senior partner, hesitated to subscribe to any step which might enlarge the area of conflict or in some other way prove detrimental to their national interests. The United States had no desire, and indeed no intention, to stand alone against the Communists in Korea. The Department of State insisted that the United States should continue to urge the United Nations to adopt a policy of bringing to bear the greatest possible collective pressure upon the Communist aggressor in Korea. This policy, it was felt, would increase the chances of reaching an honorable solution in Korea and would deter similar aggressions elsewhere.[4]

France had supported continued resistance in Korea, but was eager for a peaceful settlement if possible, and had expressed great opposition to extending hostilities outside of Korea. The Department of State therefore recommended that President Truman assure Pleven that the United States would continue to try to confine the fighting to Korea.[5]

About this time, the United States was pressing the United Nations to pass a resolution branding Communist China as an aggressor. The Department of State therefore urged the President to assure Pleven that in the American view the passage of this resolution would not constitute authorization for the extension of hostilities to China, and that the United States had no intention of asking the United Nations for authority to take any measures involving operations against Chinese territory. But the United States Government, in its capacity as the Unified Command, reserved the right to take action essential for protecting the United Nations forces under its command. Consequently, the Department of State felt that Pleven should know that in the event of large-scale air attacks against U.N. troops from Manchurian bases, the United States would bomb the bases from which the attacks originated, and that if the Chinese Communists attacked American forces outside of Korea, the United States would take counteraction. The Joint Chiefs of Staff thought that the Department of State was right and placed their seal of approval on these views.[6]

President Truman and Prime Minister Pleven, on the first day of their talks, concentrated on the situation in Asia.

---

[2] Memo, Gen Duff, Dep ACofS G–3, DA, for CofS, USA, 23 Jan 51, Incl 3, in G–3, DA file 320.2, Case 60.

[3] (1) JCS 1776/187, 26 Jan 51. (2) JCS 1776/186, 24 Jan 51. Both in G–3, DA file 091 Korea, Case 151.

[4] *Ibid.*

[5] *Ibid.*

[6] *Ibid.*

The President told the French Prime Minister substantially what his advisers had recommended. He stated forcefully that he saw no way for the United States to recognize the Communist regime in Peiping, and that he was convinced that the Communists had moved into Korea because they feared the progress being made by the western powers in the Far East. He assured Pleven that the United States was striving for world peace, but that only collective security could bring this about. The United States therefore would not negotiate with the Chinese to restore peace in Korea at the price of collective security and national self-respect.[7]

Since the first intervention by the Chinese, the United States, through Ambassador Austin, had championed a resolution before the General Assembly of the United Nations which would brand China an aggressor nation and at the same time provide a method of bringing about a cease-fire. Many member nations of the United Nations, fearing that such a step would only increase the scope of the fighting and widen the breach between Communist China and themselves, hesitated to support the resolution. But President Truman urged passage of the measure in line with his determination that "For my part, I believe in calling an aggressor an aggressor." [8] Finally, on 1 February, after much hesitancy on the part of some member nations and complete opposition by the USSR and its satellite member nations, the United Nations General Assembly adopted a resolution naming the People's Republic of China an aggressor nation and calling for the achievement of United Nations objectives in Korea by peaceful means.[9] India and Burma voted against calling Communist China an aggressor. Seven nations of the non-Communist world and Yugoslavia did not participate in the voting.[10]

The Combat Scene

Regardless of political efforts to find common ground for negotiation, the issue between the Communists and the United Nations in Korea continued to be decided at this stage of the war on the battlefield. The success or failure of the United Nations political efforts would, it appeared, depend on the success or failure of the United Nations military measures. During late January and February, General Ridgway concentrated on means of exploiting to the very limit the capabilities of the forces under his command. Conferring with his I and IX Corps commanders on 21 January, he ordered them to mount a strong combat reconnaissance into the area bounded by the Suwon-Ich'on-Yoju road and the Han River to develop enemy dispositions, disrupt hostile concentrations, and inflict maximum destruction on the enemy.[11] This reconnaissance, designated Operation THUNDERBOLT, jumped off on 25 January and made consistent progress against generally light resistance. (*Map*

[7] Truman, *Memoirs*, II, 437–38.

[8] (1) Statement by the President, 25 Jan 51. (2) *MacArthur Hearings*, p. 3513.

[9] UN Doc A/1771, quoted in Department of State Pub 4263, *U.S. Policy in the Korean Conflict, July 1950–February 1951*, p. 37.

[10] Memo, Dr. Ralph J. Watkins for Maj Gen Maxwell D. Taylor, 1 Mar 51, in G–3, DA file 381 China, Case 8/8.

[11] Memo, Notes on Conference with CG I Corps, CG IX Corps, and Vice Comdr 5th AF, Gen Ridgway, 21 Jan 51, in GHQ, UNC SGS.

U.S. INFANTRYMEN *enter the walled city of Suwon.*

VII) By the end of January, the enemy's main line of resistance still had not been developed; but Ridgway's forces had reached a line four to six miles north of the line Suwon–Kumyangjang-ni–Ich'on and were continuing their advance.

Understandably encouraged by the January advances, General Ridgway in early February outlined plans for the immediate future and his ideas on longer-range moves. Ridgway informed MacArthur that the Eighth Army was inflicting maximum losses upon the enemy and delaying to the utmost enemy attempts to push farther into South Korea. This was being done, Ridgway claimed, at the same time that complete co-ordination within and between the corps was insuring the integrity of all major units. General Ridgway reported that his forces in the western sector were moving forward in phased, closely co-ordinated advances to develop the enemy dispositions on that front and to kill as many of the enemy as possible with a minimum of friendly losses. Ridgway told MacArthu that if it proved tactically sound and militarily possible, he would send h troops as far as the Han River where they would hold. He also planned a co-ordi-

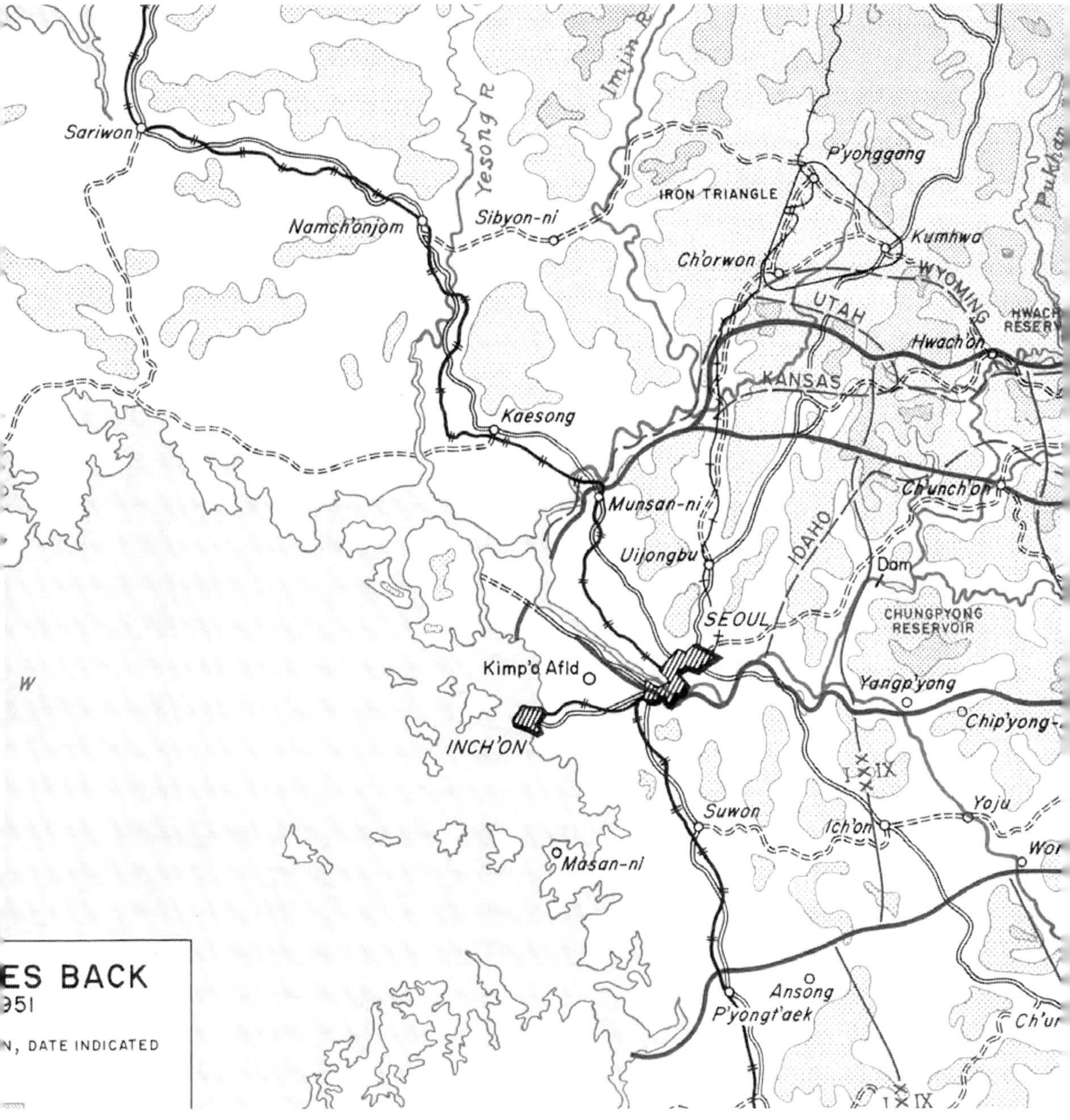

Sariwon
Yesong R
Imjin R
P'yonggang
IRON TRIANGLE
Namch'onjom
Sibyon-ni
Kumhwa
Ch'orwon
WYOMING
UTAH
HWACH
RESERV
Hwach'on
KANSAS
Kaesong
Ch'unch'on
Munsan-ni
IDAHO
Uijongbu
Dam
CHUNGPYONG
RESERVOIR
SEOUL
Kimp'o Afld
Yangp'yong
Chip'yong-
INCH'ON
Yoju
Suwon
Ich'on
Masan-ni
ES BACK
951
N, DATE INDICATED
Ansong
P'yongt'aek
Ch'u

nated attack on the central front in the very near future to reach and hold the general line Yongp'yong-Hoengsong-Yangnung.[12]

In General Ridgway's opinion, the advance to the Han, at least as far east as Yongp'yong, was a sound operation with potentially high results so long as enemy resistance did not stiffen to the point where United Nations losses canceled out military gains. But from Yongp'yong eastward to the Sea of Japan, a distance of ninety airline miles, the Korean front ran through a rugged, wooded area lacking roads and any natural terrain line on which to base defensive positions that could be easily or profitably held. General Ridgway had no illusions about setting up a static defense or making substantial advances in that sector. With regard to this part of Korea, he said, "Assuming as I do, a continuation of a major effort to destroy us, prolonged efforts to hold any such line, would in my view, require far greater forces than are now available, and entail a heavy attrition, with little or no commensurate gain." Ridgway made it plain to MacArthur that he saw little wisdom in a general advance beyond the line of the Han River in view of the great risks. He would recommend such an advance only if the Chinese forces should voluntarily withdraw north of the 38th Parallel. As to the 38th Parallel, General Ridgway told MacArthur that he considered it indefensible with his present forces. If Eighth Army tried to hold any part of the former boundary too many men would be lost.[13]

Likewise, any attempt at the moment to retake the South Korean capital would, in General Ridgway's opinion, be foolish from a purely military viewpoint. To occupy Seoul would place an unfordable river through or behind his defensive positions. Therefore, unless a sudden opportunity arose to trap and destroy a major enemy force, in which the retaking of Seoul was incidental, he would leave Seoul to the Chinese.[14]

Ridgway set forth five major assumptions: the enemy would continue a major effort to force the Eighth Army from Korea or to destroy it in place; there would be no major reinforcement of the Eighth Army; the basic plan and directives under which the Eighth Army was operating were sound and required no present modification; the 38th Parallel could not be defended with forces then available; and elsewhere north of the Han, terrain lines across the peninsula were not good enough to justify the losses required to take and defend them. He asked General MacArthur for his views on these concepts.[15]

General Ridgway's analysis and plans presented MacArthur, according to General Whitney, ". . . with a dilemma." Whitney explained later that MacArthur placed far greater stress on the factor of supply ". . . than Ridgway apparently did." MacArthur had not changed his opinion that Seoul was a vital supply hub which had to be seized if enemy supply was to be effectively curtailed and that sound psychological advantages lay in capturing the ancient capital city. "Therefore," Whitney recalls, "he had no intention of holding the Eighth Army

[12] Ltr, Gen Ridgway to Gen MacArthur, 3 Feb 51, in GHQ, FEC SGS files.
[13] *Ibid.*
[14] *Ibid.*
[15] *Ibid.*

south of the Han River. Yet he understood fully that Ridgway's caution was natural because of the heavy blow the army had sustained when he had sought to hold the Seoul area before. . . . MacArthur worded his reply carefully." [16]

MacArthur did, indeed, word his reply carefully, so carefully that he seemed to agree with all of Ridgway's proposals and ideas. MacArthur informed Ridgway on 4 February, "I interpret your objective to be such advance with concomitant pressure by your own forces as will develop the enemy's main line of resistance." If this line developed south of the Han, MacArthur agreed that no attempt should be made to push farther. If, on the other hand, the Eighth Army reached the Han without serious resistance, MacArthur believed that Ridgway should drive forward until either the enemy line had been developed or the fact established that the enemy had no such line. Ridgway had said essentially the same thing when he recommended an advance only if the Chinese had voluntarily withdrawn north of the parallel.[17]

MacArthur's thinking on retaking Seoul seemed to parallel that of the Eighth Army commander. He agreed that the military usefulness of Seoul was practically nil, but that its occupation by Ridgway's forces would yield certain valuable diplomatic and psychological advantages. More tangible advantages would accrue from taking the Inch'on port facilities and Kimp'o Airfield. Use of these would greatly reduce the supply difficulties and increase the power of Ridgway's air support in forward areas. He said that if Inch'on and Kimp'o presented "easy prey," they should be taken. In closing, MacArthur commended Ridgway highly, saying, "Your performance of the last two weeks, in concept and in execution, has been splendid and worthy of the highest traditions of a great captain." [18]

The enemy had not forgotten the September landing at Inch'on. American intelligence agencies learned that the Chinese were very worried over the possibility of another amphibious landing either at Inch'on or at the narrow waist of Korea. Ridgway asked MacArthur to consider exploiting these enemy fears by naval feints to simulate possible landings. MacArthur ordered these diversionary actions, which, at Inch'on, forced the employment of at least one North Korean division to guard the port against the threatened attack. Later on, this enemy division was pulled out of Inch'on and thrown against Ridgway's advance from the south in Operation THUNDERBOLT. Noting this, Ridgway asked MacArthur to resume the naval demonstrations to draw enemy forces away from the front. Hence, high-speed amphibious ships sailed from Pusan and again simulated actual landing operations in the Inch'on area.[19]

On 5 February, General Ridgway or-

[16] Whitney, *MacArthur: His Rendezvous with History*, pp. 460–61.

[17] Rad, C 54811, CINCUNC to CG Army Eight (Personal) for Ridgway, 4 Feb 51.

[18] (1) *Ibid.* (2) Comenting on this exchange, General Whitney noted, "Here was one of the marks of leadership; MacArthur thus got his conflicting views across to Ridgway without doing violence to sensibilities which had been suffering acutely in the difficult campaign." See Whitney, *MacArthur: His Rendezvous with History*, p. 461.

[19] (1) Comd Rpt, G–3 UNC, Jan 51, App. 4, Part III, Incl 48. (2) Rad, CG–2–621 KGOO, CG Army Eight to CINCFE, 6 Feb 51. (3) Rad, COMNAVFE to CTF 90 and CTF 95, 6 Feb 51.

GENERAL REEDER *(left)* AND GENERAL TAYLOR *(center) arrive at Taegu Airstrip to begin their tour of inspection of the fighting front. They are greeted by Maj. Gen. Henry I. Hodes, Deputy Chief of Staff, U.S. Eighth Army.*

dered the X Corps to attack in the central sector. For several days, Almond's troops made good progress in their advance, known as Operation ROUND-UP; but enemy resistance increased steadily as the X Corps approached the main enemy positions. In the west, meanwhile, other Eighth Army units drove ahead, piercing the enemy's defenses south of Seoul and forcing the Chinese back across the Han River in the Seoul area on 10 February.

In reporting these successes to the Department of the Army on 10 February, General Willoughby, MacArthur's G-2, adopted a justifiably optimistic tone. He claimed that the enemy was not voluntarily withdrawing from any of his positions but was being forced to do so. He pointed out that once Ridgway seized the Han River, the enemy could find no defensive positions short of the old North Korean defense line along the 38th Parallel. Any enemy withdrawal to the 38th Parallel would have to be viewed,

Willoughby claimed, as a decision forced on them by a series of defeats in the field with an accompanying attrition in men and matériel. He told staff officers in Washington that the enemy had definitely been kicked back and so long as United Nations pressure could keep the enemy off-balance, the initiative would remain with MacArthur.[20]

But just what MacArthur intended to do next in Korea was not clear to officials of the Department of the Army. In an effort to get abreast of the situation, Maj. Gen. Maxwell D. Taylor, General Bolté's successor as Department of the Army G–3, asked General MacArthur to explain his plans. High-level talks in which Army officials were to take part in the very near future made it necessary that MacArthur submit his views on probable short-range military developments. Taylor asked MacArthur specifically if he intended to push north and, if so, if he also intended to move his prepared defense base forward.[21]

MacArthur based his reply on the exchange of views with General Ridgway earlier in the month. In fact, had he transmitted Ridgway's letter to General Taylor without change, the effect would have been the same. "It is my purpose," MacArthur told Taylor, "to continue the ground advance until I develop the enemy's main line of resistance or the fact that there is no such line south of the 38th Parallel." A constant advance, MacArthur pointed out, would keep the Chinese and North Koreans off-balance. This would allow his own forces to take full advantage of their superior artillery firepower and armor and, as MacArthur expressed it, "to flush the enemy from concealment where he may have escaped air attack." MacArthur promised that if his attacks showed that the enemy was not present in strength south of the 38th Parallel, he would immediately notify the Joint Chiefs of Staff and request instructions before moving farther north. He was obviously preparing well in advance to refute any possible charges that he intended to make another crossing of the parallel without full authority from higher headquarters.[22]

Although the measures he had recommended against the Chinese outside of Korea had not been taken and no reinforcement had arrived, MacArthur's command had not been driven out of Korea as he had predicted. Yet he still insisted that he be allowed to bomb Manchuria. "It can be accepted as a basic fact," he told Taylor, "that, unless authority is given to strike enemy bases in Manchuria, our ground forces as presently constituted cannot with safety attempt major operations in North Korea." Where previously he had insisted that bombing China was necessary to permit his forces to stay in Korea, General MacArthur had modified this view to the extent that he now felt it was necessary to bomb China in order to operate in North Korea.[23]

MacArthur said that it was evident to him that the enemy had lost his chance of a decisive military decision in Korea. But he still considered that the Chinese retained the potential, so long as their base of operations in Manchuria was immune from air attack, of resuming the

[20] Telecon, TT 4364, GHQ and DA, 10 Feb 51.

[21] Rad, DA 83262, DA to CINCFE, 11 Feb 51.

[22] Rad, C 55315, MacArthur (Personal) for Taylor, 11 Feb 51.

[23] *Ibid.*

offensive and forcing a withdrawal upon his command. "We intend to hold the line of the Han up to the point of a major and decisive engagement," the U.N. commander claimed. He no longer feared forced evacuation, but he did anticipate being forced back from the Han, how far he could not say. Once forced back from the river, however, his forces would be able to stabilize the line because logistic difficulties would prevent the enemy from full exploitation of the initial advantage. "The capability of the enemy is inversely and geometrically proportionate to his distance from the Yalu," is the way MacArthur expressed it. MacArthur also made a plea for security of his plans by telling Taylor, "It is earnestly requested that no estimate of the situation be released in Washington." He claimed that in the past his plans had been jeopardized by leaks to the press of his secret reports and by "injudicious speculation which has emanated from more or less authoritative sources." He reminded Taylor that reports and estimates should be released only at the discretion of the field commanders.[24]

Secretary of the Army Pace grew concerned lest the success of early February probing attacks lead to overoptimism on the part of the American public. He feared this might give rise to speculation as to whether United Nations forces might not again advance to the Yalu River. He notified MacArthur on 13 February that the actual situation was being depicted for press officials, and that they were being cautioned against putting out unjustifiably glowing reports on progress in Korea. He hoped that this step would curb any wrong trend in public thinking on the matter.[25]

General MacArthur visited the battlefront in mid-February for a first-hand look at the situation in the field. When he returned to Tokyo, he issued a public statement which warned that, in spite of recent advances by his forces, the future of the Korean fighting depended upon international considerations and upon high-level decisions which had not yet been received by his headquarters. It is obvious that he still had not abandoned hope that his recommendations to bomb China, use Nationalist troops, and blockade Chinese ports would be approved. Or, if he had abandoned hope, he at least wanted to keep these recommendations alive in the public mind.[26]

With reference to the battlefield situation of the past several months, MacArthur credited his strategy of rapid withdrawal before the Chinese in December with lengthening the enemy's supply lines and "pyramiding his supply difficulties." But he insisted that recent tactical successes by the Eighth Army under Ridgway not lead to overoptimism. The Chinese, MacArthur cautioned, still retained a tremendous potential for further offensive operations. In this last contention MacArthur was correct.[27]

Chip'yong-ni

While the successes were being achieved in the west, Operation ROUND-

[24] *Ibid.*

[25] Rad, W 83277, DA to CINCFE, 13 Feb 51.

[26] This public statement was duly noted by Washington authorities and later cited as an example of MacArthur's violation of the 6 December directive. See *MacArthur Hearings*, p. 3539.

[27] Statement, General MacArthur in *Pacific Stars and Stripes*, February 14, 1951.

UP was beating itself out against strong resistance and enemy counterattacks in the central sector north of Hoengsong. The U.S. X Corps and the ROK III Corps met increasingly heavy enemy concentrations in their attempts to advance. American intelligence already had warned that the enemy was shifting most of his forces from the west to the central zone. Unable to hold in the west, the enemy apparently was massing his strength against the relatively weak center of the United Nations line.

On the night of 11–12 February, two Chinese armies and a North Korean corps struck the central front, scattered three ROK divisions, and forced the troops in this sector to abandon Hoengsong and withdraw southward toward Wonju. The enemy was obviously aiming his attack at the communications centers of Wonju, near the center, and Chip'-yong-ni, near the west flank of the X Corps sector, whose seizure would assist further advances to the south and west. General Ridgway therefore resolved that Wonju and Chip'yong-ni would be held.

At nightfall on 13 February, three Chinese divisions opened attacks against Chip'yong-ni. For three days, the 23d Regiment of the U.S. 2d Division, with the French battalion attached, staved off all efforts by the Chinese to overrun the town and killed thousands of the enemy.

Although the Chinese were stopped in the Chip'yong-ni area, enemy forces farther east bypassed Wonju and advanced south almost to Chech'on before the Eighth Army could halt them. Weakened by great losses in men and ham-strung by an inadequate logistic system, the Chinese and North Koreans then called off their attack and withdrew.

For the Eighth Army, there was no resting on laurels. Even before the front lines stabilized, General Ridgway opened Operation KILLER to destroy the enemy east of the Han River line and south of the general line Yangp'yong–Hyonch'on-ni–Haanmi-ri. The main effort was directed along the Wonju-Hoengsong and Yongwol-P'yongch'ang axes. Ridgway issued strict orders that this would be no runaway drive north. He demanded close lateral co-ordination within and between the two attacking corps, the IXth and Xth, and emphasized that his purpose was to kill enemy troops.

General Ridgway took American newsmen in Korea into his confidence on Operation KILLER, but with the strict provision that they would not disclose the plan until after the attack had started. By so doing, Ridgway hoped to prevent publication of rumor or conjecture from compromising his intentions to attack. But on 20 February, General MacArthur flew into Korea and during a press conference at Wonju that day announced to the newsmen that he was going to launch an offensive in a day or two. This startled and amazed General Ridgway, not only because MacArthur had disclosed Ridgway's intentions to the enemy, but also because the planned offensive had originated with the Eighth Army, not General MacArthur. The premature announcement, however, had no effect on the outcome of Operation KILLER. When the Eighth Army jumped off on 21 February, the enemy faded away; and within eight days, Ridgway's troops reached their assigned objective.[28]

---

28 Ridgway, The Korean War, Issues and Policies, p. 403.

Supply Problems

Despite the fading resistance, the Eighth Army's move northward had not been easy. For heavy rains and thawing in the combat area had severely hampered the movement of vehicles and transport of supplies. Besides the harmful effect of bad weather, supply operations were also inhibited by the heavy battle damage to overland supply routes. This combination of weather and battle destruction prompted Ridgway to call upon General Stratemeyer for greater effort to support his advance units by air. He asked that Stratemeyer's planes airlift a daily minimum of 200 tons of cargo by C–47 aircraft to forward airfields.[29]

Stratemeyer, believing that Ridgway was asking too much, appealed to General MacArthur, pointing out that the forward airfields were too small and dangerous to risk landing planes, even C–47 cargo transports. He claimed that these forward airstrips would have to be lengthened to at least 4,000 feet before they would be safe enough to use in support of Eighth Army front-line divisions, and asked that the job, if these airfields were to be lengthened, go to the Army Engineers.[30]

But MacArthur supported Ridgway. He did want the airfields improved, but he did not want to take engineer troops away from the Eighth Army to do it. In reply to Stratemeyer's appeal, MacArthur declared that army engineers were scarce even though the maximum number possible had been shipped from the United States. Units in Korea had been committed and were being used full time to repair the Eighth Army's overland supply lines. Therefore, no army engineers would be made available to the Air Force from Ridgway's scanty resources. Instead, Stratemeyer was directed to bring up Air Force engineer troops from Okinawa. These troops would be replaced on Okinawa by native laborers. General MacArthur noted that his experience in Japan had shown him that Japanese contractors were capable of handling the type of construction then being done by Air Force engineer troops on Okinawa.[31] Stratemeyer asked MacArthur to reconsider, but was again overruled and directed to carry out this order. He was told, however, that if he so desired, he might accomplish the airstrip improvements without bringing up Air Force engineers from Okinawa.[32]

This action, of course, did not solve the immediate problem. General Ridgway, apparently unaware of MacArthur's action, called on General Hickey, the acting chief of staff at GHQ, for his support. "In the northward advance through the mountainous terrain of central and western Korea," Ridgway explained:

it had become increasingly difficult to maintain pressure sufficient to inflict maximum casualties upon the enemy. As you know, we have experienced some difficulties arising from recent rains here—we are seeking every possible means to anticipate and provide for adverse conditions while operating in extremely rugged terrain over lines of

[29] Rad, CG–2–1399 KGLO, CG EUSAK to CG FEAF, 26 Feb 51.

[30] Rad, AX 2844 DO, CG FEAF to CINCFE, 16 Feb 51.

[31] Rad, CINCFE to CG FEAF, 16 Feb 51, in G–4, GHQ, UNC file AG370.5.

[32] (1) Staff Sec Rpt, G–4 GHQ, UNC, Feb 51, ch. II, p. 3. (2) USAF Hist Study No. 72, USAF Operations in the Korean Conflict, 1 November 1950–30 June 1952, pp. 151–52 and 155.

communication rendered undependable by the approaching rainy season.[33]

Ridgway told Hickey of his request to Stratemeyer and pointed out that in his opinion some airfields in the forward area could be developed rapidly to meet his needs with a minimum of engineer effort. "Inasmuch as the support of these additional aircraft is of vital importance to current and planned operations, I would appreciate it very much if you would give your personal attention to my recommendations on this subject." [34]

On 6 March, General Stratemeyer called on Air Force headquarters in Washington to send him five engineer aviation battalions and other specialized engineer units to construct forward airfields. This request was turned down and Stratemeyer was told by his own service, as he had originally been told by MacArthur, to bring up his engineers from Okinawa. Eventually, in April and May, Stratemeyer transferred an engineer aviation group and three engineer aviation battalions to Korea from Okinawa.[35]

Combat Strength

The seemingly insoluble problem of the lack of fighting men continued to plague the United Nations Command throughout the winter of 1951. Within the Eighth Army, General Ridgway made drastic cuts in the service units and transferred the excess men to combat units. He recommended that the percentage of service troops be trimmed down in Japan as well and he asked that the maximum possible number of men already in or arriving in the theater be diverted from service-type duty to combat. He insisted that the percentage of service personnel to combat personnel being sent to Korea was too high for the needs of the Eighth Army, and he recommended reclassification if necessary to meet the combat requirements. At this time, 57,000 men were performing service support duties in Korea and almost 35,000 in Japan.[36] But even though the greatest possible number of men was transferred, Ridgway's divisions remained well below their authorized strengths in infantry and artillery.

General MacArthur realized by late January that he could expect no major reinforcements for Korea from the United States. But such successes as Ridgway's troops had scored during the month had convinced him that the Eighth Army was going to stay and make a fight of it. As anxious as the field commander to see the divisions in Korea made as strong as possible, MacArthur took the problem to the Army Chief of Staff on 29 January. "The continuous lack of combat replacements for the seven months of combat is a matter of grave concern to me," MacArthur complained to Collins. He anticipated a suggestion to convert his local service support personnel to combat soldiers, by telling Collins that this had already been done. He had integrated ROK soldiers into American units also. But, MacArthur told General Collins, there was no acceptable

[33] Rad, G-3287 KCG, CG Army Eight to CINCFE (Hickey), 2 Mar 51.

[34] *Ibid.*

[35] USAF Operations in the Korean Conflict, 1 November 1950-30 June 1952, pp. 151-52.

[36] (1) Rad, CG1-1033 KAGCP, CG Army Eight to CO Japanese Replacement Center, Info CINCFE, 14 Jan 51. (2) Rpt, Strength of Service Unit, tab 169A, in G-3, DA files.

substitute for trained combat replacements.[37]

"To date," MacArthur pointed out, "Army divisions have been fighting from twenty to fifty percent below authorized strength in infantry and artillery units." He expressed particular concern because this lack of replacements would not let him rotate combat-weary soldiers. General MacArthur also pointed out that the shortage of combat soldiers within Ridgway's divisions necessitated extended frontages which were susceptible to infiltration and which exposed not only combat elements but supply and communications lines as well. This latter condition had resulted in abnormally high casualties in rear areas. Only the Marine division in Korea did not suffer from lack of men. MacArthur pointed out to Collins that this division was at full strength and had been for some time.[38]

The shortage of replacements within the Far East Command amounted to approximately 40,000 men. General Collins decided that the only way to take care of this deficiency, at least temporarily, was to pull more men away from the General Reserve. He did not intend to repeat the experience of the previous autumn when the General Reserve had been stripped to virtual ineffectiveness, but he did approve a levy of 14,300 men in February. Most of this levy was to be placed on the four National Guard divisions and the two RCT's called to active duty the previous September. General Collins directed the Department of the Army G–1 to start action at once to ship these men to MacArthur within the next four to six weeks. Collins expressed the hope that MacArthur's needs could be met almost in full and that, ultimately, the Army could send him 27,000 men above and beyond the normal flow of replacements.[39]

On 30 January Collins sent to MacArthur a full explanation of the steps the Army was taking to give him sufficient soldiers. "You can rest assured that we here are aware of your personnel shortages and the effect upon your operations," Collins said. He explained that since the number of men being trained and shipped to the Far East from the replacement training centers in the United States had proved inadequate, he was calling on the General Reserve. Of the 14,300 men to be taken from the Reserve, the great majority would be infantry and would arrive in the Far East during the last week of February. This, while not meeting the needs entirely, would at least bring the fighting divisions closer to authorized strength. Collins reminded MacArthur of their talks in Tokyo earlier in the month. At that time MacArthur had agreed with Collins that any sizable levy upon the General Reserve would cause a great delay in bringing units within the United States, including those destined for Europe, to a satisfactory state of combat readiness, and had further agreed that this would be a most serious mistake.[40]

Collins also informed MacArthur that the shortages of trained combat troops throughout the entire Army were such that he had directed an Army-wide reduc-

---

[37] Rad, C 54360, MacArthur (Personal) for Collins, 29 Jan 51.

[38] *Ibid.*

[39] Min, 59th mtg Army Policy Council, 30 Jan 51.

[40] Rad, DA 82320, Collins (Personal) for MacArthur, 30 Jan 51.

tion of 3 percent in service personnel. He felt that this could be done without seriously affecting the fighting efficiency of combat units.[41]

Although General Collins had told MacArthur on 29 December that it would not be practicable to obtain more troops from other members of the United Nations, the Joint Chiefs took certain steps in that direction in late January. In their opinion, increased active participation in the Korean fighting by other member states of the United Nations would not only bolster MacArthur's forces, but would serve as well to bind those states more closely together in opposition to Communist aggression, wherever it might occur. The Joint Chiefs accordingly recommended to Secretary of Defense Marshall on 24 January that he ask the Secretary of State to exert renewed pressure on other member states to furnish ground forces to MacArthur's command. They asked that the general criteria established for such forces in August 1950 be observed and that Great Britain and the NATO countries on the European continent not be solicited.[42]

When on 6 February the Assistant Secretary of the Army, Earl D. Johnson, visited the Far East Command, General Beiderlinden, MacArthur's G–1, took advantage of Johnson's presence to register additional complaints on the manpower shortages. "The failure to provide adequate replacement support has had a deleterious effect on the entire Korean operation," Beiderlinden told Johnson.

> Every expedient was employed to close the gap and maintain combat units without retraining. Wounded men were returned to the front lines again and again without sufficient recuperation to assure full recovery. Combat units were combined, stripping personnel from one to fill another. ROK, United Nations forces, indigenous personnel and incapacitated limited service, all were exploited to the maximum. . . . The end result of such personnel planning must inevitably be reflected in extended frontages, inability to develop full combat effectiveness, all resulting in adjustments in tactical planning combined with abnormal casualties.[43]

But this reclama could have had little effect, for the Army was already exerting the greatest practicable effort to meet MacArthur's needs. In fact, by 27 February, the effort had been so successful that Army officials were able to promise MacArthur twice-monthly increments of replacements totaling over 55,000 by the end of April. The majority of these would be trained in replacement training centers in the United States.[44]

National Guard Divisions

During December, General MacArthur had made several unsuccessful attempts to secure divisions of the National Guard for his theater. On 18 December, he had asked that all four of the National Guard divisions called to active service in September be moved to Japan at once. The Joint Chiefs of Staff had turned him down, "pending continued consideration at a governmental level as to the future United States courses of action in Korea. . . ."[45] Again on 30 December,

[41] *Ibid.*

[42] Memo, Gen Bradley for Secy Defense, 24 Jan 51, sub: Military Assistance for Korea.

[43] Memo, Gen Beiderlinden for Hon Mr. Johnson, 6 Feb 51, sub: Status of FEC Enlisted Replacements During FY 51, in G–1, GHQ, UNC files.

[44] Rad, DA 84382, DA to CINCFE, 27 Feb 51.

[45] (1) Rad, C 51599, CINCUNC to DA, 18 Dec 51. (2) Rad, JCS 99616, DA (JCS) to CINCFE, 23 Dec 50.

MacArthur insisted that these divisions be shipped to Japan for the defense of Japan. "Indeed," he told Washington, "it was my understanding, in which I may have been in error, that the four National Guard divisions called to active duty in September were for ultimate employment here should the necessity arise. . . ." In early January, while the issue of forced evacuation was still in serious doubt, the Joint Chiefs of Staff hinted that, if Eighth Army's line could be stabilized with the forces already in Korea, two partly trained National Guard divisions could be sent to defend Japan. But if evacuation took place, Japan would have to be defended by troops removed from Korea.[46]

After General Collins returned to the United States from his mid-January visit to the Far East, he reviewed again the possibility of furnishing National Guard divisions to MacArthur. The Eighth Army was giving a good account of itself in Korea and was stabilizing its position with the forces already available to it. Under these conditions it seemed appropriate to carry out the half-promise of 9 January to send two partly trained National Guard divisions to Japan. On 23 January, he told General MacArthur that if things in Korea kept going as well as at present and the Chinese could be contained, two divisions might be sent him to increase the security of Japan.[47] A week of continued successes in the field followed, and on 30 January Collins recommended to the other members of the Joint Chiefs of Staff that the U.S. 40th and 45th Divisions, both National Guard, be ordered to Japan to bolster the defenses there. There were, at that time, no American divisions in Japan. The Joint Chiefs of Staff agreed and forwarded the recommendation to the Secretary of Defense at once. MacArthur was informed of this development as soon as it took place.[48] After relatively lengthy consideration which involved weighing the interests of the European theater against those of the Far East, the Secretary of Defense consented to the transfer of the two major units. On 25 February, the Joint Chiefs of Staff notified the U.N. commander that the 40th and 45th Divisions would reach his command sometime in April. He was specifically ordered to leave these divisions in Japan and not to employ them against the enemy in Korea.[49]

Bombing of Rashin

In the early months of the Korean fighting, General MacArthur had been furnished a list of targets in North Korea which the Joint Chiefs of Staff thought suitable for destruction by strategic bombing. Among these key targets was the port city of Rashin. Rashin, lying only nineteen air miles south of the Soviet border on Korea's east coast, housed a major port and extensive rail yards. At the time of its selection as a bombing target, General Ridgway, then on the Department of the Army staff, had noted Rashin's proximity to the Russian

---

[46] (1) Rad, C 52391, MacArthur (Personal) for JCS, 30 Dec 50. (2) Rad, JCS 80680, JCS (Personal) for MacArthur, 9 Jan 51.

[47] Rad, DA 1706, DA to CINCFE, 23 Jan 51.

[48] (1) Rad, DA 82320, Collins (Personal) for MacArthur, 30 Jan 51. (2) Min, 59th mtg Army Policy Council, 30 Jan 51.

[49] Rad, DA 84232, CofS USA to CINCFE, 25 Feb 51.

border and had questioned its selection.[50] Nevertheless, Rashin remained on the target list and was bombed effectively on one occasion. Another bombing strike mounted on the port was diverted because of weather conditions. When reports of the Rashin bombing reached the Department of State, officials there expressed concern over the possibility of violations of the Soviet border, and asked that targets close to that border no longer be bombed.[51] The Joint Chiefs of Staff on 8 September 1950 had directed General MacArthur to make no further aerial attacks against Rashin and asked his views on the matter. On 10 September, General MacArthur replied that he concurred in the restriction on Rashin and had ordered suspension of all further attacks against the port.[52]

The matter lay more or less dormant until mid-February 1951 when the Commanding General, Far East Air Forces, General Stratemeyer, appealed to General MacArthur for permission to attack Rashin once more. Stratemeyer pointed out that the enemy continued to receive reinforcements and supplies in spite of his severely crippled transportation system, while Rashin, an important link in the enemy's supply system, remained immune from attack. Aerial reconnaissance of the Rashin area indicated a high level of activity in the city. To disrupt and destroy the North Korean coast transportation and supply system effectively, Rashin would have to be attacked and destroyed. According to Stratemeyer, the month of February offered the best weather for visual attack on Rashin. Thereafter, the weather would progressively deteriorate. Stratemeyer assured MacArthur that his aircraft could attack and demolish the city without violating the international border, and reiterated that he considered it almost mandatory that his forces be allowed to attack.[53]

General MacArthur, in a switch from his previous stand, agreed with his air commander that Rashin should be hit. He felt that the city was the keystone of the enemy's logistic system in the east and was being used at peak activity since the enemy had correctly guessed that Rashin might be immune. Accordingly, he recommended to the Joint Chiefs of Staff on 15 February that their restrictions against bombing Rashin be removed. He pointed out that specific targets in the Rashin area—large marshaling yards, extensive storage facilities, and dock areas—were particularly susceptible to visual bombing during February; and he assured them that his bombers could, without question, destroy Rashin without violating the international border.[54]

Before making a decision on the matter, the Joint Chiefs thought they should know more about Rashin and asked MacArthur to send more specific information with regard to the types and degree of military activity in the city as well as the exact location of installations.[55] Mac-

---

50 Note, handwritten, signed AMG (Gruenther), 2 Aug 50, in G–3, DA file 091 Korea, Case 22/2.

51 Memo (Tel Call), Mr. Matthews, State Dept, for OSD, in G–3, DA file 091 Korea, Case 103.

52 (1) Rad, JCS 90943, JCS to CINCFE, 9 Sept 50. (2) Memo, G–3 DA for CofS USA, 15 Feb 51, sub: Instructions to the Commander in Chief, FEC, Respecting the Bombing of Rashin, in G–3, DA file 091 Korea, Case 159.

53 Ltr, CG FEAF to CINCFE, 14 Feb 51, sub: Attack on Rashin, in GHQ, UNC AG 13686.

54 Rad, CX 55601, CINCFE to JCS, 15 Feb 51.

55 Rad, JCS 83773, JCS to CINCFE, 17 Feb 51.

Arthur admitted that he could not list the precise nature or quantity of supplies in the Rashin area, but he insisted that great depot accumulations of various types were located there. "Rashin is the last major profitable strategic target in North Korea and has remained virtually untouched," he maintained. If he could destroy this last vital link in the enemy's east coast transportation system the enemy would have suffered a major loss. Conversely, Rashin's immunity from attack remained a major threat to MacArthur's forces.[56]

General Taylor, the Department of the Army G–3, supported MacArthur and told the Chief of Staff that he felt it was operationally essential that MacArthur's request be granted. He admitted that the major risk was that USSR aircraft, maintaining surveillance in the area, might, as a local defense measure, investigate U.N. aircraft involved and set off an air battle. There was also a lesser risk that attacks upon Rashin, even if the border were not violated, might provide a basis for Russian propaganda claims alleging violations of their border.[57] There were other reasons why Washington felt it not advisable to bomb Rashin at this time. The Department of State still objected vigorously to the possibility of border violations or USSR claims of border violations. Too, there was excellent chance that USSR shipping, which used the harbor freely, might be destroyed and cause a serious international incident. Militarily, the Joint Chiefs of Staff did not feel that Rashin was as vital as MacArthur claimed. The railroad leading south from Rashin down the coast was not completed, greatly lessening Rashin's value as a supply point. Furthermore, even though bombing might accomplish the immediate destruction of the particular stores then located in Rashin, the enemy need only transfer his logistic activities a few miles north of the Russian border, to Vladivostok for example, and enjoy the same advantages afforded by Rashin.[58] The Joint Chiefs of Staff therefore turned down MacArthur's appeal for permission to bomb Rashin.[59]

Bombing of Power Installations

In late December, before the Chinese had attacked across the parallel, the Joint Chiefs of Staff had asked for MacArthur's advice on whether to bomb and destroy the power installations on the Korean side of the Yalu and the power dams on the river itself. They were contemplating such destruction only if the Chinese did drive south across the 38th Parallel. MacArthur had informed his superiors that the hydroelectric installations in northeast Korea were inactive. These, of course, had been physically inspected by X Corps units. The power plants in northwest Korea, which his forces had never seized, were an unknown quantity insofar as power output was concerned. But MacArthur felt that none of these installations had any further military significance. Their destruction, if accomplished, would necessarily therefore be a political matter. He pointed out that he had been instructed firmly in the

[56] Rad, C 55830, CINCFE (Personal) for JCS, 18 Feb 51.

[57] Memo, G–3 (Taylor) for CofS USA, 15 Feb 51.

[58] *MacArthur Hearings*, pp. 356, 431, 750, 1063, and 1331.

[59] *Ibid.*, p. 17.

past to refrain from destroying these installations and that these instructions had been widely publicized as evidence of the United Nations peaceful intent toward countries north of the Yalu. He stated that his medium bombers could certainly destroy these power plants and dams, but reminded his superiors that this change in policy "involves considerations far beyond those of the immediate tactical campaign in Korea." [60] The Joint Chiefs of Staff, understandably, let the matter drop since there were no apparent political advantages to be gained.

But on 26 February, General MacArthur once more brought the bombing of the power installations to the attention of the Joint Chiefs, telling them that General Stratemeyer had urgently requested permission to destroy the entire North Korean power complex, including those plants on the Yalu River. Stratemeyer believed that by so doing he could slow down Communist support of their war effort, undermine the enemy's morale, and cut down any surplus power going to Manchuria. MacArthur suggested that the political considerations which prevented earlier bombings might have changed. He requested instructions as to what he was to do. The Department of the Army G–3 recommended to the Chief of Staff that MacArthur be told to carry on without destroying these power plants since political considerations had certainly not changed to any great degree. On 1 March, the Joint Chiefs advised MacArthur that, in view of his previous statement that the preservation or destruction of the power installations was predominantly a political rather than a military matter, they did not believe he should bomb the power facilities.[61]

[60] (1) Rad, JCS 99713, JCS to CINCFE, 26 Dec 50. (2) Rad, C 52125, CINCFE to DA for JCS, sgd MacArthur, 27 Dec 50.

The Improved Outlook

By the end of February, limited but very real combat successes had dissipated the last traces of the specter of forced evacuation under Chinese Communist pressure. General MacArthur cheerfully reported that he was ". . . entirely satisfied with the situation at the front, where the enemy has suffered a tactical reverse of measurable proportions. His losses have been among the bloodiest of modern times." Visibly pleased by the northward progress of his forces in the field, the United Nations commander noted, "The enemy is finding it an entirely different problem fighting 350 miles from his base than when he had this 'sanctuary' in his immediate rear, with our air and naval forces practically zeroed out." [62]

The soundness of General Ridgway's tactics was praised by MacArthur.

> Our strategic plan, notwithstanding the enemy's great numerical superiority, is indeed working well, and I have just directed a resumption of the initiative by our forces. All ranks of this international force are covering themselves with distinction and I again wish to especially commend the outstanding teamwork of the three services under the skillful direction of their able field commanders, General Ridgway, Admiral Struble, and General Partridge.[63]

[61] (1) Rad, CX 56453, CINCFE to DA for JCS, 26 Feb 51. (2) Memo, G–3 DA for CofS USA, 26 Feb 51, sub: Bombing of Power Plants in North Korea, in G–3, DA file 091 Korea, Case 147/4. (3) Rad, JCS 84577, JCS to CINCFE, 1 Mar 51.

[62] Rad, C 56709, CINCFE to DA, 1 Mar 51.

[63] *Ibid.*

CHAPTER XIX

The Crossroads

Although the war had again shifted in favor of United Nations forces, the Eighth Army successes through the end of February 1951 could not be considered an indication of eventual victory. The most that could be predicted was that the enemy forces then arrayed in Korea would be incapable of forcing the Eighth Army from the peninsula.

For anyone committed to the viewpoint that a war offers its participants only the alternatives of victory or defeat, the current situation was intolerable. General MacArthur represented this viewpoint in his suggested counteractions to the Chinese intervention. The rejection of his proposals, he maintained, would lead to disaster, their acceptance to victory. He neither sought nor suggested any middle course.

President Truman, on the other hand, recognized other alternatives and was willing to examine them. Consequently, by the close of February, he had not yet granted any of MacArthur's calls for increased action against Communist China. Furthermore, the resurgence and stiffening of the Eighth Army under Ridgway had created an atmosphere in which the next course of action did not have to be decided in haste or out of a feeling of desperate weakness.

The original purpose of United Nations military operations in Korea—to repel the aggression and to restore peace and security in the area—of course remained unchanged. So did the longer range and long-standing objective of the United Nations, and particularly of the United States since Cairo, "to bring about the establishment of a unified, independent and democratic Korea." The achievement of these goals, particularly the longer range objective, by military means, however, had become less likely after the impact of Chinese intervention and the American decision in December not to commit additional forces to Korea. The alternatives consequently narrowed to some sort of accommodation that would provide a halt or at least a lull in the fighting during which diplomatic negotiation might salvage the prestige of the United States and the United Nations and at the same time bring some result not too far short of the basic objectives.

Through March 1951, the United States, as the Unified Command of the United Nations, continued to fight without having elected any new political or military courses of action. Neither the Department of State, responsible for advising the President on political matters, nor the Joint Chiefs of Staff, his principal military advisers, seemed willing to state definitely a proposed course of action until the other party had done so. Frequent meetings took place between State and Defense representatives, but each Department deferred to the other for a clear statement of what should be done in Korea.[1]

During exploratory talks on 6 February, the representatives of the Department of State had listed five courses of action which the United Nations might follow: an all-out military effort to conquer all Korea and unify the country by force; complete abandonment of Korea to the Communists; extension of hostilities to China, thus removing pressure on Korea; an indefinite military stalemate at approximately the present battle line; or a peaceful settlement through negotiation. The initiative in the first three courses would have to be taken by the United Nations, but in the fourth, stalemate, neither side would have to take the initiative. In view of the Communist rejection of overtures by the United Nations, the initiative for bringing about a peaceful settlement, the fifth step, now lay primarily with the Communists.

The Joint Chiefs maintained that they could not intelligently choose any one of these steps without knowing what political course the United States meant to follow; and since future political moves by the United States remained obscure, the Joint Chiefs recommended no military course of action other than a continuation of an aggressive defense.[2] The Department of State nevertheless informed American allies participating in Korea of the five alternative courses of action that the United Nations might consider.[3]

The Secretary of State took the position on 23 February 1951 that neither the United Nations nor the United States had assumed any obligation to unify Korea by military means. The 7 October 1950 resolution of the General Assembly was permissive but not mandatory on this point. Secretary Acheson believed that most governments having troops in Korea, including the principal allies of the United States, would not support unification as a war aim but would continue to support it as a political objective.[4]

The Joint Chiefs of Staff were not sure that the political objectives were still valid and recommended to the Secretary of Defense on 27 February that these objectives be reviewed for either reaffirmation or modification. Once these objectives were firmly established, the Department of State should be able at least to develop some short-range political courses leading toward those political objectives. The Joint Chiefs of Staff felt they would then be able to analyze the military capabilities of the United Nations and recommend military courses of action to be taken in conjunc-

[1] (1) *MacArthur Hearings*, pp. 920–21. (2) Summary of Notes on JCS-State Mtg, 13 Feb 51, JSSC Rpt to JCS, p. 123.

[2] (1) *MacArthur Hearings*, pp. 920–21. (2) JSSC Rpt to JCS, p. 119.

[3] Rad, DA–IN 3983, 21 Feb 51, in G–3, DA files.

[4] JCS 1776/192, Incl B, App. to Annex A.

tion with and in furtherance of these political courses of action.[5]

The 38th Parallel Again

The 38th Parallel assumed an ominous significance in the eyes of some United Nations members in February and March as MacArthur's forces again drove northward and it appeared that the Eighth Army, after pushing forward slowly, would soon be able to cross over. Many officials, allied and American, viewing the thrice-crossed parallel as a symbolic barrier beyond which MacArthur's men should not again venture lest the enemy strike even harder in retaliation, became greatly concerned.

General MacArthur had fended off newsmen's questions on the subject in mid-February by telling them that for the time being any talk of crossing the parallel except by patrol actions was purely academic. MacArthur took this opportunity to reaffirm his belief that the Chinese should be attacked on their own soil, holding that the existing superiority of the Chinese Communist enemy must be materially reduced before he could seriously consider conducting major operations north of the 38th Parallel.[6]

This was merely a public airing of the view he had already expressed to General Taylor on 11 February, when he had pointed out that unless he received authority to strike enemy bases in Manchuria, his ground forces as then constituted could not safely attempt major operations in North Korea. He had at the same time, of course, told Taylor that even if he found it possible to cross the parallel in force, he still would not do it until he had received instructions from Washington.[7]

General MacArthur's directives with regard to the 38th Parallel had not changed. He still possessed the authority to cross granted him on 27 September by the United States and tacitly confirmed on 7 October by the United Nations General Assembly. But the Department of State was keenly aware of the concern felt by some of the members of the United Nations over the advisability of re-entering North Korea. To allay this concern, and in anticipation of the arrival of United Nations forces at the parallel, Secretary Acheson on 23 February asked Secretary Marshall to consider revising the 27 September directive so as to limit MacArthur's advance. Acheson added that any subsequent decision to move substantial forces above the parallel would require preliminary discussions with other governments having troops in Korea.[8]

Acheson enclosed a memorandum the tenor of which was generally pessimistic and which he suggested Marshall send the President. In it, Acheson pointed out that any decision to press for the unification of Korea by military action would mean a vast increase in United States military commitments; would almost certainly require the extension of hostilities to Communist China; would greatly increase the risk of direct Soviet

[5] Memo, JCS, sgd Bradley, for Secy Defense, 27 Feb 51, sub: Action to be Taken by U.N. Forces With Respect to the 38th Parallel.

[6] Statement, Gen MacArthur, 13 Feb 51, in *MacArthur Hearings*, p. 3539.

[7] Rad, C 55315, MacArthur (Personal) for Gen Taylor, 11 Feb 51.

[8] JCS 1776/192, Incl B, App. to Annex A.

intervention; and would require a major political effort to obtain the agreement of other directly interested nations to take such action.[9]

Acheson judged that virtually all members of the United Nations, including most of those actively participating in Korea, strongly opposed any general advance across the 38th Parallel. This opposition was based on the belief that once the enemy had been driven out of South Korea the primary objective of repelling the aggression had been accomplished; that an advance in North Korea would make an early negotiated settlement of the Korean fighting impossible, since the enemy would accept nothing less than the *status quo ante bellum*; that crossing the parallel would greatly increase the pressure for extending the hostilities into China and in turn would involve American military resources to an increased extent in indecisive operations in Asia; and that a crossing would greatly increase the risk of Soviet involvement and general war.[10]

A major advance across the parallel, Acheson claimed, would require full consultation with major allies and their agreement, which under current circumstances would be extremely difficult to obtain. Any unilateral re-entry into North Korea by the United States, on the other hand, would create a severe crisis within the free world and could lead to the withdrawal of certain allies from the Korean War. Acheson did concede that all of South Korea must be captured, claiming that such would constitute a major victory for United Nations forces since it would deny the enemy their main objective. Nor did Acheson propose to forbid MacArthur's men to set foot across the parallel; rather, he proposed that no major crossing should be made. He recognized that so long as fighting in Korea continued, MacArthur must be free to attack with naval and air power across the parallel and to take such ground action in North Korea as was required to interrupt enemy offensive preparations.[11]

It is evident that the Department of State officials were looking forward to a possible settlement of the Korean crisis by negotiation. They considered it important that United Nations military action produce a desire on the part of the enemy to negotiate rather than to fight, and at the same time not create a situation in which he would balk at a negotiated settlement. In other words, MacArthur's forces should inflict so many casualties on the enemy that he would be anxious to negotiate, but on the other hand, this punishment should take place in the vicinity of the parallel and not in the course of pushing the enemy so far back that he would refuse to accept a settlement at the line where the fighting ended.[12]

Secretary of the Army Pace, Secretary of the Air Force Finletter, and Secretary of the Navy Dan A. Kimball examined the Acheson proposals and found them reasonable. All three agreed that MacArthur should not attempt a general advance north of the 38th Parallel except to take advantage of favorable terrain for defense. Secretaries Pace and Finletter wanted the United States to adopt this policy of restraint and to an-

9 *Ibid.*
10 *Ibid.*
11 *Ibid.*
12 *Ibid.*

nounce it to the world "as a matter of principle." But on this point Secretary Kimball dissented on grounds that discussing such a decision with other governments or publicly announcing that MacArthur was more or less bound to the 38th Parallel would have a bad effect from a military standpoint.[13]

The Joint Chiefs of Staff, however, took hearty exception to the Department of State proposals. They pointed out that so long as the political objectives of the United Nations remained unchanged, its military forces should not be forbidden, for political reasons, to advance north of the 38th Parallel. Such a prohibition would be wholly inconsistent with the political objectives.[14]

The Joint Chiefs of Staff believed, along with Generals Ridgway and MacArthur, that any directive halting MacArthur at the parallel would permit the enemy to build up in North Korea such a concentration of military forces that MacArthur's own forces would be jeopardized. Nor would a United Nations prohibition against crossing the 38th Parallel impose a comparable restriction on enemy forces. The Joint Chiefs told the Secretary of Defense that their own combined military experience convinced them it would be impracticable to undertake aggressive defensive operations to keep the numerically superior enemy off-balance and to disrupt his preparations for new offensives if the 38th Parallel became a limiting feature of military operations. In sum, MacArthur had to have freedom of maneuver if for no other reason than to insure the safety of his forces.[15]

The Joint Chiefs considered it premature even to make a preliminary determination of MacArthur's action when he reached the parallel. They reminded Secretary Marshall of MacArthur's announced intention to apply to them for instructions if he found no major enemy strength disposed south of the parallel. Until MacArthur reported his findings, the Joint Chiefs considered any decision on crossing the parallel to be militarily unsound. For any decision to restrain United Nations forces made on the political level and in consultation with other nations would inevitably be disclosed to the Chinese and North Koreans who then could base their own courses of action upon known intentions of friendly forces. Pressing once again for a decision by the Department of State as to the course of action to be taken to reach United States political objectives in Korea, the Joint Chiefs of Staff told Secretary Marshall, "Until this governmental decision is reached there should be no change in that part of the directive to General MacArthur which now permits him so to dispose his forces either north or south of the 38th Parallel as best to provide for their security." [16]

Because of these strong objections, Secretary Marshall told Secretary Acheson that he did not believe the memorandum opposing a general advance across the 38th Parallel should be sent to President

[13] Memo, Secys Army, Navy, and Air Force for Secy Defense, 26 Feb 51, sub: State Dept Draft Memo for the President on the 38th Parallel, in G–3, DA file 381 Korea, Case 3/3.

[14] Memo, JCS for Secy Defense, 27 Feb 51, sub: Action to be Taken With Respect to the 38th Parallel.

[15] *Ibid.*

[16] *Ibid.*

Truman. Marshall himself agreed with the Joint Chiefs of Staff that there was a risk in disclosing to the enemy a United States military decision, that freedom of action and freedom of maneuver had to be maintained for United Nations ground forces, and that it was, in any event, too early from a military point of view to reach a final determination on crossing the parallel.[17]

On 2 March, General MacArthur submitted through channels a proposed report to the United Nations for the period 15–28 February which concluded with the statement:

> While President Truman has indicated that the crossing of the parallel is a military matter to be resolved in accordance with my best judgment as a theater commander, I want to make it quite clear that if and when the issue actually arises, I shall not arbitrarily exercise that authority if cogent political reasons against crossing are then advanced and there is any reasonable possibility that a limitation is to be placed thereon.

But Washington authorities saw no profit in unnecessarily calling the attention of the United Nations to the 38th Parallel and asked MacArthur to delete this portion of his report. Both the Department of State and the Department of Defense agreed that references to the 38th Parallel from the military point of view should be avoided whenever possible; and General MacArthur subsequently agreed to the excision of this part of his report.[18]

Advances in Korea

The United Nations surge up the Korean peninsula had slackened somewhat in late February. But by 2 March, General Ridgway completed plans for Operation RIPPER in which all corps would move northward through successive phase lines to seize Hongch'on and Ch'unch'on in the central sector and to destroy all enemy forces, material, and supplies in the path of the advance.[19] (*See Map VII.*) Ridgway's troops opened Operation RIPPER on 7 March. Stubborn delaying actions permitted only short gains during the first week, but by 13 March, enemy resistance began to diminish. By 16 March, the enemy was attempting to disengage and withdraw, and by the 18th Seoul was once again in United Nations hands and all other objectives were generally attained. The enemy's decreasing effort to contest Eighth Army's advances, observations of sizable enemy groups moving northward out of the battle areas, and statements by captured soldiers, all pointed to an enemy decision to fall back on prepared positions north of the 38th Parallel. Enemy reserve forces had been located close to the parallel for some time, and MacArthur's intelligence officers therefore reasoned that the enemy had had time to prepare strong defenses on or near this line of latitude.[20]

The Department of the Army G–3, General Taylor, was somewhat displeased because he was not given the details of Operation RIPPER in advance, nor even

---

[17] Ltr, Secy Defense Marshall to Secy State Acheson, 1 Mar 51, in G–3, DA file 381 Korea, Case 3/4.

[18] (1) Rad, C 56709, CINCFE to DA, 2 Mar 51. (2) Telecon, TT 4477, 7 Mar 51 (3) Telecon, TT 4479, 9 Mar 51.

[19] Ltr, CG EUSAK to CINCFE, 2 Mar 51, sub: Operation RIPPER, with Incls, in G–3, GHQ, UNC files.

[20] Telecon, TT 4498, DA to GHQ, 15 Mar 51.

TROOPS OF THE 1ST BATTALION, 35TH REGIMENT, 25TH DIVISION, *pile into assault boats to cross the Han River, south of Seoul, 7 March 1951.*

told that it was taking place. On 17 March, he asked MacArthur to send him the details embodied in operational directives issued by General Headquarters and/or Eighth Army. He suggested further that in the future the Department of the Army be an information addressee for all operational directives, to include those of the Eighth Army. MacArthur's headquarters had not planned Operation RIPPER nor had it issued any operational directives or orders. Ridgway had taken care of the whole thing, merely advising MacArthur of his plans. General MacArthur brushed aside General Taylor's request by telling him that Operation RIPPER was merely a development of the constant interchange between his headquarters and the commanders in Korea, and that no formal orders had been issued by him. MacArthur insisted that Taylor was being kept fully informed of all operations of the command but that it was impracticable to give Taylor every detail of intercommand arrangements.[21]

[21] (1) Rad, DA 86022, DA to CINCFE, 17 Mar 51. (2) Rad, CINCFE to DA Mar 51.

Seeking More Forces

As one means of increasing MacArthur's ground strength in Korea, the Joint Chiefs of Staff meanwhile had recommended to the Secretary of Defense in late January that the Department of State be asked to seek additional forces from U.N. members who they believed were not contributing all they could. Secretary Marshall had asked the Department of State to do this on 30 January. On 23 February, Secretary Acheson told Marshall that some action would be taken to carry out this proposal. Australia and New Zealand would be pressed to furnish an additional infantry battalion each. Canada would be asked to increase its commitment to brigade size, according to the original plan which had been canceled in October. Certain Latin American countries also would be asked to send ground forces to Korea. The Department of State thought that it would be unwise, however, to ask Turkey and Greece for more ground forces and also that there were no other countries capable of sending forces to Korea at that time.

General Taylor pointed out several factors which he felt should be taken into consideration by the Army Chief of Staff with regard to forces from other U.N. countries. He told General Collins on 14 March that in light of the need for redeploying United States ground combat units to more strategic areas as soon as possible, it would be a good idea to remove them from Korea if they could be replaced. Too, a United Nations force composed of complete divisions from several different countries other than the United States would provide a means of testing certain organizational and operational methods under study in NATO. Furthermore, it would be a long time, according to General Taylor, before ROK Army units could be strengthened to a point where they might relieve some United States combat troops.

It appeared to Taylor that Turkey, Greece, Great Britain, the Philippines, Argentina, Brazil, Mexico, Australia, and New Zealand were capable of furnishing sizable forces up to division strength for service in Korea. The difficulty would be in providing logistic support to these forces since the total matériel resources of the United States were needed for current U.S. programs. Hence, any equipment furnished for training or employing new units in Korea could be furnished by the United States only if its programs were reduced accordingly.[22]

American military authorities, despite possible logistic headaches, continued to insist that other nations could and should contribute more heavily to the United Nations fighting team in Korea. Robert Lovett, Acting Secretary of Defense, notified Secretary of State Acheson on 31 March that the Department of Defense was not satisfied that everything possible had been done to induce these other nations to furnish more forces, and that the current situation in Korea presented an opportunity to renew requests for fuller participation by other members of the United Nations. Lovett charged that the heavy commitment of

---

[22] Memo, Gen Taylor, G–3 DA, for CofS USA, 14 Mar 51, sub: Increasing Foreign Contingents in Korea to Div Size, in G–3, DA file 320.2 Pac, Case 63.

United States ground troops, the high casualties suffered, the long months of unrelieved combat duty, and the desirability of reassigning experienced soldiers to form cadres for mobilization of new units in the United States and of redeploying battle-tested units to other strategic areas made it all the more imperative that U.S. units in Korea be relieved.[23]

Mr. Lovett wanted real, not token, assistance from these other nations, and requested that the Department of State once again prevail on such countries having trained manpower resources to provide contingents of worthwhile size and to equip these units and support them themselves. He particularly had in mind the Commonwealth nations of Australia, New Zealand, and Canada, all of which could, he felt, well afford to increase the size of their contingents in Korea. He thought also that Great Britain might be able to furnish a full division, while Latin American countries such as Brazil and Mexico appeared to have the military manpower necessary to send sizable units to Korea.[24]

President Truman Is Displeased

Almost by default, a political course of action began to emerge in mid-March. Encouraged by the results of Operation RIPPER, which proved that the military initiative in Korea no longer lay with the enemy, U.S. policy planners decided that efforts to achieve a negotiated settlement should be renewed. These planners, from both Defense and State, now believed that Ridgway's punishing attacks might have rendered the Chinese more amenable to a political settlement.

To both Departments, it appeared that the most logical beginning of a negotiated settlement was for President Truman to appeal directly to the Chinese Communists. For while earlier attempts to bring about negotiations had failed, President Truman had in none of these instances been the one to suggest opening negotiations. Furthermore, the situation seemed particularly propitious because enemy forces were being pushed back into North Korea and could therefore negotiate on the basis of their prewar status.[25]

The Department of State drafted such a Presidential declaration and after obtaining the Joint Chiefs' approval of its content, began to clear it with the other United Nations members having troops in Korea.[26] In substance, the President was to point out that the aggressors in Korea had been driven back to the general vicinity from which their unlawful attack had first been launched and that, therefore, the principal objective of repelling North Korean and Chinese Communist aggression against the Republic of Korea had been achieved. He would assert further that United Nations objectives, such as unification and the establishment of a free government in all of Korea, could and should be accomplished without more fighting and bloodshed. The Chinese Communists were, in effect, to be invited to cease fire and to negotiate a settlement of the outstanding issues. They were also to be

[23] Ltr, Actg Secy Defense (Lovett) to Secy State, 31 Mar 51, in G-3, DA file 091 Korea, Case 148/19.
[24] *Ibid.*

[25] *MacArthur Hearings*, p. 343.
[26] *Ibid.*, pp. 343-44.

warned that if they refused to negotiate, the United Nations would be forced to continue the fighting.[27]

On 20 March, the Joint Chiefs of Staff alerted General MacArthur to the planned Presidential announcement. He was also informed that some nations consulted believed that an advance by major forces of the United Nations Command across the 38th Parallel would endanger further diplomatic efforts, and was reminded that time would be needed to determine the reactions of all concerned, including the Communist governments. The Joint Chiefs of Staff had already told the Department of State that the 38th Parallel had no tactical significance, a judgment with which the Department of State now agreed. But State had asked the military advisers just what freedom of action MacArthur should have for the next few weeks in order for him to maintain contact with the enemy and at the same time insure the safety of his forces; and the Joint Chiefs, in turn, asked MacArthur to make his own recommendations as to what latitude he required.[28]

MacArthur interpreted the latter request as a possible prelude to an order forbidding him to cross the 38th Parallel and immediately recommended that no further military restrictions be imposed upon his command. He explained that since he was forbidden to use his naval and air arms as he had suggested, and since the enemy's ground potential so far exceeded his, it remained completely impractical anyway to attempt to clear North Korea. In sum, MacArthur felt that his current directives were adequate and should not be changed.[29]

The proposed Presidential announcement was never made. For while it was still being prepared, General MacArthur issued a public statement on 24 March that in the eyes of Washington officials completely vitiated the contemplated political move. In his statement, MacArthur declared that the tactical successes of his forces clearly showed Communist China to be a vastly overrated military power weak in everything but human resources. Continuing, he said, "Even under the inhibitions which now restrict the activity of the United Nations forces and the corresponding military advantages which accrue to Red China, it has shown its complete inability to accomplish by force of arms the conquest of Korea." The confident tone of this statement contrasted sharply with MacArthur's reports to Washington two months earlier. He also reiterated his oft-aired contention that ". . . the fundamental questions continue to be political in nature and must find their answer in the diplomatic sphere." [30]

Unmindful of the President's scheduled call on the enemy for negotiation, MacArthur then declared:

> Within the area of my authority as the military commander, however, it should be needless to say that I stand ready at any time to confer in the field with the Commander-in-Chief of the enemy forces in the earnest effort to find any military means whereby realization of the political objectives of the United Nations in Korea, to which no nation may justly take exception,

---

27 JSSC Rpt to the JCS, p. 131.

28 Rad, JCS 86276, JCS to CINCFE, 20 Mar 51.

29 Rad, C 58203, CINCUNC (MacArthur) to DA for JCS, 21 Mar 51.

30 MacArthur, *Reminiscences*, pp. 387–88.

might be accomplished without further bloodshed.

President Truman was angered by MacArthur's statement since it tacitly preempted the President's prerogatives and criticized, by implication at least, the national policy. Besides infuriating the President, MacArthur's announcement brought down upon Washington a rash of inquiries from allies of the United States as to whether MacArthur's words were the precursor of a drastic change in national policy.[31]

President Truman, on the same day he heard MacArthur's statement, called in Acheson, Rusk, and Lovett to discuss what response to MacArthur's act would be appropriate. They agreed that the 6 December directive to MacArthur and the other commanders made plain what they could and could not say without prior clearance. They further agreed that MacArthur had violated this directive. But MacArthur was not censured for this violation, only reminded once again of the directive itself. In an immediate dispatch, the Joint Chiefs notified MacArthur, "In view of the information given you 20 March 1951 any further statements by you must be coordinated as prescribed in the order of 6 December. The President has also directed that in the event Communist military leaders request an armistice in the field, you immediately report that fact to the JCS." [32]

General MacArthur had not known the contents of the proposed Presidential declaration. The information he received from the Joint Chiefs on 20 March did little more than tell him that some sort of Presidential announcement was to be made. Also, in his own offer to confer in the field with the enemy commander, MacArthur had stressed the terms, "Within the area of my authority as a military commander . . . ," and ". . . to find any military means. . . ." Evidence that such a move would have been quite proper is available. Shortly after the Inch'on landing, when it was thought the North Koreans might sue for peace terms, the Deputy Under Secretary of State told the Department of Defense, "A cease-fire should be a purely military matter and accordingly they (the North Koreans) should communicate their offer to the Commanding General of the unified command . . . who is the appropriate representative to negotiate any armistice or cease-fire agreement." [33]

On the related issue of recrossing the 38th Parallel, General MacArthur did not intend to hold the Eighth Army below the line unless so ordered by Washington. Nevertheless, he instructed General Ridgway on 22 March not to move above the parallel in force until specifically authorized to do so. To any press inquiries on the probability of a crossing, Ridgway was to reply that the decision would have to be made by Mac-

31 (1) *Ibid.* (2) The President interpreted MacArthur's action as threatening the enemy with an ultimatum, implying that the United States and its Allies might attack China without restraint. This had implications far greater than usurpation of a prepared statement which the President had intended to make. In the President's mind, MacArthur had once again openly defied the policy of his Commander in Chief. See Truman, *Memoirs*, II, 442–43.

32 *MacArthur Hearings*, pp. 344, 3542.

33 Ltr, Dept of State (Deputy Under Secy Mathews) to OSD (Gen Burns), 15 Sep 50, in G–3 DA file 091 Korea, Case 99, App. to JCS 1776/105.

Arthur himself. MacArthur informed Ridgway that a new directive for operations in Korea was expected from Washington shortly, apparently in the belief that the Joint Chiefs of Staff, by asking what freedom he needed in his future operations, meant to send him new instructions, including orders on crossing the 38th Parallel. But Washington authorities sent him no new directive on either the parallel or the conduct of future operations.[34]

As to the immediate future, MacArthur told Ridgway, "My present intention is to continue current type of action north of the parallel, but not to proceed further than your logistics would support a major operation." MacArthur evidently was more concerned with the logistical than with the political implications of re-entering North Korea. He continued, "At that time to pass from the present tactics which you have so ably conducted to ranger-type probing by battalions or companies from divisional fronts operating for ten-day periods with self-contained supplies supplemented by guerrilla type activities. If you have any suggestions, let me have them." General Ridgway replied that he would issue all necessary instructions to insure compliance. He interpreted MacArthur's term "in force" as permitting at least one reinforced infantry battalion per corps to cross the parallel if a potentially fruitful opportunity should present itself.[35]

[34] Rad, C 58292, MacArthur (Personal) for Ridgway, 22 Mar 51.

[35] (1) *Ibid.* (2) Rad, G-3 412 KCG, Ridgway (Personal) for MacArthur, 22 Mar 51.

Bevin and the British View

In spite of the intransigence thus far shown by the Communists toward every United Nations suggestion of settling the Korean problem by talking instead of fighting, the British Government remained hopeful that the Communists would eventually agree to negotiate. On 30 March, British Foreign Secretary Bevin proposed a new attempt at negotiation. Bevin suggested the issuance of a clear statement of Korean policy, agreed to by all countries having forces in Korea and specifically endorsed by the unified command. This, he thought, would provide a basis for approaching the Peiping and USSR governments in order to explore Chinese Communist readiness to negotiate a settlement by some procedure other than that of the Committee of Good Offices of the United Nations, which he felt could not by itself obtain the co-operation of the Chinese. Specifically, Bevin recommended a joint declaration by all nations having forces in Korea—expressing their desire to see an independent and unified Korea, their agreement to the withdrawal of all foreign troops, and their readiness to achieve these objectives by other than military means. At the same time, President Truman, in his capacity as Chief Executive of the state providing the unified command, would announce that the unified command fully endorsed the military implications of the joint declaration. Following these two statements of policy, the Chinese and the Russians would be asked to express their views as to the best means of bringing about a peaceful settlement in Korea. With reference to the Presidential dec-

DIRECTING ARTILLERY FIRE ON CHINESE POSITIONS *as U.N. forces advance against the communists near the 38th Parallel.*

laration proposed earlier and General MacArthur's statement of 24 March, the British foreign secretary noted that MacArthur's action was further reason for considering an entirely new procedure since it was now unlikely that any further statement by the unified command alone would be taken seriously by the Communists.[36]

[36] Rad, 825, Dept of State to USUN NY, 30 Mar 51, in G–3, DA file 091 Korea, Case 167/2.

Ridgway Re-enters North Korea

Lacking specific instructions to the contrary, General MacArthur meanwhile approved plans developed by General Ridgway for advancing above the 38th Parallel. On 22 March, Ridgway informed MacArthur that he had prepared plans to advance, if MacArthur approved, to a line that, except for a short stretch in the west, lay just above the parallel, generally between the con-

GENERAL MACARTHUR *is greeted by General Ridgway upon MacArthur's arrival at Kangnung, 3 April 1951.*

fluence of the Han and Yesong Rivers on the west coast and the town of Yangyang on the Sea of Japan. Ridgway explained that operations to reach this line would have as their objective not the seizure of terrain but the maximum destruction of enemy troops and matériel, and that they would be conducted with particular care to maintain major units intact and to keep casualties to a minimum. He assured MacArthur further that he had no intention of outrunning his logistical support. MacArthur approved Ridgway's plan without hesitation, and without referring it to Washington.[37]

Ridgway opened the first phase of this advance (Operation COURAGEOUS) on 22 March, moved steadily forward all along the front, and attained positions generally along the 38th Parallel by the 30th.

[37] Rad, G-3 412 KCG, Ridgway (Personal) for MacArthur, 22 Mar 51.

Except for a small area in the west, South Korea thus was cleared of organized enemy forces. The latter suffered enormous casualties, although as a result of the relatively slow Eighth Army advance compelled by Ridgway's insistence on careful co-ordination and the preservation of lateral security, the enemy units themselves managed to withdraw intact.[38]

By the time the Eighth Army regained the parallel, Ridgway, on 29 March, completed the details of instructions for the next forward step, which he called Operation RUGGED. The new objective, Line KANSAS, differed slightly from Ridgway's 22 March concept by starting at the junction of the Han and Imjin Rivers, not the Han and Yesong, then running northeastward and eastward to Yangyang.[39]

General MacArthur flew into Korea on 3 April to discuss this next step northward with General Ridgway and to look at the ground situation. At that time, Ridgway explained that when he had sought approval for an advance to the Yesong-Yangyang line, he had believed he would find good hunting in the western area between the Imjin and the Yesong. But recent intelligence had revealed very few enemy forces in that region and Ridgway therefore had decided not to advance as far as the Yesong. Ridgway told MacArthur that the strongest possible line he could seize was the one toward which he was now aiming, Line KANSAS. MacArthur agreed and told Ridgway he wanted him to make a very strong fight for this line, with any advance beyond it carefully limited and controlled.[40]

On 5 April, MacArthur notified the Joint Chiefs of Staff that the Eighth Army had jumped off in its drive across the parallel to destroy enemy forces and supplies south of Line KANSAS. MacArthur also informed the Joint Chiefs that Ridgway intended to follow Operation RUGGED with Operation DAUNTLESS which would take the Eighth Army twenty miles farther into North Korea, in the west central zone, and enable it to seize Line WYOMING and thereby gain control of an area known to be a point of concentration for enemy troops and supplies. MacArthur explained that, once Lines KANSAS and WYOMING had been seized, he intended to maintain contact with the enemy only by patrols of battalion size. The existing logistical limitations, combined with the terrain, weather conditions, and intelligence of enemy dispositions, had convinced him that a further advance in force beyond the present objective lines was not feasible.[41]

The enemy did not strongly resist the crossing of the parallel. By 9 April, all units in the U.S. I and IX Corps and ROK I Corps had fought their way forward to positions on Line KANSAS; and although the U.S. X and ROK III Corps in the central and east central sectors had been slowed down by rugged terrain and inadequate supply routes, these two corps by the same date were drawing near their KANSAS objectives. Throughout this early April advance, Ridgway and MacArthur were aware that the

---

[38] Comd Rpt, Eighth Army, Narrative, Mar 51.

[39] *Ibid.*

[40] Ridgway, The Korean War: Issues and Policies, p. 420.

[41] Rad, C 59397, CINCFE to DA, 5 Apr 51.

enemy, particularly the Chinese, was building up in rear areas and was daily increasing his capability to launch an offensive. The enemy build-up was especially notable in the Ch'orwon-P'yong-gang-Hwach'on triangle in the west central area, which in turn accounted for Ridgway's plan to seize this area by advancing to Line WYOMING. On 31 March, General MacArthur had reported to the Department of the Army that an enemy offensive of great strength might be expected at any time after 1 April. He estimated that the Chinese had 274,-000 troops in Korea and 478,000 regular troops in Manchuria. The North Koreans were believed to have approximately 198,000 men, including guerrillas, available for an attack.[42]

Keeping the enemy's offensive capability constantly in mind, Ridgway made plans to contain the expected offensive by rolling to the rear with the enemy's punch. On 12 April, he issued Operation Plan AUDACIOUS, which called for an orderly, fighting withdrawal through successive phase lines. This withdrawal would be made only on Ridgway's order and would be conducted in such a manner as to inflict maximum losses on the enemy and to preserve all friendly units intact.[43]

About the time Ridgway issued this plan, he became aware that publishing it would be one of his last acts as the Eighth Army commander. As a result of a decision made by President Truman two days earlier, the general who had revitalized the Eighth Army was about to be elevated to higher command, not primarily because of Ridgway's accomplishments, but more because of the President's exasperation with General MacArthur.

---

42 (1) Rad, CX 59065, CINCFE to DA for G-2, 31 Mar 51. (2) Telecon, TT 4597, DA and GHQ, 13 Apr 51. (3) Telecon, TT 4603, DA and GHQ, 15 Apr 51.

43 Comd Rpt, Eighth Army, Apr 51, Narrative, pp. 11-13.

CHAPTER XX

The Relief of MacArthur

The focus shifted from military operations after President Truman suddenly relieved General MacArthur of all his military commands. The President took this step following five days of consultation with his chief military and civilian advisers. The culmination came on 10 April when he directed General Bradley to send General MacArthur a message stating:

> I deeply regret that it becomes my duty as President and Commander in Chief of the United States Military Forces to replace you as Supreme Commander, Allied Powers; Commander in Chief, United Nations Command; Commander in Chief, Far East; and Commanding General United States Army, Far East. You will turn over your commands, effective at once, to Lieutenant General Matthew B. Ridgway.[1]

The abrupt dismissal of so distinguished a soldier as General MacArthur aroused considerable furor in the United States and elsewhere. Charges of "cavalier treatment" and "foreign pressure" as well as broad hints of political machination followed his dismissal. The entire matter was aired extensively between May and August 1951 before the Committee on Armed Services and the Committee on Foreign Relations of the United States Senate. No definite conclusions were drawn, but testimony given the committees provided some indication of the reasons which impelled President Truman's decision.[2] Charges that MacArthur's removal was fostered, and actually engineered, by certain nations allied with the United States in Korea, particularly the British, were not well founded. While these nations, through press media and even through official channels, criticized General MacArthur's conduct of the campaign and expressed

---

[1] Rad, JCS 88180, Bradley (Personal) for MacArthur, 11 Apr 51.

[2] The Joint Committee on Armed Services and Foreign Relations which conducted these hearings was composed of Senator Richard B. Russell, Chairman; Senator Styles Bridges; Senator Alexander Wiley; Senator H. Alexander Smith; Senator Bourke B. Hickenlooper; Senator William F. Knowland; Senator Harry P. Cain; Senator Owen Brewster; and Senator Ralph E. Flanders. Witnesses appearing before the committee included General MacArthur; Secretary of Defense Marshall; General Bradley; General Collins; General Vandenberg; Admiral Sherman; Secretary of State Acheson; General Wedemeyer; and former Secretary of Defense Johnson. Among its indefinite conclusions the committee reached the following: "The removal of General MacArthur was within the constitutional powers of the President but the circumstances were a shock to the national pride," and "There was no serious disagreement between General MacArthur and the Joint Chiefs of Staff as to military strategy in Korea." See *MacArthur Hearings*, pp. 3601–02.

grave fears that his recommendations might lead to a general war, no evidence existed that any of these nations petitioned President Truman for MacArthur's dismissal.

For months prior to his relief, General MacArthur had, according to Washington officials, expressed opinions on matters beyond his purview in a manner not befitting a military commander. These opinions had not only embarrassed President Truman and his advisers, but threatened, these same officials claimed, to have a profound effect upon international public opinion and to jeopardize the relationships of the United States with its allies. Until President Truman's order forbidding expressions of personal opinion on political and military policy, such expressions were more acts of military impropriety than of misconduct. But after the order, any public expression of opinion contrary to established national policy violated a Presidential directive.

Although it may have been a contributing factor, General MacArthur's conduct of the campaign, from a purely military standpoint, did not bring about the President's decision. His inability to anticipate the Chinese attack in late November and the subsequent withdrawal of the United Nations forces in December apparently did not cause his dismissal from command.[3]

Immediately after MacArthur's relief, President Truman stated publicly that MacArthur was unable to give his wholehearted support to the policies of the U.S. Government in matters pertaining to his official duties. He pointed out that while full debate on matters of national policy was a vital element of any true democracy, military commanders had to be governed by the policies and directives issued to them in the manner provided by U.S. laws and the Constitution.[4] Hence, General MacArthur's removal from command seems to have stemmed from his official protestations and public expressions of dissatisfaction with United States Far Eastern military and political policies made by him between August 1950 and April 1951.

The Formosa Issue

The first occasion after the outbreak of the Korean War on which General MacArthur ran afoul of President Truman developed not over Korea, but over the general issue of American policy toward Formosa. This problem had been under discussion by officials of the Department of Defense and the Department of State for some time before the Korean situation developed.[5]

---

[3] President Truman later stated that he did not blame General MacArthur for the failure of his November offensive. The President felt that MacArthur was no more to be blamed for the fact that he was outnumbered than was General Eisenhower for his heavy losses in the Battle of the Bulge. The difference, as the President saw it, between Eisenhower in 1944 and MacArthur in 1950 was the manner in which MacArthur tried to excuse his failure. See Truman, *Memoirs*, II, 381–82.

[4] During conversations with the author in June 1961, former President Truman declined to elaborate on statements already made in his memoirs as to his reasons for dismissing General MacArthur. He stated that General MacArthur had been, and remained, a "great American and a great general" and that he had no desire to tarnish MacArthur's public image. Truman did, however, assert emphatically that his course of action had been the only one open to him and that, faced again with the same situation, he would do the same thing. To do otherwise would have been to abdicate his great responsibility as Commander in Chief.

[5] Memo, Burns for Rusk, Asst Secy State for Far

MACARTHUR RECEIVES WARM WELCOME ON HIS RETURN TO THE UNITED STATES.

During the extraordinary conferences at Blair House after the North Korean invasion of South Korea, General Bradley had read to the assembled high officials a memorandum MacArthur had given Secretary of Defense Johnson during the latter's Tokyo visit. This paper, which Secretary Johnson thought brilliant and to the point, set forth in cogent terms the reasons why Formosa should not be allowed to pass to the control of Communist China, but should instead be fully protected by the United States.[6] President Truman, on 27 June 1950, ordered General MacArthur to deploy the Seventh Fleet to prevent attacks on Formosa by the Chinese Communists and, conversely, attacks by the Formosan garrison on the Chinese mainland.[7] In a public announcement on the same date, President Truman explained that

Eastern Affairs, 29 May 50, sub: Notes on State-Defense Conference Held 25 May 1950, cited in Hoare, *The JCS and National Policy*, vol. IV. ch. IV, "The Knotty Problem of Formosa."

6 *MacArthur Hearings*, p. 2579.

7 DA, TT 3426, 27 Jun 50.

he had taken this action because, "the occupation of Formosa by Communist forces would be a direct threat to the security of the Pacific area and to United States forces performing their lawful and necessary functions in that area." He also fended off any charge that the United States intended to seize the island stronghold by declaring, "The determination of the future status of Formosa must await the restoration of security in the Pacific, a peace settlement with Japan, or consideration by the United Nations." [8]

When visited by General Collins in mid-July, General MacArthur had told the Army Chief of Staff that as soon as the Korean situation had become sufficiently stabilized he intended to visit Formosa for talks with Chiang Kai-shek. The Joint Chiefs on 28 July 1950 informed MacArthur that the Chinese Communists had announced their intention of capturing Formosa and would probably succeed unless the Chinese Nationalists made timely efforts to defend the island. They had recommended to the Secretary of Defense, they stated, that the Nationalists be permitted to break up hostile concentrations through military action, even if it meant attacks on the mainland.[9]

MacArthur gave full concurrence to this proposal, and informed the Joint Chiefs of Staff that he and a selected staff would visit Formosa about 31 July to survey the situation.[10] In reply the Joint Chiefs of Staff suggested that, pending new instructions on certain policy matters being considered by the Departments of State and Defense, MacArthur might prefer to send a senior officer to Formosa on 31 July, and to proceed later himself. They added, however, that if he felt it necessary, he should feel free to go since the responsibility was his own.[11] MacArthur chose to make the initial Formosa visit in person so that he could resolve uncertainties arising out of conflicting reports from the island about the status of Chiang's government and its armed forces.[12]

MacArthur, accompanied by Admiral Struble, flew to Taipeh on 31 July where for two days he conferred with Chiang Kai-shek and his generals. But not until five days after his return to Tokyo did MacArthur report to the Joint Chiefs of Staff.[13] Meanwhile, press reports speculating that MacArthur had made binding agreements and political promises to Chiang Kai-shek caused Washington officials considerable uneasiness since they could not judge the validity of these reports. In addition, the Department of State heard from its representative in Taipeh that MacArthur was about to transfer fighter squadrons to Formosa, a move not authorized by Washington, and a move which General MacArthur had not actually planned.[14] Chiang Kai-shek added fuel to the flame by issuing a public statement that could be interpreted as indicating the existence of

[8] Statement by the President of the United States, 27 Jun 50, *MacArthur Hearings*, p. 3369.

[9] Rad, JCS 87401, JCS to CINCFE, 28 Jul 50.

[10] Rad, C 58994, CINCFE to JCS, 29 Jul 50.

[11] Rad, JCS 87492, JCS to CINCFE, 29 Jul 50.

[12] Rad, C 59032, CINCFE to JCS, 30 Jul 50.

[13] MacArthur considered his report to be timely. He stated later, "Full reports on the results of the visit were promptly made to Washington." See MacArthur, *Reminiscences*, p. 340.

[14] (1) Rad, JCS 87878, JCS to CINCFE, 3 Aug 50. (2) Hoare, *The JCS and National Policy*, vol. IV, pp. 20-21.

extensive secret agreements between himself and MacArthur.[15] There was also an erroneous but widespread belief that MacArthur had made the trip to Formosa without the knowledge or approval of the nation's leaders.

Nevertheless, the uninformed speculation in the press and the lack of real knowledge as to what MacArthur had done on Formosa, coming at a time when the United States was trying to convince Communist China that there were no ulterior motives lurking behind President Truman's action toward Formosa, caused the President, in a sternly worded message over Secretary of Defense Johnson's signature, to caution MacArthur. On 4 August, MacArthur was reminded in no uncertain terms, "No one other than the President as Commander-in-Chief has the authority to order or authorize preventive action against concentrations on the mainland. The most vital national interest requires that no action of ours precipitate general war or give excuse to others to do so." [16]

MacArthur replied the next day that he fully understood and was operating meticulously in accordance with the President's decision of 27 June.[17] Then, on 7 August, he submitted a full report of his conference with Chiang Kai-shek. He indicated Chiang's willingness to co-operate and that there was a real potential in the armed forces on Formosa, although substantial improvements would be necessary. He explained that he had directed periodic sweeps of the Formosa Strait by elements of the Seventh Fleet, periodic reconnaissance flights over certain of the coastal areas of China, and familiarization flights by small groups of United States aircraft to include temporary and refueling landings on Formosa.[18]

President Truman subsequently sent Averell Harriman to Tokyo, reputedly to caution MacArthur not only to keep the President better informed, but, on other than military matters, to make recommendations, not decisions. Afterward, Harriman stated that General MacArthur had not overstepped his military bounds in making the trip to Formosa; President Truman announced his satisfaction with General MacArthur's performance; and General MacArthur declared that anyone who hinted of friction between himself and the President was guilty of "sly insinuations, brash speculations and bold misstatements." [19]

Communist nations made much of President Truman's order neutralizing Formosa and charged that the United States intended to take over and occupy the island. In August, the Peiping regime accused the United States of aggression against Formosa and asked the United Nations Security Council to order the withdrawal of ". . . all of the United States armed invading forces from Taiwan. . . ." [20] In refutation of this charge, President Truman on 25 August directed United States Ambassador Austin to address the Secretary-General, Trygve Lie, on the matter. Austin sent

[15] *MacArthur Hearings*, pp. 3383–84.

[16] Rad, WAR 88014, JCS to CINCFE, 4 Aug 50.

[17] Rad, C 59418, CINCFE to JCS, 5 Aug 50.

[18] Rad, C 59569, CINCFE to JCS, 7 Aug 50.

[19] (1) *Time*, August 21, 1950. (2) For Harriman's report to President Truman on this visit, see Truman, *Memoirs*, II, 349–53; for MacArthur's reaction to Harriman's visit, see MacArthur, *Reminiscences*, p. 341.

[20] State Dept *Bulletin*, XXIII (4 September 1950), p. 396.

Lie a complete account of the official American attitude toward Formosa, including a 19 July statement to the Congress by President Truman in which he declared ". . . that the United States has no territorial ambitions whatever concerning that island, nor do we seek for ourselves any special position or privilege on Formosa." On 28 August, the President sent Austin more ammunition with which to demolish the Communist charges concerning Formosa by telling him that the United States would welcome United Nations consideration of the case of Formosa.[21]

A week earlier, General MacArthur, who had been invited to speak at the Fifty-First National Encampment of the Veterans of Foreign Wars (VFW) in Chicago, sent instead a paper which he proposed be read at the meeting. In this paper, MacArthur stressed the strategic importance of Formosa and insisted that the United States must, at any cost, retain control of that island. He strongly hinted that the United States would be able to use Formosa as a base in any future operations against the Asiatic mainland. He pointed out also that Formosa would be a formidable threat to American security if controlled by an unfriendly power, terming it an "unsinkable aircraft carrier and submarine tender." "Nothing could be more fallacious," he charged, "than the threadbare argument by those who advocate appeasement and defeatism in the Pacific that if we defend Formosa we alienate continental Asia. Those who speak thus do not understand the Orient." [22]

This strong statement evoked an equally strong reaction from President Truman when he was informed of it by Harriman on 26 August before its publication. The President read MacArthur's paper to Harriman, General Bradley, and Secretary Johnson, then directed that MacArthur withdraw the statement. Secretary Johnson immediately cabled MacArthur that the President directed him to withdraw the message ". . . because various features with respect to Formosa are in conflict with the policy of the United States and its position in the United Nations." [23]

According to Johnson, when the President learned that MacArthur's rather lengthy statement to the VFW had been transmitted through Army communications facilities from Japan, he was particularly indignant. Johnson testified before the Senate committee, later investigating the relief of General MacArthur, that on 26 August President Truman discussed with him the advisability of relieving MacArthur as the commander in Korea, but decided to take no such action at that time.[24]

General MacArthur's immediate response to the Presidential order was to fire a protest to the Secretary of Defense, claiming that his VFW message had been carefully prepared to support fully the President's policy decision of 27 June with respect to Formosa and pointing out that the subject of Formosa had been freely discussed in all circles, "Govern-

[21] (1) Ltr, Ambassador Austin to Secy-Gen Lie, 25 Aug 50. (2) Ltr, President Truman to Ambassador Austin, 28 Aug 50. (3) *MacArthur Hearings*, pp. 3473–76.

[22] Quotation in Whitney, *MacArthur: His Rendezvous with History*, pp. 378–79.

[23] *MacArthur Hearings*, pp. 2586, 3480.

[24] *MacArthur Hearings*, pp. 2587, 3665.

mental and private, both at home and abroad." MacArthur obviously felt he could separate his views as a private citizen from those as the commander in chief, United Nations Command. For he observed that the views embodied in his statements to the VFW were "purely my personal ones." Noting that the VFW undoubtedly had given wide distribution to his speech in advance press releases, MacArthur advised Johnson that suppressing his message under these conditions would be a grave mistake.[25] General MacArthur nevertheless was ordered to withdraw his message to the veterans' group. President Truman later softened the blow to MacArthur's feelings by transmitting to him the text of Austin's message on Formosa to Secretary-General Lie and his own letter to Austin on the same subject with the statement, "I am sure that when you examine this letter . . . you will understand why my action of the 26th in directing the withdrawal of your message to the Veterans of Foreign Wars was necessary." [26]

Although General MacArthur informed the national commander of the VFW that he had been directed to withdraw his message, MacArthur's statement appeared in print in *U.S. News and World Report* on 1 September 1950,[27] and thereby provided excellent grist for the USSR propaganda mill. Andrei Vishinsky, Soviet Ambassador to the United Nations, for instance, twisted MacArthur's comments on Formosa to his own purposes when, in a speech, he stated, "None other than General MacArthur recently informed, with cynical candor, the whole world about the decision of the ruling circles of the United States of America at all costs to turn Taiwan [Formosa] into an American base in the Far East." [28]

Even so, it is probable that the Formosa incident, had it been the last instance of friction, would have been forgotten, or at least overlooked, by Mr. Truman. But it was to be only the beginning of a series of incidents that progressively weakened President Truman's patience with General MacArthur.

MacArthur's Disagreement With United States Policy in Korea

During the first five months of the Korean fighting, General MacArthur did not openly criticize the directives under which he operated in Korea. From time to time he did complain through official channels about shortages of men and equipment. But at the Wake Island Conference, the United Nations commander indicated that he understood the problems Washington faced in supporting him when he told the assembled officials that no commander in the history of war had ever had more complete and adequate support from all the agencies in Washington.[29] When, in late October, he found the terms of his current directive a little too restrictive, he did not hesitate to ignore it by ordering all his forces north to the border instead of using only ROK forces in that area, as he

[25] (1) Rad, DEF 89880, DEF to CINCFE, Johnson (Personal) for MacArthur, 26 Aug 50. (2) Rad, C 61325, CINCFE to DA for Secy Defense, MacArthur (Personal) for Johnson, 27 Aug 50. (3) MacArthur, *Reminiscences*, pp. 385–86 and 389.

[26] *MacArthur Hearings*, p. 3480.

[27] JSSC Rpt to JCS, p. 45.

[28] *MacArthur Hearings*, p. 2002.

[29] *MacArthur Hearings*, p. 213.

had been instructed. But he did not openly criticize that directive.

During November, as signs that the Chinese were intervening in Korea began to appear, signs of differences between MacArthur and Washington officials also began to develop. MacArthur's order for bombing the Yalu River bridges on 6 November, for instance, aroused consternation in Washington. More than that, the order was not in accordance with the instructions issued him on 29 June, warning him to stay well clear of the USSR and Manchurian borders in conducting his air operations. When three days later the Joint Chiefs of Staff suggested that MacArthur's mission in Korea might have to be changed in view of Chinese intervention, he hotly proclaimed his disagreement [30] and his mission was not changed. The 11th-hour suggestion by the Joint Chiefs of Staff, prompted by the conviction of other nations participating in the Korean fighting that MacArthur should check his attack along a line short of the Yalu, was likewise roundly condemned by General MacArthur.[31] Yet at no time in this period did General MacArthur argue his case publicly, although he did deny that he had received any suggestion from "any authoritative source" that he should stop at any line short of the international boundary.[32]

With the intervention by Chinese Communist armies, however, General MacArthur's attitude toward public disclosures changed abruptly. Smarting from the defeat his forces had suffered, MacArthur spoke out sharply in his own defense, and in published statements in early December charged that the limitations upon his operations were an enormous handicap "without precedent in military history," and intimated that selfish interests in Europe were causing support to be withheld from his forces.[33]

These statements were widely assessed as a criticism of the United Nations policy of limited war in Korea and as an oblique criticism of the Truman administration in its conduct of the war. They were probably not so intended. General MacArthur pointed out quite rightly that at no time had he asked for authority to retaliate beyond the inviolate northern boundary of Korea. His statements, issued at a time when the administration was trying earnestly to reassure uneasy allies, were nonetheless of great concern to President Truman and his advisers. British fears that MacArthur might involve the west in a large-scale war with Communist China made his pronouncements especially regrettable. Too, his insistence on blaming operational restrictions for the situation in Korea was taken in Washington as a reflection on the judgment of the man who had decided to impose those restrictions. This inference was unfortunate, even if not intended as such, for the Joint Chiefs of Staff believed that MacArthur was partly responsible for his own predicament. They had been persuaded to approve his plans for the No-

[30] Rad, C 68572, CINCFE to DA for JCS, 9 Nov 50.

[31] Rad, C 69808, MacArthur to DA for JCS, 25 Nov 50.

[32] (1) Statement, Gen MacArthur (in reply to question from Arthur Krock), New York *Times*, December 1, 1950. (2) *MacArthur Hearings*, p. 3496.

[33] (1) Statements, Gen MacArthur, 2 Dec 50, in *U.S. News and World Report*, December 8, 1950, pp. 16–22, and in the New York *Times*, December 2, 1950. (2) *MacArthur Hearings*, pp. 3532–34.

vember advance by his great confidence that the Chinese either would not, or could not, intervene effectively.[34]

President Truman found MacArthur's statements at this time particularly objectionable. He was irked by MacArthur's publicizing the fact that because Washington had not let him do things his own way his 24 November attack had been defeated and the Chinese had intervened. President Truman charged that he should have relieved MacArthur at that point. He had not done so because he did not wish it to appear that he was being relieved because the offensive had failed. Truman had never gone back on people when luck was against them and he did not intend to do so in this case.[35]

Commenting later, President Truman stated that there was no excuse for the way in which MacArthur began to publicize his views following the failure of his offensive. Within four days, in four different ways, according to President Truman, MacArthur blamed his troubles publicly on Washington's order to limit the hostilities in Korea. He spoke of extraordinary inhibitions without precedent in military history and made it quite plain that no blame should be attached to him or his staff.[36]

As a result of MacArthur's statements President Truman had sent to the heads of all executive departments his 5 December memorandum ordering government officials to clear all public statements concerning foreign policy with the Department of State and all concerning military affairs with the Department of Defense. Although this memorandum was addressed to all executive branches, it was directed specifically at General MacArthur.[37]

Meanwhile, MacArthur had proposed to General Collins, then visiting in the Far East, that the United States should carry the war to China, through bombing, blockade, and other measures. Throughout December and January and into February, he insisted upon these measures, always through official media albeit quite forcefully. Not until mid-February did General MacArthur bring his views to the attention of the public. On 13 February, in a statement to the press, he contended that unless he was allowed to reduce materially the superiority of the Chinese, ostensibly through attacks upon their "sanctuary," he could not seriously consider conducting major operations north of the 38th Parallel; and on 7 March, in another such statement, he held that vital decisions, yet to be made, must be provided on the highest international levels. These statements may have been perfectly true, and there is considerable evidence that the latter opinion especially was valid; but both were considered by Washington authorities to be comments affecting both foreign and military policies of the United States Government that were not cleared by the Department of Defense and were, therefore, in violation of the President's directive of 5 December.[38]

---

[34] Hoare, *The JCS and National Policy*, vol. IV, ch. VII, pp. 32-33.

[35] Truman, *Memoirs*, II, 384; see also pp. 444-50.

[36] *Ibid.*, 382.

[37] (1) Rad, JCS 98134, JCS to CINCFE, CINCEUR, CINCAL, *et al.*, 6 Dec 50. (2) *MacArthur Hearings*, p. 880.

[38] (1) *MacArthur Hearings*, pp. 3539-40. (2) Ltr, Secy Marshall to Hon Richard B. Russell, 17 Aug 51. (3) *MacArthur Hearings*, pp. 3665-66.

General Marshall later said that MacArthur's statement of 24 March offering to negotiate with enemy leaders in the field was the culminating factor in President Truman's decision to relieve MacArthur. Marshall charged that this statement contained a thinly veiled hint that the enemy should either negotiate or the war would be carried to the Chinese mainland. He based this view on that portion of MacArthur's statement which said, "The enemy therefore must by now be painfully aware that a decision of the United Nations to depart from its tolerant effort to contain the war to the area of Korea through expansion of our military operations to his coastal areas and interior bases would doom Red China to the risk of imminent military collapse." [39]

The President found this to be a most extraordinary statement for a military commander of the United Nations to issue on his own responsibility. He construed MacArthur's statement as defiance of his orders as Commander in Chief and a challenge to his authority. Additionally MacArthur was, in the President's view, flouting the policy established by the United Nations. By this act General MacArthur had left the President no choice. He felt he could no longer tolerate MacArthur's insubordination.[40] President Truman nonetheless delayed a final decision; neither the general public nor General MacArthur was aware of President Truman's steadily mounting dissatisfaction.

But the stage had already been set for the final act. On 20 March General MacArthur, in reply to a personal letter from Joseph W. Martin, House of Representatives Minority Leader, sent a relatively mild commentary to the Congressman on American foreign policy. There was nothing new in what he said, nor was it said in a particularly inflammatory manner. MacArthur merely reiterated his views that the Asian theater was fully as important as the European and that the United States must prosecute the Asian war until victory was achieved. MacArthur later testified that the letter to Congressman Martin was, in his mind, so trifling a matter that he could scarcely recall it.[41] But it was no trifling matter to President Truman; and when Martin chose to make public the contents of the letter on 5 April 1951, the President reacted strongly and quickly. On 6 April, he called together his special assistant, Averell Harriman, Secretary of State Acheson, Secretary of Defense Marshall, and the chairman of the Joint Chiefs of Staff, General Bradley, and put the matter squarely before them. What should be done with General MacArthur? Harriman told Truman that MacArthur should have been fired two years ago.[42] But Secretary Marshall appreciated the possible repercussions that

[39] (1) Ltr, Secy Marshall to Sen Russell, 17 Aug 51, in *MacArthur Hearings*, pp. 3655–66. (2) Statement of Gen MacArthur, 24 Mar 51, in *MacArthur Hearings*, p. 3541.

[40] Truman, *Memoirs*, II, 441–42.

[41] (1) *MacArthur Hearings*, p. 113. (2) See also MacArthur, *Reminiscences*, pp. 385–86 and 389.

[42] Harriman was referring to MacArthur's plea in 1949 that he was too busy when requested to return to the United States to discuss matters concerning the occupation of Japan, and to subsequent difficulty in convincing MacArthur to withhold approval from a bill of the Japanese Diet which was contrary to the approved economic policy for the occupation. See Truman, *Memoirs*, II, 447.

GENERAL RIDGWAY

might come with MacArthur's dismissal and advised caution. He observed that if the President relieved General MacArthur it might be difficult to get pending military appropriations through Congress. General Bradley approached the question from the point of view of military discipline. He believed that MacArthur had acted in an insubordinate manner and that, consequently, he deserved to be relieved of his command. But he told Truman that he wished to talk with General Collins before making a final recommendation. Secretary Acheson believed that MacArthur should be relieved, but only after a unanimous decision to do so by the Joint Chiefs of Staff.

President Truman had already decided to relieve General MacArthur but he kept this decision to himself for the time being. He asked all four of his advisers to return again the next day for more discussion, and directed Secretary Marshall to restudy all messages exchanged with General MacArthur in the past two years. The next morning, 7 April, the same group met again in President Truman's office. Secretary Marshall told the President that after

GENERAL VAN FLEET

reading all the messages he agreed with Harriman that MacArthur should have been relieved two years earlier. Before this brief meeting ended, President Truman directed General Bradley to obtain the views of the remaining Joint Chiefs of Staff and to be prepared to make a final recommendation on 9 April.[43]

General Bradley, General Collins, General Vandenberg, and Admiral Sherman met on the afternoon of 8 April in the Pentagon and discussed the military aspects of MacArthur's relief thoroughly. At the conclusion of this conference, these officers conferred briefly with Secretary Marshall. All of the Joint Chiefs of Staff agreed that from the military viewpoint the relief of General MacArthur should be carried out. General Collins, the Army Chief of Staff, later testified that he based his belief on two principal factors: first, he was convinced that General MacArthur was not in sympathy with the basic policies governing the operation of United Nations forces in Korea. "I felt," Collins stated, "that the President, as our Commander in Chief, was entitled to have as a commander in the field a man who was more in sympathy with the basic policies and more responsive to the will of the President as Commander in Chief." Collins felt, secondly, that MacArthur had failed to comply with the instructions given him directing him to clear any public statement that he made which involved matters of policy, particularly of foreign policy.[44]

After receiving the views of his principal advisers on 9 April, President Truman announced his decision to relieve General MacArthur of his commands in the Far East. General Ridgway was to replace MacArthur and Lt. Gen. James A. Van Fleet was to become the new commanding general of the Eighth Army.

The President originally intended to notify General MacArthur of his relief at 2000 Washington time on 11 April (1000 Tokyo time, 12 April). A message was sent to Secretary of the Army Pace, then visiting in Korea, telling him to deliver the relief message to MacArthur at his residence, the American Embassy in Tokyo, at the time indicated. But Secretary Pace failed to receive these instructions because of a breakdown in a

[43] Truman *Memoirs*, II, 447–48.

[44] *MacArthur Hearings*, p. 1187.

communications power unit in Pusan.[45] In the meantime, late on 10 April, indications appeared that the action to be taken had become known publicly. It was then decided by President Truman to accelerate the transmission of the official notification to General MacArthur by approximately twenty hours.[46]

According to General MacArthur, his first inkling that he had been relieved came from his wife. One of MacArthur's aides had heard the news on a radio broadcast and told Mrs. MacArthur who then informed her husband. The official notification, MacArthur claimed, did not reach him until half an hour later.[47] MacArthur immediately ordered General Hickey to telephone General Ridgway in Korea and to notify him of the change in command. He turned over to Hickey the functions of command until General Ridgway could leave the front and fly to Tokyo to take over in person.[48]

[45] MacArthur called the communications failure "incredible. . . . He [Pace] was in Korea at the moment in immediate message contact with my headquarters, which had similar contact with Washington." See MacArthur, *Reminiscences*, p. 395.

[46] *MacArthur Hearings*, p. 345.

[47] (1) *Ibid.*, p. 155. (2) See also MacArthur, *Reminiscences*, p. 395.

[48] *MacArthur Hearings*, p. 155.

CHAPTER XXI

New Direction, New Policy

As the controversy within the national defense structure heightened and finally culminated in General MacArthur's relief, tension also mounted steadily at the battlefront. Pushed back by Ridgway in March and April, enemy forces had compressed like a huge spring, and apparently were about ready to burst forth again in a major attack. Just before MacArthur's removal, Ridgway had called his American corps commanders together and warned them that, if the Chinese struck in full strength, the Eighth Army might be in for the worst period since it had entered Korea.[1]

General Ridgway learned of the startling command change late on 11 April. The next day he flew to Tokyo, where he found MacArthur "amazed" by the President's action in relieving him but not outwardly bitter or resentful. MacArthur briefed Ridgway on some of the key problems of the command, then wished Ridgway well in his new assignment.[2]

The change of commanders was attended by some confusion as to General Van Fleet's role. The Secretary of Defense, in notifying Ridgway of his own appointment as United Nations commander, had added that Van Fleet was being sent to Korea for "such duties as you may direct." Immediately after meeting with MacArthur, Ridgway telephoned Secretary of the Army Pace and, expressing uncertainty as to his authority in choosing a commander for the Eighth Army, asked how he was expected to use General Van Fleet. But Secretary Pace knew no more about the matter than did Ridgway and could only refer the question to Secretary Marshall.[3]

The Joint Chiefs of Staff promptly set the matter straight. They notified Ridgway that the President had personally approved Van Fleet as the successor to the command of the Eighth Army. They themselves suggested the phrase, "for such duties as you may direct," since they realized the imminence of a major Chinese offensive and felt that General Ridgway might wish to retain direct command in the field for the time being, presumably until the threatening enemy offensive had been turned back. Until that time, they suggested, Van Fleet

---

[1] Ridgway, The Korean War, Issues and Policies, pp. 418–19.

[2] *Ibid.*, p. 160.

[3] Comd Rpt, Eighth Army, Apr 51, sec. II, Book 1, OCG.

could serve as General Ridgway's deputy commander.[4]

But General Ridgway saw no reason to delay turning the Eighth Army over to its new commander. He already had made plans to meet any Chinese attack; and the army staff and subordinate commanders could be counted on to carry their new chief safely past the difficult familiarization period. General Van Fleet arrived in Korea at 1230 on 14 April and immediately assumed command of the Eighth Army. General Ridgway, who had returned to Korea to receive Van Fleet, left for Tokyo at 1900 the same day.[5] Two days later, on 16 April, General MacArthur and his family left Japan for the United States.

On 16 April, Ridgway reported to Washington, "Although the enemy has remained on the defensive since mid-February, only partially exploiting his enormous potential of more than sixty divisions, he retains the capability of assuming the offensive at any time." [6] Three days later, Ridgway placed definite restrictions on the advance of Van Fleet's divisions out of conviction that the Chinese would attack soon and that for Van Fleet to overextend his own lines to attack could be dangerous. Accordingly, Van Fleet was not to send any strong force farther than Line WYOMING without Ridgway's prior approval.[7]

Responding to Ridgway's instructions on the same day, the new Eighth Army commander informed his superior that he planned to jump off in a general advance toward Line WYOMING on 21 April. He also asked Ridgway to approve a secondary limited objective attack in the east, to be opened on 23 April, to seize the Kansong-Inje road by securing what he called Line ALABAMA from Yanggu to Songhyon-ni.[8]

General Ridgway authorized Van Fleet to secure the Kansong-Inje road, but designated a modified Line ALABAMA which ran somewhat farther east and west than the line proposed by Van Fleet. Seizure of this line would bring the eastern quarter of the Eighth Army front well north of the 38th Parallel. Ridgway specified that no large enemy groups were to be bypassed and that lateral co-ordination within and between corps was to be maintained.[9]

In briefing his staff on 19 April, Ridgway described the current enemy situation as much more favorable for United Nations forces than might appear. Existing North Korean units, in his judgment, could operate successfully only against South Korean troops; and while he expected a Chinese offensive soon, he was certain that the Eighth Army could defeat it.[10]

The Chinese Offensive—22 April

United Nations forces were still edging toward Lines WYOMING and ALABAMA when the enemy launched their expected offensive. The opening attack on the night of 22 April halted Van

[4] Rad, JCS 88374, JCS to CINCFE (Personal) for Gen Ridgway, 12 Apr 51.

[5] Comd Rpt, Eighth Army, Apr 51, sec. II, Book 1, OCG.

[6] Rad, C 60250, CINCUNC to DA, 16 Apr 51.

[7] Rad, CX 60388, CINCFE to CG Eighth Army, 19 Apr 51.

[8] Rad, G-4 3901 KCG, Gen Van Fleet (Personal) for Ridgway, 19 Apr 51.

[9] Rad, C 60684, CINCFE to CG Army Eight, 21 Apr 51.

[10] MFR, sgd Surles, 19 Apr 51, in SGS, GHQ, FEC files.

Fleet's central forces just short of Line WYOMING when three Chinese armies struck toward the Yonch'on-Hwach'on area. After this initial blow, actually a secondary effort, the Chinese delivered the main attack against I and IX Corps in the west in an attempt at a double envelopment of Seoul. A tertiary drive developed simultaneously in the east near Inje. The Eighth Army held firm against the assaults everywhere except in the central sector, where one ROK division crumpled and fell back in considerable disorder for almost twenty miles. This breakthrough prompted Van Fleet to withdraw, as previously planned by Ridgway, to Line KANSAS.

By 24 April, it appeared that 337,000 Chinese were driving toward Seoul in the main enemy effort, and that the enemy's secondary effort in the central zone was being pressed by about 149,000 troops.[11] In both the west and central sectors, these enemy forces followed the Eighth Army withdrawal closely despite high losses. (*Map VIII*) By the end of April, additional withdrawals took the Eighth Army back an average of thirty-five miles to a new defense line arching no more than five miles north of Seoul in the west and running generally northeastward across the peninsula to a point slightly above Yangyang on the east coast. But Van Fleet's forces, after again inflicting tremendous casualties on the enemy, fell back no farther and the Communist offensive beat itself out along this line. General Ridgway's 19 April evaluation had been well based.[12]

[11] Telecon, TT 4635, GHQ and DA, 24 Apr 51.
[12] Comd Rpt, Eighth Army, Narrative, Apr 51.

Missions and Responsibilities—CINCUNC

Ridgway applied a similar searching analysis to all of his new duties as commander in chief, United Nations Command. One of his first acts was to obtain from his staff a complete recapitulation of the missions and authorities he had inherited from General MacArthur.

In summing up, Ridgway's staff described specific positive missions, inferred positive missions, restrictive missions, and complementary authorities. As commander in chief of the United Nations Command, Ridgway was responsible for maintaining the integrity of United Nations forces; continuing to fight in Korea so long as, in his judgment, such action offered a reasonable chance of success; maintaining a blockade of the entire Korean coast; stabilizing the situation in Korea or evacuating to Japan if forced out of Korea; taking all appropriate steps to ensure stability throughout Korea; and taking actions under the auspices of the United Nations to establish a unified, independent, and democratic government in the sovereign state of Korea.[13]

He was forbidden to take military action against Chinese territory without authority from Washington; to use non-Korean forces in areas bordering on Manchuria or Russia; to allow any of his forces to cross those borders; to take air or naval action against Rashin; or to attack the hydroelectric installations in North Korea in the vicinity of the Yalu

[13] Study, CINCUNC and CINCFE, Current Military Missions as Derived From JCS Communications, no date, in Comd Rpt, GHQ UNC, Apr 51, Annex IV, Part III, Doc. 35.

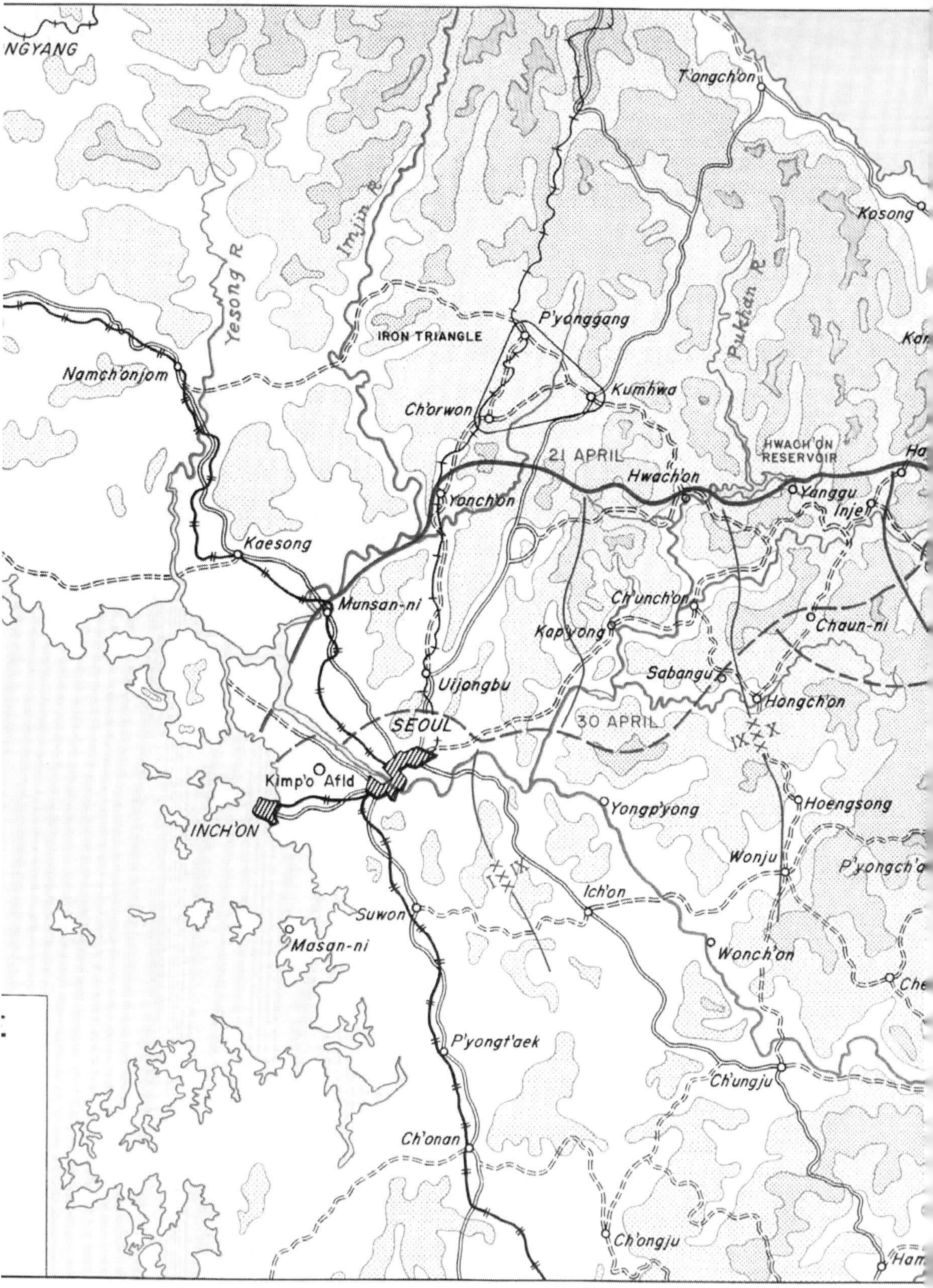
NGYANG
T'ongch'on
Kosong
Imjin R
Yesong R
Pukhan R
P'yanggang
IRON TRIANGLE
Namch'onjom
Kumhwa
Ch'orwon
21 APRIL
HWACH'ON RESERVOIR
Hwach'on
Yanggu
Inje
Yonch'on
Kaesong
Munsan-ni
Ch'unch'on
Chaun-ni
Kap'yong
Sabangu
Uijongbu
Hongch'on
SEOUL
30 APRIL
IX
Kimp'o Afld
INCH'ON
Yongp'yong
Hoengsong
Wonju
P'yongch'a
Ich'on
Suwon
Masan-ni
Wonch'on
P'yongt'aek
Ch'ungju
Ch'onan
Ch'ongju

River. Nor could he consider the region above the waist of Korea as a general objective area, but he could, subject to the restrictions on troops and targets, operate north of the 38th Parallel.[14]

Finally, he had authority to command all military forces assisting the Republic of Korea which were placed under the unified command of the United States; to use ROK soldiers and civilians in North Korea so long as they were designated as United Nations instrumentalities under his control; and to dispose of prisoners of war in such a manner as would least interfere with military operations, providing such disposition was in consonance with the provisions of the Geneva Convention.[15]

Once aware of what was expected of him by the Joint Chiefs of Staff, General Ridgway sent his major commanders written instructions telling them exactly what he expected of them. As became apparent to these subordinates, Ridgway's approach to the Korean conflict differed completely from that of General MacArthur—Ridgway's greatest concern seemed to be that some rash act of his command might cause the conflict to spread to other areas.

General Ridgway warned his principal commanders, "The grave and ever present danger that the conduct of our current operations may result in an extension of hostilities, and so lead to a world-wide conflagration, places a heavy responsibility upon all elements of this command, but particularly upon those capable of offensive action. "In accomplishing our assigned missions," Ridgway emphasized,

> this responsibility is ever present. It is a responsibility not only to superior authority in the direct command chain, but inescapably to the American people. It can be discharged only if every commander is fully alive to the possible consequences of his acts; if every commander has imbued his command with a like sense of responsibility for its acts; has set up, and by frequent tests has satisfied himself of the effectiveness of, his machinery for insuring his control of the offensive actions of his command and of its reactions to enemy action; and, in final analysis, is himself determined that no act of his command shall bring about an extension of the present conflict, except when such act is taken in full accordance with the spirit of the accompanying letter of instructions.[16]

Ridgway emphatically pointed out that international tensions within and bearing upon the Korean theater had created acute danger of World War III, and that the instructions from Washington reflected the intense determination of the American people, as well as of all free peoples of the world, to prevent World War III, if it could be done without appeasement or sacrifice of principle. "In the day to day, in fact the hour to hour, performance of his duties," Ridgway concluded, "I therefore desire that every responsible commander, regardless of rank, bear constantly in mind that the discharge of his responsibilities in this respect is a sacred duty." [17]

In his instructions to General Van Fleet, General Ridgway warned that until American intelligence agencies had determined otherwise, Van Fleet was to assume that the enemy forces were determined to drive the Eighth Army from Korea or to destroy it in place, and that

[14] *Ibid.*
[15] *Ibid.*

[16] Rad, C 60965, CINCFE to JCS, 25 Apr 51, quoting memo from Ridgway to All Comdrs, 22 Apr 51.
[17] *Ibid.*

Russia might at any time attack the United Nations Command. "You will further base your operations," he told Van Fleet,

> on the assumptions that your own forces will be brought to and maintained at approximately TO&E strength, but that you will receive no major reinforcements in combat organizations or service support units; that the duration of your operations cannot now be predicted; that you may, at any time, be directed by competent authority to initiate a withdrawal to a defensive position and there be directed to defend indefinitely; that you may at any time be directed by competent authority to initiate a retirement designed to culminate in an early evacuation of the Korean peninsula.[18]

General Ridgway then charged Van Fleet with a mission which appears to have been Ridgway's own idea, a balance between what the Joint Chiefs of Staff had directed him to do and the military capabilities which he felt he possessed. "Your mission," he told Van Fleet, "is to repel aggression against so much of the territory (and the people therein) of the Republic of Korea, as you now occupy and, in collaboration with the Government of the Republic of Korea, to establish and maintain order in said territory." [19]

Keeping in mind constantly the restrictions upon his own authority, General Ridgway cautioned Van Fleet that, while he could operate north of the 38th Parallel, he could only use Korean troops in the areas bordering on Manchuria and Russia and that he must prevent any border-crossing by these forces.[20] "You will direct the efforts of your forces," he ordered Van Fleet, "toward inflicting maximum personnel casualties and material losses on hostile forces in Korea, consistent with the maintenance intact of all your major units and the safety of your troops." Indicating that he was satisfied with the way things were going in Korea, Ridgway further stated:

> The continued piecemeal destruction of the offensive potential of the Chinese Communist and North Korean armies contributes materially to this objective, while concurrently destroying Communist China's military prestige. . . . Acquisition of terrain in itself is of little or no value.[21]

Although General Ridgway had not yet been so directed, he fixed a line beyond which the Eighth Army could not advance without his permission. Clearly, Ridgway intended to keep a much tighter rein on the new field commander than MacArthur had kept upon him. In similar instructions to Admiral Joy and General Stratemeyer, Ridgway gave strict warnings against violating the borders of China and the USSR, and forbade these commanders to employ their forces except in support of United Nations operations within a 20-mile range of USSR territory or within a 3-mile range of Chinese Communist territory.

New Orders for CINCUNC

Ridgway's responsibilities as commander in chief, Far East Command, and commander in chief, United Nations Command, while nominally separate, were nonetheless closely related, and on

[18] Ltr of Instructions, Gen Ridgway to Gen Van Fleet, 22 Apr 51.
[19] *Ibid.*
[20] *Ibid.*

[21] *Ibid.*

occasion became so intertwined as to interfere with each other. This had been one of the major causes of General MacArthur's professed inability to understand or reconcile his instructions.

This conflict between roles became especially noticeable when General Ridgway on 17 April asked that the Joint Chiefs authorize him, at his own discretion, to withdraw United Nations forces from Korea in the event of a USSR attack and to use them to defend Japan. He must have known that such a unilateral procedure would certainly exceed his authority as a United Nations commander, and probably made the request to emphasize not only the threat of a Russian attack on Japan but also the need for making his missions and authorities clearer by distinguishing between the United Nations in Korea and the United States in the Far East. In this purpose, if such it was, he was eminently, but not immediately, successful.[22]

The immediate response of the Joint Chiefs of Staff was to turn Ridgway down. They agreed in principle to a withdrawal of United Nations forces from Korea in the event of Russian attack, but retained the right to control such a withdrawal. They notified Ridgway that "Subject . . . to the immediate security of your forces both in Korea and Japan, you will initiate major withdrawal from Korea only upon instructions furnished you after receipt of information from you as to conditions obtaining." Since Ridgway's first concern in the event of a Russian attack undoubtedly would be the security of his forces, these instructions were, to a degree, unclear.[23] On the related matter of employing other than American and ROK forces against a Russian attack, Ridgway, as he probably expected, was told that he would not plan on using such forces for that purpose "pending further instructions." [24]

Prompted by Ridgway's request, the Joint Chiefs of Staff ordered the Army Chief of Staff to prepare for General Ridgway a directive governing the conduct of Korean operations. Exhaustively analyzing all outstanding JCS directives to CINCFE and CINCUNC, the Army G–3, General Taylor, furnished General Collins with a proposed new set of instructions for General Ridgway. Collins presented it to the Joint Chiefs who passed it on to the Secretary of Defense and the President for final approval.[25]

In the meantime, General Ridgway had been doing his own analyzing and on 30 April sent to the Joint Chiefs of Staff a draft directive to himself, which he proposed they approve and return to him as his authority for operating in Korea. But General Ridgway had waited too long. The President had already approved the set of instructions prepared for Ridgway by the Army G–3, and on 1 May these instructions were sent to him.[26]

But even this newest directive did not clearly separate Ridgway's responsibilities as CINCUNC from those as CINCFE. The over-all mission assigned to Ridgway as CINCUNC was "to assist

[22] Rad, C 60308, CINCFE to JCS, 17 Apr 51.

[23] Rad, JCS 88950, JSC to CINCFE, 19 Apr 51.

[24] *Ibid.*

[25] Memo, ACofS G–3 for CofS, 21 Apr 51, sub: Directive to CINCFE for Opns in Korea, in G–3, DA file 381 Korea, Case 6.

[26] Rad, JCS 90000, JCS to CINCFE, 1 May 51.

the ROK in repelling aggression and to restore peace and security in the area," and his military objective was "to destroy the armed forces of North Korea and Communist China operating within the geographic boundaries of Korea and waters adjacent thereto." But in pursuit of his objectives in Korea, the security of forces under his command and his basic CINCFE mission of defending Japan were still overriding.[27] Here, then, the Joint Chiefs of Staff again had failed to differentiate between the needs and powers of CINCFE and CINCUNC and had defined these unclear relationships with insufficient precision.

The usual injunctions against violating Communist China or Russian territory were passed on to Ridgway. If the Chinese attacked United States forces outside of Korea by sea or air, in Japan for instance, retaliation against China's mainland would be permitted, but no retaliatory step would be taken without the specific approval of the Joint Chiefs of Staff.[28]

For the first time, the Joint Chiefs of Staff limited the advance of United Nations forces. Ridgway was to make no general advance beyond the original trace of the Lines KANSAS-WYOMING without prior approval. They did authorize deeper limited movements. Ridgway could, at his own discretion, move north of the line in limited operations, including amphibious and airborne operations, to keep the enemy off-balance and to maintain contact, or for the omnibus and elastic purpose of "insuring the safety of your command." [29]

The restriction on general advances reflected a growing conviction that the Korean problem would not be solved by military action alone.[30] Later critics claimed that this restriction interfered with the field commander, and attached various sinister motives to its proponents. The truth was that the Joint Chiefs of Staff, and General Ridgway as well, had realistically faced the military facts. They knew that no major reinforcement was available or would become available from United States forces. They were aware of the great reluctance of member nations of the United Nations to send more fighting men and equipment to the United Nations Command and of the desire of some of these nations to get out of the Korean situation gracefully and without unduly aggravating the Chinese. They were fairly certain that some sort of armistice negotiations could be developed if the U.N. advance into North Korea was so modest that the enemy could profitably agree to a demarcation line along the line of contact. They were aware that each mile of advance into North Korea changed the balance of logistic capability in favor of the enemy and brought the United Nations forces within ever closer range of the increasingly strong enemy air force based in Manchuria. They were reluctant to accept the numerous casualties that a further advance would cost. But most of all, both the Joint Chiefs of Staff and General Ridgway knew that the

27 *Ibid:*

28 *Ibid.*

29 *Ibid.*

30 In this connection see: (1) Rad, C 59397, CINCFE to JCS, 5 Apr 51. (2) Rad, CX 60388, CINCFE to CG Eighth Army, 19 Apr 51. MFR, sgd Surles, 19 Apr 51, SGS, in GHQ, UNC, SGS files. (3) Rad, C 60965, CINCFE to JCS, 25 Apr 51, quoting Memo for Ridgway to All Comdrs, 22 Apr 51.

Chinese had ground forces available in North Korea and Manchuria that had not been tapped and that far outnumbered those of the United Nations Command in the area. Furthermore, the Joint Chiefs of Staff evidently expected some sort of political effort toward stopping the military operations. They told Ridgway that, if Communist military leaders requested an armistice in the field, he was to report that fact to them immediately and await instructions.[31]

Meanwhile, the Joint Chiefs of Staff reminded Ridgway, the prohibition against the use of the 40th and 45th Infantry Divisions had not been lifted. These divisions were to be kept in Japan and the integrity of their men and units preserved. With regard to the ROK Army, Ridgway could use such forces as were already available, but the United States would not furnish logistic support for any other major ROK units at that time.[32]

A New Directive for CINCFE

When General Ridgway examined his new instructions, he found that they did not coincide with the version which he had recommended on 30 April, nor did he completely understand them. He therefore notified the Joint Chiefs of Staff on 7 May that he was sending to Washington "an officer of my staff thoroughly familiar with our problems and points of view." "I request," he continued, "that he be permitted to consult with your planners on current ambiguities and conflicting instructions." [33] He asked, at the same time, that he be allowed to operate under the instructions he himself had proposed to them "until you approve new instructions as a result of these consultations." Such authorization would, he claimed, remove all doubt as to his missions and as to means and methods permitted in their accomplishment.[34]

General Collins was puzzled as to what constituted the ambiguities and conflicting instructions to which Ridgway referred and directed his G–3 to examine and report on the two sets of instructions. After a comparison, the Army G–3 reported some conflicts. General Ridgway, for instance, had recommended that he be allowed to order his forces across the Manchurian and USSR borders if he felt it necessary, whereas the Joint Chiefs of Staff had told him that under no circumstances would United Nations forces cross those borders. Ridgway also felt that he, as CINCUNC, should command all military forces placed under the unified command by United Nations members, but the Joint Chiefs of Staff had specifically excluded the U.S. 40th and 45th Divisions from CINCUNC control.[35]

The Joint Chiefs agreed to receive the liaison officer from General Ridgway's headquarters but refused to cancel their current directive in the interim. They would clarify any pertinent portions, where necessary, by radio.[36]

The Joint Chiefs received a better idea

---

[31] (1) Rad, JCS 90000, JCS to CINCFE, 1 May 51. (2) Memo, JCS for Secy Defense, 27 Mar 51, Incl to JCS 1776/201 forwarded to CINCFE on 4 Apr 51.

[32] *Ibid.*

[33] Rad, C 61932, CINCFE to DA for JCS, 7 May 51.

[34] *Ibid.*

[35] Memo, ACofS G–3 (sgd Duff) for CofS USA, 7 May 51, in G–3, DA file 381 Korea, Case 6/15.

[36] Rad, JCS 90687, JCS to CINCFE, 9 May 51.

of what General Ridgway considered wrong with their directive when on 9 May he registered three major objections. Ridgway objected most strongly to the mission itself, claiming that it was impossible, with the forces he had, to destroy the armed forces of North Korea and Communist China within Korea, especially since his forces could make no general advance beyond Lines KANSAS-WYOMING. Second, he charged that the two concepts embodied in the phrases "security of forces under your command" and "your basic mission as CINCFE of the defense of Japan," were antithetical, and that these two conflicting concepts, when further complicated by the Joint Chiefs of Staff order that he would not evacuate Korea except with their permission, "constitute a serious abridgement of the authority and freedom of action I believe you intend me to have in order to discharge assigned responsibilities." [37]

A successful defense of Japan, Ridgway declared, would be almost impossible if the United States military force in Korea were destroyed by Russian attack. The only forces then remaining in Japan would likely be the two National Guard divisions, such provisional units as could be organized from the support forces in Japan, and the comparatively new and ill-prepared Japanese National Police Reserve. Ridgway felt that since his CINCFE mission of defending Japan took precedence over his CINCUNC mission of clearing Korea, he should at least be able to say when his troops in Korea would withdraw and begin the defense of Japan. If this decision were left to Ridgway, he could withhold such an announcement until the proper time; and the United States Government could announce simultaneously a decision to evacuate ROK soldiers and governmental officials along with the Eighth Army. But, Ridgway claimed, "If authority is retained in Washington, I believe the difficulty of avoiding premature disclosure, with consequent risk of disaster, will be very great if not insurmountable." [38]

The third major objection to the Joint Chiefs' instructions sprang from the permission granted to attack, under certain conditions, enemy air bases in Manchuria and China. Ridgway saw need also for authority to make prior reconnaissance of these areas.[39]

These specific objections prompted no immediate changes in Ridgway's directive. The Joint Chiefs of Staff refused to grant Ridgway permission to pull his forces out of Korea on his own initiative. Without being specific, they told him that strategic considerations demanded that they retain authority for ordering any such withdrawal. Their latest instructions, they believed, gave sufficient latitude to plan a withdrawal and to take any preliminary action required. They did not share Ridgway's fear of a premature disclosure and assured him that there would be plenty of time for them to delegate withdrawal authority to him and that they would not make any premature disclosure of intentions in Washington.[40]

The Joint Chiefs of Staff fully realized

[37] Rad, C 62088, CINCFE to DA for JCS, 9 May 51.

[38] *Ibid.*

[39] *Ibid.*

[40] Rad, JCS 90999, JCS to CINCFE, 12 May 51.

that the forces available to Ridgway were too weak to destroy the North Korean and Chinese Armies. But, they pointed out optimistically, if the Chinese withdrew, this condition might change. In any event, they considered the mission currently assigned to Ridgway to be in keeping with existing national objectives. These objectives were at the time, of course, under intensive review. "Depending on actions of the President on recommendations of the National Security Council," the JCS informed Ridgway, "your mission will be made to accord therewith." But, Washington authorities bluntly told Ridgway, for the time being, "Your mission remains unchanged." [41]

The Chinese Attack Again

The Chinese and North Korean Armies fell back to recover in early May. But it was clear that they had not been decisively defeated, and equally plain that they were replacing losses and rebuilding their offensive strength for another try at destroying the United Nations forces. Prisoner reports and other intelligence sources, in fact, had long indicated that the enemy spring offensive would come in two installments.[42]

Despite great losses in the first installment, the enemy still had a numerical advantage over Van Fleet. The Chinese ground forces totaled about 542,000 men, according to intelligence estimates on 1 May. Smaller in numbers, but still dangerous, the North Korean Army numbered more than 197,000. Across the Yalu in Manchuria, an additional Chinese force of almost 750,000 stood waiting. Van Fleet, on the other hand, commanded 269,772 U.S. Army, U.S. Marine, and allied troops and 234,993 ROK Army troops.[43]

In view of the likelihood of another enemy offensive, Van Fleet's most profitable tactics were to keep the enemy off-balance, break up enemy attack formations, and reconnoiter to discover enemy dispositions and plan of attack. Besides, General Ridgway had ordered Van Fleet on 25 April to "maintain the offensive spirit of your Army and retain the initiative." General Van Fleet therefore directed on 4 May that each of his front-line divisions establish a patrol base, manned by a complete RCT supported by corps armored units, in advance of its main position. These regimental patrol bases were set up seven or eight miles in front of the main line of resistance, and from them armor-supported patrols ranged ten to twelve miles into enemy territory to carry the fight to North Korean and Chinese screening units. Other elements of Van Fleet's forces meanwhile cleared the Kimp'o Peninsula west of Seoul and made substantial advances up the main arterials leading north and east

41 *Ibid.*

42 (1) Comd Rpt, GHQ UNC, May 51, Annex IV, Part 1, p. 3. (2) Telecons TT 4668, DA and GHQ, 3 May 51; TT 4680, 5 May 51; and TT 4682, 7 May 51. (3) Comd Rpt, Eighth Army, Narrative, May 51.

43 (1) Enemy statistics from Order of Battle, CCF and NKA 1 May 1951, DIS, FEC, UNC 3159, 4 May 51. (2) Data on U.N. forces from Comd Rpt, Eighth U.S. Army, Narrative, May 1951, p. 29 and Plate No. 12, sec. II, Book 2.

BUILDINGS BURNING *in village north of Ch'unch'on as a result of artillery fire of the 1st Cavalry Division.*

from Seoul, recapturing Uijongbu and Ch'unch'on.[44]

At the same time, when General Ridgway visited Korea on 3–4 May, Van Fleet announced that if the enemy did not attack soon, he intended to open a general offensive himself. Ridgway agreed.[45]

Accordingly, Van Fleet began planning an advance and tentatively set 12 May as the opening date.[46] But in the week following, intelligence sources provided clear signs that the enemy was all but ready to move out once more in a full-scale offensive. On 12 May, General Ridgway reported to Washington, "It appears the enemy is again in the advanced stages of preparation for the resumption of the offensive which he can

[44] (1) Comd Rpt, Eighth Army, Narrative, May 1951. (2) Comd Rpt, GHQ UNC, Introduction, May 51.

[45] Rad, C 61848, CINCFE to DA, 5 May 51.

[46] Rad, CX-1483, CG Eighth Army to All Comdrs, 9 May 51.

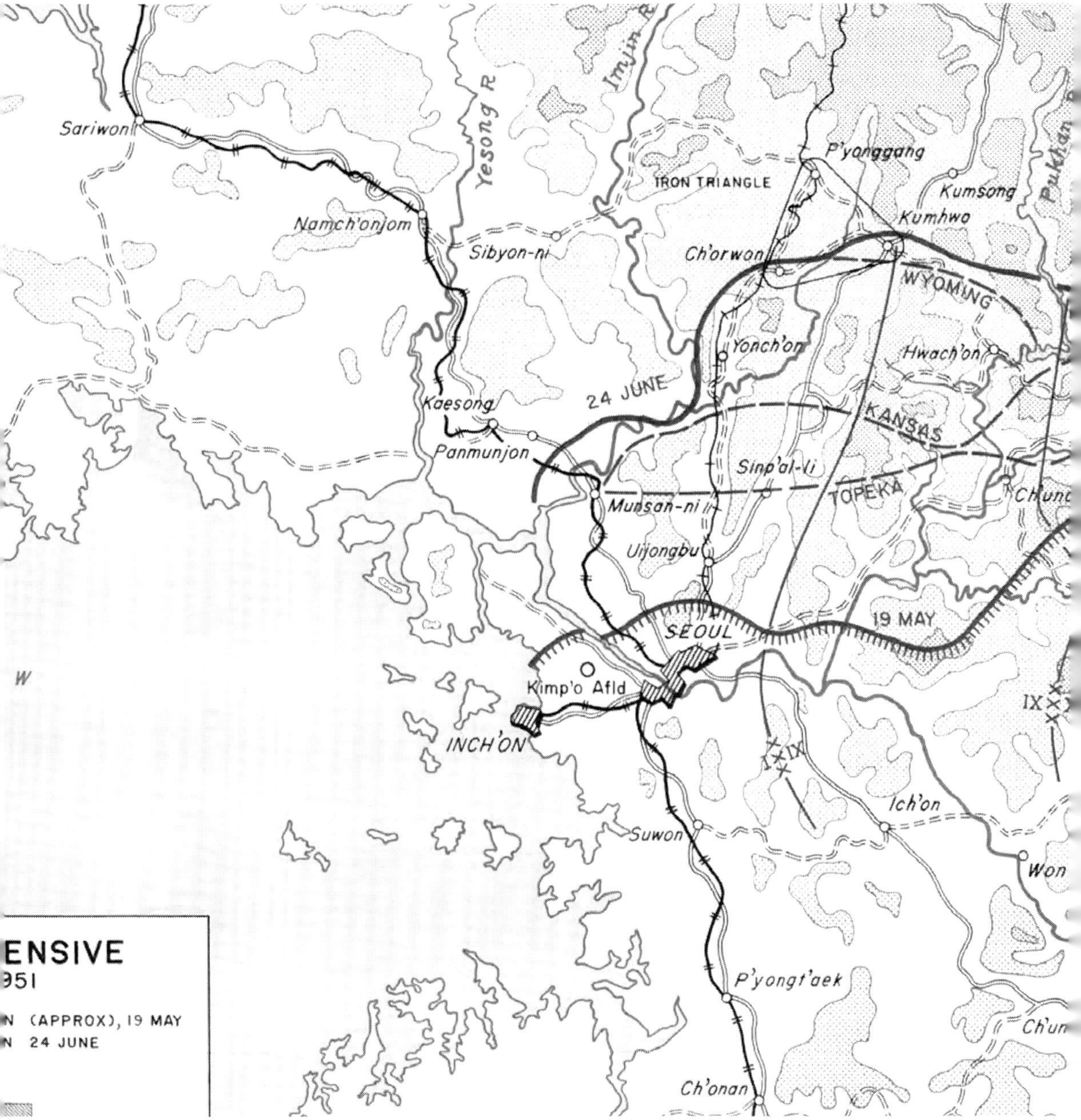

Sariwon
Yesong R
Imjin R
Namch'onjom
Sibyon-ni
IRON TRIANGLE
P'yonggang
Kumsong
Kumhwa
Pukhan R
Ch'orwon
WYOMING
Yonch'on
Hwach'on
24 JUNE
Kaesong
KANSAS
Panmunjon
Sinp'al-li
TOPEKA
Munsan-ni
Ch'unc
Uijongbu
SEOUL
19 MAY
Kimp'o Afld
INCH'ON
IX
Ich'on
Suwon
Won
P'yongt'aek
Ch'un
Ch'onan
W
ENSIVE
951
N (APPROX), 19 MAY
N 24 JUNE

launch at any time, probably within the next 72–96 hours." [47] By that date, Van Fleet already had postponed his own advance indefinitely.[48]

From the disposition of enemy forces Ridgway predicted that five Chinese armies would make the main effort down the west central sector toward the lower Han River corridor, along with a secondary drive on Seoul by three Chinese armies and one North Korean corps. He expected a lesser attack farther east by two Chinese armies and three North Korean corps striking down the Ch'unch'on-Hongch'on axis.[49] But the enemy mass shifted eastward during the period 10–16 May; and late in the afternoon of the 16th, five Chinese armies launched the main enemy effort down the Ch'unch'on-Hongch'on axis. In the west the enemy made only strong probing attacks.[50] (*See Map VIII.*)

To General Ridgway it appeared that the Chinese had concentrated their principal strength of seven armies on a 25-mile front from Ch'unch'on northeastward to the Hwach'on Reservoir, and that only four enemy armies remained in the 40-mile sector to the west. Ridgway estimated it would take the Chinese a full week, probably longer, to shift their mass again to the Ch'orwon-Seoul axis in the western sector, and therefore saw an opportunity ". . . for the Eighth Army to deliver strong attack on the Uijongbu axis, using at least two U.S. divisions with the objective of relieving the pressure on IX and X Corps by threatening vital enemy lateral communications through Chorwon." He held high hopes for this strategy and told General Van Fleet that if the contemplated counterattack were successful, "unlimited opportunities for major exploitation would result." [51]

After reconnoitering the front on 19 May, Ridgway ordered Van Fleet to attack immediately not only up the Uijongbu-Ch'orwon corridor, but across the entire front. The enemy had obviously overextended and Ridgway hoped to catch him off-guard.[52]

Consequently, even as the enemy was still attempting to move south along a 40-mile front in the east central sector, Van Fleet ordered his forces forward. (*Map IX*) The sudden reversal of direction caught the enemy by surprise. As a result, not only were substantial ground gains registered, but also in a single day the Eighth Army claimed to have killed 21,000 enemy and wounded 14,000.[53]

Along the entire front, U.N. troops continued their counterattack against moderate to weak resistance. By the end of May, Van Fleet's forces had just about made their way back to Line KANSAS, and perhaps more important, had killed the enemy at a rate higher than ever previously achieved by Eighth Army.

With the enemy's much-vaunted offensive transmuted into rout and confusion and the Chinese and North Korean forces reeling back into North Korea,

[47] Telecons, TT 4682, DA and GHQ, 7 May 51 and TT 4704, 12 May 51.

[48] Rad, GX-5-1176 KGOO, CG Eighth Army to All Comdrs, 11 May 51.

[49] Rad, CINCFE to DA, 12 May 51.

[50] Rad, G (TAG) 172 KCG, Ridgway (Personal) for Collins, 20 May 51.

[51] Rad, C 62789, Ridgway (Personal) for Van Fleet, 18 May 51.

[52] Rad, G (TAG) 172 KCG, Ridgway (Personal) for Collins, 20 May 51.

[53] (1) Rad, GS-5-3290, CG Eighth Army to All Comdrs, 19 May 51. (2) Rad, BCX 6355DI, CG FEAF BOMBCOM to All Comdrs, 22 May 51.

General Ridgway was justifiably confident in reporting to the Joint Chiefs of Staff on 30 May that the enemy had suffered a major defeat in Korea. The estimate by field commanders of the total number of enemy soldiers killed in late May was so high that General Ridgway could not, he told the Joint Chiefs, accept it as credible. Nevertheless, he assured his superiors in Washington that the Communist casualties far exceeded those they had suffered during the 22 April offensive; and since the majority of enemy casualties were infantrymen, the loss of combat effectiveness by major enemy tactical units was much greater than a mere reduction in numbers would indicate. Also, the relatively primitive nature of enemy medical and evacuation facilities would reduce the number of wounded returned to duty, and would thus compound enemy losses. Moreover, nearly 10,000 prisoners, the vast majority of whom were Chinese, had been taken by the Eighth Army.[54]

Huge quantities of enemy matériel, Ridgway reported, had been and were still being captured. Artillery, mortars, and automatic weapons were seized in amounts exceeding anything previously taken in the Korean fighting. Accompanying these enemy losses, understandably, was a noticeable drop in the fighting spirit of Chinese Communist forces. A shortage of food was also lowering enemy morale. Captured Chinese reported that their units had had to eat grass and roots because of the exhaustion of ration supplies. In sum, Ridgway judged that "A plainly evident disorganization now exists among both the Chinese Communist forces and the North Korean Peoples Army forces." [55]

The outlook for the United Nations Command, in contrast to that for the enemy, was comparatively bright. "Eighth Army," General Ridgway told his superiors, was "at near full strength with morale excellent and logistic capabilities little affected to date by deteriorating weather. . . ." [56] He concluded his report to the JCS with a significant prognosis: "I, therefore, believe that for the next sixty days the United States Government should be able to count with reasonable assurance upon a military situation in Korea offering optimum advantages in support of its diplomatic negotiations." [57]

Efforts at Political Settlement

The recent U.N. success in blunting two major enemy drives took place against a background of continuing reexamination of the nation's goals and the laying of plans for achieving those goals. The opposition of American allies to increased involvement in Asia, the apparent reluctance of the American public to increase operations in Korea, and the uncommitted war potential of the Chinese Communists all emphasized the wisdom of negotiating a settlement. Hence, the thinking of most of the nation's officials tended strongly in that direction.

In the course of a meeting on 19 March 1951 of the Joint Chiefs of Staff,

---

[54] Rad, C 63744, CINCFE to DA (for JCS), 30 May 51.

[55] *Ibid.*

[56] *Ibid.*

[57] *Ibid.*

Secretary of State, and Secretary of Defense, a possible cease-fire leading to an armistice in Korea was thoroughly discussed. The Joint Chiefs of Staff were specifically directed to re-examine the proposed armistice terms they had submitted on 12 December 1950 and to determine whether these terms were now valid.[58] But encouraged and emboldened by the fine showing of U.N. military units in Korea under direction of General Ridgway, the Joint Chiefs of Staff, from the military point of view, did not then see many advantages in an armistice. For even in late March, the Chinese were suffering heavy losses, and the Joint Chiefs were not averse to seeing those losses continue. "Any arrangement which did not prejudice their [the Chinese Communists'] position in Korea but which would end the infliction of large losses on the Communists would be greatly to their advantage," the Joint Chiefs told the Secretary of Defense. Conversely, any armistice arrangement which would keep United Nations forces in Korea, and which did not prejudice the position of the Communist forces there, would be greatly to the disadvantage of the United States. "Such an arrangement," the Joint Chiefs held, "would in all probability, jeopardize the security of our forces, constitute an unwarranted drain on our military resources, and tie down our forces in Korea almost as effectively as if they were engaged in combat." Consequently, from a purely military point of view, the Joint Chiefs judged that "an armistice arrangement of itself would not, even temporarily, constitute an acceptable solution of the Korea situation." [59]

On the other hand, the Joint Chiefs realized that the losses being incurred by the Communists would probably tend to make them more conducive to a political settlement than heretofore. Indeed, it might be possible, in light of the military situation, to take political action to end the aggression, conclude the fighting, and insure against its resumption. But the nation's military leaders insisted that any such political resolution had to provide for a settlement under circumstances which would permit the ultimate attainment of United States objectives without forfeiture of or prejudice to the nation's general position with respect to Russia, to Formosa, and to the seating of the Chinese Communists in the United Nations.[60]

On 5 April, the Joint Chiefs of Staff furnished the Secretary of Defense, for transmittal to the President and to the National Security Council, their views on the military outlook for Korea and on the military position which the United States should maintain. If Russia intervened, either with "volunteers" or as part of a general war, United Nations forces should be withdrawn from Korea. If Russia did not start a general war prior to settlement of the Korean problem, there were two ways to look at the world situation. The Joint Chiefs felt that if the immediate objective of the Russian strategy lay in western Europe, it would be to Russia's advantage to keep the maximum number of United Nations

[58] Memo for Secy Defense, 27 Mar 51, sub: U.S. Position Regarding an Armistice in Korea.

[59] *Ibid.*

[60] *Ibid.*

forces tied up in Korea. On the other hand, if Russia's immediate objectives were in the Far East, they would be fostered if the United Nations forces left Korea.[61]

Conversely, under either of the two conditions cited, the Communists would profit by leaving their military forces in Korea. An armistice which left Communist armies in Korea would be to the great disadvantage of the United Nations, and would place a heavy drain on U.S. military reserves by forcing the retention of troops in Korea.

But the Joint Chiefs of Staff sounded the keynote of subsequent American policy toward Korea, both military and political, when they told the Secretary of Defense, "The Korean problem cannot be resolved in a manner satisfactory to the United States by military action alone." The Korean problem was a symptom of world tension which could only be relieved in a manner satisfactory to the United States when, and if, there was a general relaxation of the world tensions.[62]

They concluded with four significant recommendations, recommendations later integrated into the basic policy developed by the National Security Council as a statement of American objectives and procedures in facing the Communist threat in the Far East. These were: (1) The United States forces in Korea must pursue their current military course of action there until a political objective for that country appeared attainable without jeopardizing United States positions with respect to Russia, Formosa, and seating the Chinese Communists in the United Nations. (2) Dependable South Korean units should be generated as rapidly as possible and in sufficient strength to take over the major part of the burden from United Nations forces. (3) Preparations should be made immediately for action by naval and air forces against the mainland of China. (4) Action should be taken as a matter of urgency to ascertain the policies and objectives of the allies toward Korea specifically and the Far East in general, and also to discover the degree and nature of the support which the United States could expect from them if, while continuing the present military course of action in Korea, operations against the mainland of China were initiated.[63]

It grew more apparent each day, particularly in view of the much-publicized hearings on the relief of General MacArthur which started on 3 May in Washington, that the United States badly needed a clear, workable statement of its military and political objectives with regard to Asia and, particularly, to Korea. Not only had the American people become confused over the issues; they did not fully understand or appreciate the reasons why the United States had pressed for UN intervention. American military and political strategists had themselves failed to agree on the proper objectives, much less the proper ways and means of attaining them. General MacArthur's return, accompanied as it was by vituperative and bitter attacks on national policy, only served to cloud the issues further.

Finally, in mid-May, after long and

[61] Memo, JCS (Bradley) for Secy Defense, 5 Apr 51, sub: Military Action in Korea.

[62] *Ibid.*

[63] *Ibid.*

careful consideration of the views of all main advisory bodies, including the Joint Chiefs of Staff, the National Security Council submitted to the President a statement of policy which the council believed the U.S. Government should now follow in facing the Communists in Korea and throughout Asia. This statement of policy was approved by President Truman on 17 May. In accepting the advice of the National Security Council, the President decided that the United States would retain as an ultimate objective a political, not military, solution which would provide for a united, independent, and democratic Korea. At the same time, he directed all governmental agencies to do whatever required to put the policy in effect at once.[64]

The significance of this blueprint for American action in Asia can hardly be overstated insofar as its effect on the Korean problem is concerned. For it was this blueprint, with some modification, that the United States followed from that day forward in bringing a conclusion to the fighting in Korea. There was nothing startlingly new in this policy. Seemingly by instinct, the United States had been following most of the precepts right along. But by setting forth in a single statement the best possible answers to all the questions which had been repeatedly asked for many months, the National Security Council took a firm step toward the immediate goal of stabilizing the situation in Korea. The statement implied no hope of military victory in Korea; but it did bespeak a certain confidence that Communist designs could be thwarted even though United States aims could not be fully accomplished.[65]

The significant portion of the new policy was, of course, the American Government's intention to seek through United Nations machinery a settlement acceptable to the United States which would as a minimum terminate hostilities under appropriate armistice arrangements. But until such an armistice could be brought about, the United States would "continue to oppose and penalize the aggressor." In the absence of a negotiated settlement, ". . . recognizing that there is no other acceptable alternative . . . ," the United States would keep up the current military course of action ". . . without commitment to unify Korea by military force . . . ," but with the purpose of inflicting heavy losses on the enemy, preventing the seizure of South Korea, and limiting Communist capabilities for aggression elsewhere in Asia. While the United States was determined to avoid extending hostilities beyond Korea, it intended to deflate the military strength and prestige of the Chinese by inflicting heavy losses upon them in Korea at every opportunity. Furthermore, the Joint Chiefs were to prepare detailed plans for punitive action against China itself should China take aggressive action outside Korea or if United Nations forces were compelled by military action to evacuate Korea. These punitive actions would include those previously recommended by General MacArthur—blockade, military operations against China, and exploitation of Chinese Nationalist forces.[66]

---

[64] JCS 1992/82.

[65] *Ibid.*

[66] *Ibid.*

Rebuilding the ROK Army

A principal feature of the policy established on 17 May was aimed at generating strong, dependable ROK military forces to take over from U.N. forces. The ROK Government itself was fully in favor of this objective, at least of the intention to strengthen the ROK forces. General MacArthur, who in early April had turned thumbs down on proposals to arm additional South Korean military units, had been gone from the Far East Command less than a week when the Korea representative to the United Nations handed General Bradley a request that the United States arm and equip ten additional ROK divisions, and that the new divisions be commanded, if possible, by American officers.[67] On 24 April, President Rhee sent President Truman a duplicate request.[68]

Past performances of ROK battle units, most recently during the enemy's April offensive, lent little support to these requests. Yet, on 26 April, General Collins sent General Ridgway a message which seemed to indicate that a substantial increase in the number of ROK divisions was being contemplated in Washington. "It is highly desirable," Collins told Ridgway, "to develop possible ways of utilizing available Korean manpower in order to augment United Nations forces and eventually to replace some United States units." General Collins, mainly seeking information, asked Ridgway to tell him the answer to five general questions. These were: How many Korean males of military age could be put into the Army without damaging agricultural and industrial programs of the ROK? Would it be possible to expand leadership training facilities for ROK troops to meet an increased mobilization and, if so, would it require more American personnel to train them? Should Americans command Korean units and, if so, at what level of command should Americans be so employed? Was it advisable to send senior Korean officers to higher American military schools? Should the TO&E of Korean divisions be changed? General Collins concluded by asking General Ridgway if he thought the ROK Army should have ten more divisions as President Rhee had requested.[69]

General Ridgway's response to these questions was emphatically negative. Like Ridgway, General Van Fleet also frowned on the idea of activating more divisions. "If excess trained officers and non-commissioned officers are available," he stated, "they are needed in units presently constituted." [70]

Both Ridgway and Van Fleet took decided stands against placing any American officers in command of ROK Army units, because of the language barrier, for one reason. But also, command, to

---

[67] When Bradley reported this to General Marshall, the Secretary of Defense evinced great interest. He suggested that this request might be taken into consideration in the contemplated rotation plan then being set up for Korea. He thought some officers, rather than rotating, would be happy to transfer to South Korean divisions if promotion were involved. Memo, Gen Bradley for Gen Collins and Adm Davis, 19 Apr 51, in G–3, DA file 091 Korea, Case 174.

[68] Rad, 230830Z, USAMB Korea to CINCFE, Apr 51.

[69] Rad, DA 89517, Collins (Personal) for Ridgway, 26 Apr 51.

[70] (1) Rad, C 61433, Ridgway (Personal) for Collins, 1 May 51. (2) Rad, C 61589, CINCFE to DA, Ridgway (Personal) for Collins, 2 May 51.

be effective, must include complete authority to administer and to discipline; and since the Republic of Korea was a sovereign nation, American officers would have no inherent authority to discipline ROK soldiers. Van Fleet pointed out further that if American command were to be effective it would have to extend down to battalions and companies and therefore would require large numbers of trained officers and noncoms. He insisted ". . . that the basic problems with the ROK Army at this time are training and development of leadership qualities. This is a long range project, especially the development of an officer corps as would be true in any new army." Ridgway not only agreed, but also believed the "Creation of an officer corps is . . . the first and prime consideration." [71]

The nation's top authorities nonetheless had decided that the United States would "develop as rapidly as possible dependable ROK military units in sufficient strength eventually to assume the major part of the burden of United Nations forces in Korea." [72] Hardly had this new policy been approved, when President Rhee announced that if the United States would only equip his already well-trained soldiers, American troops could be withdrawn from Korea and the job left to the ROK Army. Disturbed by this announcement, General Ridgway sought, through Ambassador Muccio, to induce Rhee to make no more damaging statements. The Joint Chiefs of Staff were equally concerned and called on the Secretary of Defense to seek further action by the Department of State. They told General Marshall, "Any action of the United States will never be entirely successful without the full cooperation of the Government of the Republic of Korea. . . ." The Department of State accordingly directed Ambassador Muccio to convey to President Rhee, "in the strongest terms," the grave concern which the United States felt over the continuance of such statements.[73]

Revision of Ridgway's Directives

Two officers of General Ridgway's personal staff meanwhile had been in Washington since 11 May to explain and support Ridgway's own proposed directive of 30 April rather than the one that Ridgway had been given on 1 May. In their appeal to the Army Chief of Staff, Ridgway's representatives suggested that if the Joint Chiefs of Staff turned down Ridgway's directive as it stood, that the existing directive then be revised sufficiently to make it acceptable to Ridgway.[74] General Collins elected to begin with the latter approach, and on 23 May presented the revamped directive to the Joint Chiefs of Staff. This revised version was approved by the Joint Chiefs of Staff and was dispatched to General Ridgway on 1 June.[75]

---

[71] Rad, C 63287, CINCFE to CSUSA, Ridgway (Personal) for Haislip, 25 May 51.

[72] JCS 1992/82.

[73] (1) Memo, Gen Bradley for Secy Defense, 23 May 51, sub: President Rhee (ROK) Statements, in G-3, DA file 091 Korea, Case 178/4. (2) Ltr, Rusk to Lovett, 1 Jun 51, same file, Case 178/3.

[74] (1) MFR, 26 May 51, sub: Staff Visit, GHQ Representative to DA, sgd Hefelbower and Smith, in JSPOG GHQ, UNC files. (2) Memo, Hefelbower and Smith for CofS USA, 16 May 51, sub: Proposed Modification of CINCUNC and CINCFE Missions, in G-3, DA file 381 Korea, Case 178/3. (3) Rad, DA 91236, Hefelbower (Personal) for Hickey, 16 May 51.

[75] Rad, JCS 92831, JCS to CINCFE, 1 Jun 51.

The major differences between the revised directive and the one given Ridgway on 1 May lay in a definite division of his responsibilities as CINCFE and as CINCUNC and in a drastically changed statement of his mission. No changes were made in the restrictions upon his operations in Korea. But as a result of the National Security Council policy decision, approved by the President on 17 May, certain new instructions were given him with regard to development of ROK forces, and on planning for retaliatory action against Communist China.[76]

With regard to his duties as CINCUNC (his duties as CINCFE were stated separately), by far the most important feature of this new directive was the altered mission with which Ridgway was charged. Influenced both by Ridgway's protestations that he could not clear all of Korea, and by the recent decision to settle the Korean situation by political means, the Joint Chiefs of Staff instructed Ridgway to:

Inflict maximum personnel and materiel losses on the forces of North Korea and Communist China operating within the geographic boundaries of Korea and waters adjacent thereto, in order to create conditions favorable to a settlement of the Korean conflict which would as a minimum: (A) Terminate hostilities under appropriate armistice arrangements; (B) Establish authority of the ROK over all area south of a northern boundary so located as to facilitate, to the maximum extent possible, both administration and military defense, and in no case south of the 38th Parallel; (C) Provide for the withdrawal by appropriate stages of Non-Korean Armed Forces from Korea; (D) Permit the building of a sufficient ROK military power to deter or repel a renewed North Korean aggression.[77]

The restriction against a general advance beyond Line KANSAS-WYOMING (defined by the JCS as a line passing approximately through the Hwach'on Reservoir area) was retained in the new directive. Nor was General Ridgway granted permission to withdraw from Korea at his own discretion.

Ridgway now knew clearly what was expected of him, and what limits were set upon his authority as commander in chief, United Nations Command. These goals and restrictions would obtain throughout the remainder of his term of duty.

[76] Memo, ACofS G-3 for CofS USA, 22 May 51, sub: Proposed Changes in Directives to CINCFE, App. A.

[77] *Ibid.*

CHAPTER XXII

Signs of Armistice

General Van Fleet proposed late in May to carry the fight well behind enemy lines. He asked General Ridgway to let him mount an amphibious landing on Korea's east coast to surround and pinch off a large segment of the Chinese and North Korean Armies. Basically, Van Fleet had in mind a maneuver resembling Operation CHROMITE, a deep amphibious encirclement co-ordinated with an overland drive. His target area lay well up the east coast, nearly to Wonsan.[1]

General Ridgway opposed the landing. First, the objective area lay beyond the limiting line set by the Joint Chiefs of Staff. This line, of course, could be altered for sufficiently valid reasons. But Ridgway also objected that the advantages to be gained, even if the operation were successful, did not justify the great risks involved. For Ridgway's main mission in Korea was to destroy the greatest possible number of enemy forces with the least possible loss of his own men; and he had decided that he could best do this by a gradual advance to the Line KANSAS-WYOMING, not by an amphibious landing deep behind enemy lines. Furthermore, since it was impossible to clear all of Korea of enemy forces under conditions then obtaining, it would be unwise to risk heavy casualties merely for a chance to inflict equal casualties on a more numerous enemy. Van Fleet tried to counter these arguments, but Ridgway stood fast and the plan was shelved.[2]

Ridgway did authorize a limited advance on the east coast beyond the line set by the Joint Chiefs of Staff. He told Van Fleet that he could seize a line running from the east end of the Hwach'on Reservoir to the east coast, a move which would advance Line KANSAS well north of its position as defined in April. Ridgway did not consider this move to be a general advance because its purpose was to maintain contact and to keep the enemy off-balance. Nor did the Joint Chiefs of Staff object when Ridgway notified them of his decision.[3]

---

[1] Rad, GX-5-5099 KGOP, CG Eighth Army to CINCFE, Van Fleet (Personal) for Ridgway, 28 May 51.

[2] (1) MFR, 31 May 51, sub: Conference Between Gen Ridgway and Gen Van Fleet, copy in GHQ UNC, SGS files. (2) Hearings on Ammunition Shortages in the Armed Services, Senate Committee on Armed Services, 83d Congress, 1st Session, April 1953 (hereafter cited as *Van Fleet Hearings*), Testimony of Gen Collins, pp. 1-5.

[3] Rad, C 63730, CINCFE to JCS, 30 May 51.

Adhering to General Ridgway's concepts, Van Fleet on 1 June ordered the fortification of Line KANSAS. For the time being, the attacks toward Line WYOMING would continue, and once that line was occupied patrol bases would be established beyond it. If the enemy launched another major offensive, the forces on Line WYOMING might withdraw to Line KANSAS to defend there. Otherwise, "From positions along the line WYOMING and the patrol base line," Van Fleet ordered, "limited objective attacks, reconnaissance in force, and patrolling to the maximum capability will be conducted . . . to inflict damage on the enemy, confuse him and keep him off balance." [4]

On 9 June, General Ridgway received Van Fleet's estimate of probable developments within Korea during the next sixty days. This estimate closely paralleled his own. The enemy, despite the beating he had taken, still had numerical superiority and retained the capability to launch at least one major offensive within the next two months. Van Fleet himself fully expected the Chinese to strike again as soon as they had built up enough strength, and planned to counter this enemy threat, at least locally, by vigorous limited offensives which would, when combined with deception, keep the enemy off-balance or cause him to attack prematurely. Van Fleet had made plans for three such limited offensives, all calling for the swift seizure of objective areas, the destruction of enemy supplies in these areas, and, after short occupation, a return to Line KANSAS-WYOMING.[5]

The Eighth Army reached Line KANSAS-WYOMING by mid-June; and on the 14th, General Ridgway, basing his predictions on General Van Fleet's report, sketched for the Joint Chiefs of Staff a picture of what could be expected in Korea during the coming two months. The enemy's logistic situation was worse than that of his own forces. Enemy lines of communications were too long. Recent heavy rainfall and effective interdiction by U.N. air forces were further aggravating Chinese supply problems. The Eighth Army, on the other hand, currently enjoyed adequate logistic support. This support would remain adequate, Ridgway pointed out, provided Van Fleet made no general advance north of Line KANSAS-WYOMING during the period. To advance, Ridgway claimed, would "tend to nullify EUSAK's present logistic advantage over the enemy." [6]

Regardless of a poor supply situation, the sheer weight of superior numbers in North Korea and Manchuria made enemy forces capable of keeping the over-all initiative and of launching at least one major offensive in the next sixty days. Happily, the terrain along Line KANSAS-WYOMING offered excellent defensive positions if properly organized. Ridgway intended to hold Eighth Army along this general line, at least for the next two months, and to keep punishing the enemy by mak-

---

[4] Ltr of Instructions, CG EUSAK to All Corps Comdrs, 1 Jun 51, copy in JSPOG Staff Study, Advances North of the 38th Parallel.

[5] MFR CofS GHQ from SGS, GHQ, 13 Jun 51, in GHQ, FEC SGS files.

[6] Rad, CX 64976, CINCFE to DA for JCS, 14 Jun 51.

ing limited offensive operations with KANSAS-WYOMING as a base.[7]

Political Factors Influence the Battle Line

The new national policy, now based on a political settlement as opposed to a complete military victory, of course deterred any grander plans for a general offensive. Even if the enemy might be defeated, the cost in lives would be considerable; and certainly nothing was to be gained by paying a high price for terrain which might very possibly be returned to the enemy at the conference table. Indeed, behind the national policy lay factors and conditions beyond the power of the theater commander, or for that matter, of national leaders to control. The decision to seek a solution by political means in Korea was an outgrowth of world-wide considerations.[8]

Since General Ridgway had been directed to create conditions favorable to a settlement of the Korean conflict under appropriate armistice arrangements, he gave considerable thought to the best location of a cease-fire line. The Joint Chiefs of Staff, as of 27 March, had judged that the demilitarized zone should be an area about twenty miles in width centered at, or north of, the 38th Parallel, although they realized that the exact location would be determined on the basis of the positions of opposing ground units in combat at the time of a cease-fire.

General Ridgway took special note of this last fact and, in early June, asked for Van Fleet's views on the best location for his forces during a cease-fire. Van Fleet replied that Line KANSAS would be the most feasible location. "It is assumed," Van Fleet told Ridgway, "that the Communist forces will violate the terms of the treaty as they have in the

[7] Some officials later charged that the U.N. forces did not take sufficient advantage of the enemy's weakened condition in early June 1951, asserting that they could have destroyed the "remnants" of the Chinese and North Korean Armies in Korea.

General Van Fleet later commented on this matter in press statements in 1952 and during a hearing before the U.S. Senate Committee on Armed Services in March 1953. When asked whether he had been correctly quoted that the United Nations could have won the war in Korea, General Van Fleet admitted that while he did not believe complete victory was possible in June 1951 he had felt that at that time he had the Communist armies on the run. ". . . They were hurting badly, out of supplies, completely out of hand or control; they were in a panic, and were doing their best to fall as far back as possible, and we, stopped by order, did not finish the enemy." When asked if he had recommended the counteroffensive be resumed, Van Fleet replied, "Oh yes, I was crying for them to turn me loose." *Van Fleet Hearings*, p. 32.

[8] "Our whole policy in Korea, in fact, both military and political," Ridgway later maintained, "will be a question for historians to debate. My own conviction is that the magnificent Eighth Army could have driven the Chinese beyond the Yalu—if this country had been willing to pay the price in lives such action would have cost. Personally, I strongly doubt that such a victory would have been worth the cost—particularly in light of the fact that our Government seemed to have no firm policy on what steps to take thereafter. Seizure of the line of the Yalu and the Tumen would have been merely the seizure of more real estate. It would have greatly shortened the enemy supply lines and greatly lengthened our own. It would have widened our front from 110 miles to 420, and beyond that front would lie Manchuria and the whole mainland of Asia, in which all the wealth and manpower of this country could have been lost and dissipated. So it is useless to speculate on what might have been. I was not privy to the councils of our leaders at home when they decided to accept the Russian-sponsored overtures for a truce. But in retrospect, I do not feel constrained to quarrel with that decision." See General Matthew B. Ridgway, "My Battles in War and Peace, the Korean War," *Saturday Evening Post* (February 25, 1956), p. 130.

past by improving their potentialities for unexpected renewal of aggression." This being so, Van Fleet insisted that his forces must occupy ground suitable for strong defense even during a cease-fire; and Line KANSAS met that requirement. Furthermore, in anticipation of some type of diplomatic agreement which would require a 10-mile withdrawal from the line of contact, Van Fleet considered it essential that his forces be at least ten miles in advance of the terrain they would eventually occupy during a cease-fire.[9]

Ridgway's own Joint Strategic Plans and Operations Group had been working on the same general problem, examining various schemes of maneuver that would carry the Eighth Army above Line KANSAS so that this line would not be lost in any withdrawal required by cease-fire arrangements. On 13 June, JSPOG officers briefed Ridgway on four such schemes; and after hearing them, Ridgway concluded that Van Fleet should devise long-range plans for a general advance to the line P'yongyang-Wonsan.[10] On the 19th, he directed Van Fleet to plan the seizure of the P'yongyang-Wonsan line with a main effort over the Seoul-Wonsan axis and a secondary drive up the Seoul-P'yongyang axis. Since Van Fleet earlier had stated that he could make no general advance for at least the next sixty days, Ridgway left the target date up to him.[11]

He cautioned Van Fleet constantly to remember that the enemy might at any time choose to negotiate a political settlement, and if this happened, a 20-mile-wide demilitarized zone might be established on the basis of the locations of opposing ground units in combat at the time. "Therefore," Ridgway pointed out, "successive main lines of resistance should be selected with a suitable outpost line, and when and if negotiations appear imminent, every effort should be made to make contact with the enemy ten miles in advance of the outpost line of resistance." This line of contact would be known as the "cease-fire" line. If negotiations were successful, a demilitarized zone would probably be set up twenty miles in depth, having as its center line the cease-fire line. Within the terms of the agreement, both sides would likely withdraw at least ten miles from the cease-fire line. This would place Van Fleet's forces on the outpost line in advance of his selected main line of resistance. Ridgway of course had no information that the enemy intended to negotiate. But he directed Van Fleet to submit his operation plan by 10 July.[12] For if negotiations began in the near future, General Van Fleet's concept of using Line KANSAS as his main line of resistance to be occupied during a cease-fire would apply. Consequently, Ridgway wanted Van Fleet to be at least twenty miles in advance of KANSAS at the beginning of any negotiations. This, of course, would permit the 10-mile withdrawal and the manning of the outpost line of resistance. He assured Van Fleet that he would try to warn him of any imminent negotia-

---

[9] Ltr, Van Fleet to CINCUNC, 9 Jun 51, sub: Location of EUSAK During a Cease-Fire (Military Viewpoint).

[10] MFR, 17 Jun 51, sub: Planning Directive, sgd Lutes.

[11] Ltr, CINCFE to CG Eighth Army, 19 Jun 51, sub: Planning Directive.

[12] *Ibid.*

tions so that Van Fleet could move at least part of his troops up to a general line of contact twenty miles in advance of KANSAS.[13]

The wisdom of preparing a cease-fire line had apparently occurred to the Joint Chiefs of Staff about the same time. By mid-June, in view of their increasing conviction that political negotiations might soon develop, they had begun to doubt the wisdom of limiting Van Fleet's advance. At a meeting on 15 June, they decided that it was desirable to revise the current directives to General Ridgway; and General Collins received the task of preparing a proposed revision that would remove any restrictions on ground operations except those inherent in Ridgway's mission as CINCFE for the defense of Japan.[14]

General Collins and General Vandenberg wondered if it might not be wise to let Ridgway operate in strength as far to the north as his resources would permit. They saw the current enemy disorganization and his comparative weakness on the immediate front as an excellent opportunity to seize more terrain and to better Eighth Army's position in the event of a cease-fire. On 20 June, they asked Ridgway what he thought of a change that would remove "any undue restrictions upon your ability to exploit tactically the current situation," and that would authorize him to "conduct such tactical operations as may be necessary or desirable to support your mission . . . to insure the safety of your command; and to continue to harass the enemy." [15] In actuality, it did not really matter whether or not the Joint Chiefs of Staff removed their restriction on his advance, since the realistic restrictions imposed by terrain, logistics, troop strength, and the enemy would, in the final analysis, limit his advance anyway. But the Joint Chiefs of Staff asked for his ideas on how any future advance into North Korea would affect the target area of his air force, whether an advance would trigger enemy air attack, and what would be the effect of longer lines of communication.

General Ridgway concurred in the proposed removal of any restriction on his advance. But he asked to be allowed to defer answering the questions about operations deeper into North Korea since he had directed the Eighth Army commander to submit plans for a general advance not later than 10 July and wished to have Van Fleet's ideas before answering the Joint Chiefs of Staff.[16]

But Ridgway continued to evince great interest in the selection of a cease-fire line, explaining that it should be at least twenty miles out in front of Line KANSAS, preferably extending from the confluence of the Han and Yesong Rivers in the west, generally northeast past Ch'orwon and Kumhwa to Kosong on the east coast. Ridgway pointed out that this cease-fire line did not include the Ongjin and

[13] Ltr, Ridgway to Van Fleet, 22 Jun 51, sub: Location of EUSAK During a Cease-Fire.

[14] Memo, Gen Taylor for Gen Collins, ACofS G-3, DA, for CSUSA, 16 Jun 51, sub: Revision of Directive to CINCFE for Opns in Korea, in G-3, DA file 381 Korea. General Taylor noted that approval of such a revision would better enable Ridgway to exploit tactically the current or subsequent situation. On the other hand, the requirement that CINCFE maintain the security of his forces would serve to limit his advance.

[15] (1) Memo, Col Arns, Dep Secy JCS, for Gen Taylor, 19 Jun 51, sub: Possible Change in JCS 92831. (2) Rad, JCS 94501, JCS to CINCFE, 20 Jun 51.

[16] Rad, C 65529, CINCFE to JCS, 22 Jun 51.

Yonan peninsulas along the west coast, both originally a part of South Korea. But the value of including these two peninsulas was out of proportion to the difficulty of defending them. He asked that the Joint Chiefs of Staff modify their position regarding a demilitarized area as described in their memorandum of 27 March to the Secretary of Defense in order to conform to the description of the cease-fire line he had proposed. The Joint Chiefs of Staff did not consider this necessary. They pointed out that their instructions which called for him to create conditions favorable to an armistice did not imply that he was to gain military control of all areas south of the 38th Parallel and that such was not intended if the tactical situation did not warrant it. They did not want to make an issue out of excluding the Ongjin and Yonan peninsulas from the provisions of their directive since it "would have undesirable political implications, particularly, if such came to attention of the ROK government." [17]

After receiving Ridgway's concurrence in the decision to revise his operating directives, the Joint Chiefs of Staff took the matter up with the Secretary of Defense and the Secretary of State, and finally with the President. President Truman approved the removal of a definite limiting line on Korean operations; and on 10 July the Joint Chiefs of Staff informed Ridgway that he was ". . . authorized to conduct such tactical operations as may be necessary to or desirable to support your mission." [18]

[17] (1) Rad, C 65529, CINCFE to JCS, 22 Jun 51. (2) Rad, JCS 95125, JCS to CINCFE, 23 Jun 51.

[18] (1) JCS 1776/234, 27 Jun 51. (2) Note to Holders, 12 Jul, App. A, 27 Jun 51, in G-3, DA file 381 Korea, Case 8.

Moves Toward Negotiation

General Ridgway's concern with cease-fire lines was well timed. On 23 June appeared the first solid indication that the Communists were prepared to negotiate when during a U.N.-sponsored radio broadcast entitled "The Price of Peace," Jacob Malik, USSR delegate to the United Nations Security Council, hinted broadly that his government was in favor of such negotiations at an early date. Having assumed all along that the USSR Government was the chief instigator of the Communist aggression in Korea and that ultimate control of Communist forces in Korea rested with Russia, U.S. authorities took Malik's remarks seriously.[19]

General Ridgway, informed that cease-fire proceedings might soon develop as a result of Malik's speech, immediately took steps to forestall any letdown among his forces. "Two things should be recalled," he cautioned his commanders.

> One is the well-earned reputation for duplicity and dishonesty possessed by the USSR, the other is the slowness with which deliberative bodies such as the Security Council produce positive action. I desire that you personally assure yourself that all elements of your command are made aware of the danger of such a relaxation of effort and that you insist on an intensification rather than a diminution of the United Nation's action in this theater.[20]

Only two days earlier, Ridgway had assured Van Fleet that he would give

[19] For a detailed account of the preliminary steps leading to the opening of the armistice conference and of the negotiations themselves, see Walter G. Hermes, *Truce Tent and Fighting Front*, UNITED STATES ARMY IN THE KOREAN WAR (Washington, 1966).

[20] Rad, CX 65667, CINCFE to All Comdrs, 24 Jun 51.

timely warning of any imminent negotiations so that Van Fleet might move forces forward twenty miles above Line KANSAS. But on 26 June, after Ridgway and Van Fleet toured the battlefront and weighed the situation anew, the two generals decided against any advance beyond Line KANSAS-WYOMING. They agreed that such an advance was feasible tactically and logistically, but that the probable cost in casualties was too great a price to pay.[21]

Enemy forces meanwhile had not slackened their build-up nor tempered their reactions to the Eighth Army's probing, especially on the central front. General Ridgway's intelligence staff concluded that the enemy was, regardless of armistice moves, regrouping in preparation for further offensives. Air sightings in the last week of June indicated that enemy offensive preparations were well advanced; numerous forward supply dumps, artillery positions, and troop movements were reported in the central area; and prisoners reported the enemy's intention to launch a Sixth Phase Offensive sometime in July.[22]

The enemy build-up, however, became of secondary importance on 29 June when General Ridgway received instructions from the Joint Chiefs of Staff to approach the enemy on possible armistice negotiations. President Truman had directed that at 0800 on 30 June, Ridgway was to broadcast a message to the commander in chief, Communist Forces in Korea, saying:

> I am informed that you may wish a meeting to discuss armistice providing for the cessation of hostilities and all acts of armed forces in Korea with adequate guarantee for the maintenance of such armistice. Upon receipt of word from you that such a meeting is desired I shall be prepared to name my representative. I would also at that time suggest a date at which he could meet with your representative. I propose that such a meeting could take place aboard a Danish Hospital ship in Wonsan harbor.

Ridgway broadcast this message as directed.[23]

On the date of Ridgway's broadcast, the Joint Chiefs of Staff gave Ridgway instructions for conducting cease-fire talks with the Communists should such talks develop. They told him that the principal military interests of the United Nations in an armistice lay in the cessation of hostilities in Korea, in assuring that the fighting would not resume, and in guaranteeing the security of United Nations Command forces. Further, any cease-fire talks were to be limited strictly to military questions related to Korea.[24]

Upon receiving these guidelines, Ridgway and his Joint Strategic Plans and Operations Group developed an agenda to be proposed to the Communists at the first session, selected a delegation to represent the United Nations Command at the conference table, and worked out the physical arrangements for maintaining the United Nations Command delegation, including communications, transportation, security liaison, and other routine matters. On 3 July, the Joint

[21] (1) Ltr, Ridgway to Van Fleet, 22 Jun 51, sub: Location of EUSAK During a Cease-Fire. (2) *Van Fleet Hearings*, p. 651.

[22] Telecons, TT 4846 and TT 4884, DA and GHQ, 20 Jun 51 and 28 Jun 51.

[23] Rad, JCS 95258, JCS to CINCFE, 29 Jun 51.

[24] Rad, JCS 95354, JCS to CINCFE, (Personal) for Ridgway, 30 Jun 51.

Chiefs of Staff approved Ridgway's plans without change.[25]

Meanwhile, on 1 July, the Communist leaders replied to Ridgway's message in a broadcast sponsored jointly by Kim Il Sung, Supreme Commander of the Korean People's Army, and Peng Teh-huai, who styled himself Commander of the Chinese Volunteers. They agreed to meet with United Nations Command representatives but proposed that the place of meeting be in the Kaesong area rather than aboard the Danish ship, and that the meetings begin between 10 and 15 July.[26]

The enemy proposal to meet at Kaesong, while not entirely unacceptable to Ridgway, was interpreted as only a further demonstration of a known Communist policy never to accept a proposal *in toto*. Ridgway therefore told the Joint Chiefs of Staff that he planned to accept Kaesong as the conference site and to halt combat operations along the Munsan-Kaesong road and in the Kaesong area. On 3 July, he notified the Communists that he was prepared to meet their representatives at Kaesong on 10 July "or at an earlier date if your representatives complete their preparations before that date." He proposed, in order to insure efficient arrangement of the many details for the meetings, that three liaison officers of each side meet in Kaesong on 5 July or as soon thereafter as practicable. The Communists agreed to this procedure, but set the date for the meeting of liaison officers at 8 July.[27]

The mere promise that negotiations to end the fighting in Korea might be forthcoming had in the meantime prompted widespread speculation in the American press. Such expressions as "Let's Get the Boys Back Home" and "The War Weary Troops" were beginning to appear in the more irresponsible journals. General Ridgway took violent exception to these sentiments. "I can hardly imagine a greater tragedy for America and the free world," he informed the Joint Chiefs of Staff on 4 July, "than a repetition of the disgraceful debacle of our Armed Forces following their victorious effort in World War II. We can never efface that blot on the record of the American people on whom the responsibility squarely rests." Ridgway vowed that he would do everything within his power to eliminate such thinking among the officers and men of his command. "If this be 'thought control' then I am for it, heart and soul." [28]

Ridgway also feared that public pressure for an armistice might force him into military concessions. He told the Joint Chiefs of Staff, "I wish with great earnestness to point out the importance I attach to the retention by United Nations forces of so much of Korea as will permit occupation and defense of Kansas line with a suitable outpost zone for its protection." He reiterated the view that KANSAS was the strongest defensive line in the general area.

> It is the most advanced strong defensive terrain which the tactical situation under your directives permits me to reach, and there, logistically to support my forces. . . . Any position taken by our government which would compel me to abandon the

[25] (1) Rad, CX 66160, CINCFE to JCS, 1 Jul 51. (2) Rad, JCS 95438, JCS to CINCFE, 3 Jul 51.

[26] Rad, CX 66188, CINCFE to JCS, 2 Jul 51.

[27] (1) *Ibid.* (2) Rad, DA-IN 11098, Ridgway to JCS, 5 Jul 51.

[28] Rad, C 66323, CINCFE to JCS, 4 Jul 51.

Kansas line or deny me a reasonable outpost zone for its protection would vitally prejudice our entire military position in Korea. Request that copy of this message be furnished the Secretary of Defense personally.[29]

Next, before sending his officers to negotiate with their enemy counterparts, Ridgway explained to them at length the objectives they were to attempt to achieve and what he considered proper demeanor at the conference table. He stressed that their skillful conduct of armistice negotiations might well mark the beginning of communism's recession in Asia.[30]

The Prognosis

Following two days of stage setting by the liaison officers of both sides, the two delegations met for the first time at Kaesong on 10 July. This first meeting rang down the curtain on the war of movement. A full year of bitter fighting had served only to bring the opposing forces into balance. As armistice negotiations began, United Nations Command ground forces in Korea exceeded 550,000, the bulk of which comprised 17 divisions (7 American and 10 ROK), 4 brigades, 1 separate regiment, and 9 separate battalions. Enemy forces totaled about 459,000 divided among 13 Chinese armies and 7 North Korean corps. The significant point of difference was in available reserves. Whereas the United Nations Command had no appreciable source of reinforcement anywhere, its opponents had close at hand some 743,000 Chinese troops in Manchuria.

The willingness of the Chinese to negotiate an armistice rather than commit their large reserve to battle undoubtedly was prompted in large part by the high losses they had sustained since intervening eight months earlier. By 10 July 1951, estimates of total enemy casualties had risen above 1,200,000, divided almost evenly between the Chinese and North Koreans. The costs to United Nations Command forces also had been dear. By the end of June 1951 American combat losses stood at about 78,800, of whom approximately 21,300 were killed in action or subsequently died as a result of their combat participation. Losses among other United Nations contingents were in proportion to the Americans'; and ROK Army casualties numbered 212,554, including 21,625 dead. The ROK civilian population had paid a still higher price, suffering some 469,000 casualties, of whom at least 170,000 had been killed.

Whereas neither of the opposing forces had been able to achieve a final victory, each had made significant gains. The United Nations Command forces had at least met their objective of repelling the aggression against South Korea; for with the exception of a small area in the west, the republic had been cleared of enemy forces, and even some territory above the 38th Parallel now was under United Nations Command control. A startling gain for Communist China had resulted from battle successes during the past winter. These victories had raised the prestige of Mao Tse-tung's regime and won it a front-rank position as a military power. While offset to some degree by a lack of an atomic capability and a de-

[29] *Ibid.*

[30] Memo, Gen M. B. Ridgway for General and Flag-Officer Members of the U.N. Delegation, 7 Jul 51, in GHQ, UNC SGS files.

pendence on the USSR for industrial and technical support, Communist China's new prominence was certain to upset the political balance in Asia.

But these gains would serve neither side for the duration of the Korean War. On the battlefield, both sides might claim to have the stronger force; but the opposing commanders would recognize that the near parity of military power left them only the prospect of directing operations in a war they could not win. At the conference table, both delegations might profess to be negotiating from positions of superior strength, but each would know that the other possessed no decisive advantage. The challenge here, then, would be to achieve an armistice under favorable terms without being able to dictate those terms. Indeed, those responsible for policy and direction during the first year of the Korean War had set the scene for what could prove to be a long, tedious stalemate at two locations: in the truce tent and on the fighting front.

Bibliographical Note

The official documentary sources upon which this book is based fall logically into two categories, records seen originally in the Far East Command, and Department of the Army records seen in Washington, D.C., and Alexandria, Virginia. The former records, exploited during research from mid-1950 to early 1953, comprise a variety of record collections and papers some of which have been destroyed as part of the normal records management processes. The remainder have been returned to the United States and are now in custody of the National Archives and Records Service of the General Services Administration. Department of the Army records examined in the Washington, D.C., area from 1953 to 1956 have been relocated from original repositories as part of the decentralization of records during the late 1960's but remain in custody of the National Archives and Records Service.

Far East Command Records

Included under this grouping are official records which were maintained under control of the Adjutant General, FEC, the records maintained with the General Staff and Special Staff Sections of GHQ, FEC/UNC, the records of FEAF and COMNAVFE, and the records of Eighth Army and X Corps. In addition, certain "convenience" files were maintained for the Office of the Chief of Staff, GHQ, FEC, and for the Office of the Commander in Chief Far East Command/United Nations Command, all of which were made available for research. The latter files contained memorandums and reports not normally available in the Adjutant General's files. Within the substantial body of Far East Command records, the following were especially important in preparation of this history:

Chief of Staff, GHQ FEC/UNC Files

These files, originally located in the Dai Ichi building in Tokyo, adjacent to CINCFE/CINCUNC's offices were not designed for permanent retention. They reflect all major activities of CINCUNC and consist of personal messages and memorandums, command letters, personal letters, and miscellaneous items. Their particular value lies in the manner in which they reflect the day-to-day trends in CINCUNC's thinking and in the fact that they contain his instructions to his senior staff members.

Chief of Staff Daily Folders

Prepared by the SGS, GHQ, for presentation to the Chief of Staff each morning, these folders contained copies of all pertinent communications between Washington and Tokyo and between CINCUNC and his subordinate commanders. Memorandums for record of important telephone calls, copies of miscellaneous memorandums for record on meetings, and liaison officer reports on visits to Korea are among the particularly valuable items in these files not elsewhere available. These files, comprising twelve file drawers of material, were turned over to the Military History Section, GHQ, FEC/UNC, in October 1951.

JSPOG Files

Under control of the ACofS G–3, GHQ, FEC/UNC, the Joint Strategic Plans and Operations Group maintained separate files comprising all the joint planning files of the theater. These files, arranged in books by subject, contained detailed staff studies of contemplated operations, operations plans and operations orders, interspersed with attached handwritten comments by key officers of command.

FEC Reports

Several recurring reports prepared within the FEC have provided information and views on Korea not available elsewhere. These are:

The Annual Narrative Historical Report, GHQ FEC

At the outbreak of the Korean War, the only recurring historical report prepared by the FEC was the Annual Narrative Historical Report, required by Army regulations. The FEC had prepared such a report for 1949 and subsequently prepared a report for the period 1 January–31 October 1950. Thus the period 25 June–31 October 1950 is covered, not by a monthly Command Report (see below), but by the Annual Report which is less detailed. Nevertheless, the staff sections of GHQ/FEC, particularly the ACofS, G–3, included in this Annual Report unique information and documents on Korean planning and operations. In this connection, the Eighth Army was relieved of the requirement for submitting a historical report covering the period 25 June–31 October 1950 and such historical records of Eighth Army's activities as exist for that period consist of War Diaries. The period 1 January to 25 June 1950 is not covered by a historical report from the Eighth Army.

The Monthly Command Report, FEC/UNC

Beginning on 1 November 1950, GHQ, FEC/UNC, prepared and submitted each month to the Department of the Army a Command Report, describing in detail the operations, activities, and problems of the command. The basic narrative report is accompanied by annexes from each General and Special Staff Section of GHQ FEC/UNC. The most valuable of these annexes, from the historian's viewpoint, are those of the ACofS, G–3, the ACofS, G–2, and those prepared by the Commander in Chief and the Chief of Staff GHQ FEC/UNC. The latter annex did not appear until April 1951 when the Military History

Section, GHQ FEC/UNC, assigned an officer to prepare this report for the Chief of Staff.

Daily Intelligence Summary and Special Reports

From the beginning of the war Daily Intelligence summaries were prepared by the Theater Intelligence Division, ACofS, G–2, FEC/UNC. These summaries, in booklet form, contain detailed information on enemy dispositions, order of battle, the combat situation, and estimated enemy intentions. Each summary contains several maps illustrating enemy and friendly action for each 24-hour period. These Daily Intelligence summaries received wide distribution within the command and were sent to Washington daily. Copies are filed with the Command Report, GHQ FEC/UNC. In addition to these summaries, the ACofS, G–2, prepared in similar format special intelligence reports on such subjects as enemy order of battle, enemy LOC's, etc., copies of which were also placed with the monthly Command Report.

Daily Operations Report

The ACofS, G–3, GHQ FEC/UNC, issued a Daily Operations Report, covering friendly and enemy information, and setting forth in some detail the combat operations for each 24-hour period. In many respects this report duplicates the Daily Intelligence Summary, which, of the two, is more useful to the historian. Copies of the Daily Operations Report are also available with the Command Report, GHQ FEC/UNC.

X Corps Special Reports

The X Corps prepared special reports on the Inch'on landing, on the Wonsan landing, on the Hungnam evacuation, and on the battle of the Soyang River. These reports, although omitting derogatory information, are nonetheless useful in establishing dates, locations, and the general chain of events. Copies are in custody of the National Archives and Records Service.

Department of the Army Records

ACofS G–3 Correspondence Files

By far the most productive segment of Department of Army Records for the historian interested in Korea are the correspondence files of the ACofS G–3, DA, for the period. Bearing a file identification of G3 091 Korea, these voluminous files record the actions and recommendations of the G–3 and of the Chief of Staff, U.S. Army, and contain many important national policy papers. They contain all pertinent Joint Chiefs of Staff documents received by the Department of the Army together with the Army input to the Joint Chiefs of Staff. They are complete files in that each action is backed up by all pertinent correspondence and is carried through to the final recommendation or conclusion. Of particular interest are thirty-one notebooks prepared by the Far East and Pacific Branch, ACofS G–3, covering specific matters. These notebooks contain radios, memorandums, and charts and tables used in briefing the ACofS G–3.

JSSC Committee Report to the JCS

In preparing testimony to be presented before the Senate committee investigating the relief of General MacArthur, the Joint Chiefs of Staff instructed the Joint Strategic Survey Committee to prepare a synthesis of all matters relating to the relief of MacArthur and the conduct of the Korean War. This committee presented the Joint Chiefs of Staff with a well-documented narrative of exceptional value to the historian for the period 25 June 1950–30 April 1951. This document is filed with the Top Secret records of the ACofS G–3, DA, for 1951 in custody of the National Archives and Records Service.

Radio Files

The exchanges between Washington and Tokyo, notably between the JCS and CINCUNC, were carried on by radio communication. The messages sent and received by the JCS/Department of the Army to CINCUNC/CINCFE are on file in the Staff Communications Center, Office of the Chief of Staff, DA. Messages exchanged between CINCUNC/CINCFE and his subordinate commanders are in the Adjutant General's file, FEC, at the Federal Records Center, GSA, Kansas City, Missouri.

Teleconferences

Several important teleconferences were held in the first days of the Korean War, followed by routine teleconferences for the remainder of the war, the latter conferences usually on intelligence matters. Copies of these teleconferences are on file in the Staff Communications Center, Office of the Chief of Staff, DA.

Secondary Sources

The MacArthur Hearings

As part of the furor surrounding the relief of General MacArthur, the Senate of the United States conducted an investigation. Virtually every responsible official of the Department of Defense testified at the Senate hearings on this matter. The record of this testimony has been printed by the U.S. Government Printing Office in five volumes totaling 3,691 pages of testimony, an appendix of selected documents, and an index. The rather formidable title of this testimony is *Military Situation in the Far East, Hearings Before the Committee on Armed Services and the Committee on Foreign Relations, United States Senate, Eighty Second Congress, First Session, To Conduct an Inquiry into the Military Situation in the Far East and the Facts Surrounding the Relief of General of the Army Douglas MacArthur From His Assignments in that Area.* This material is cited in the present volume as the *MacArthur Hearings.*

General Smith's Chronicles

Major General Oliver Prince Smith, USMC, commander of the 1st Marine Division during the first part of the Korean War, maintained a diary which he furnished to the Historical Branch, G–3 Division, USMC. Capt. Nicholas Canzona and Mr. Lynn Montross of that branch made available selected portions of that diary to the author. The designation of this document is "Notes by Lt. Gen. Oliver P. Smith on the Operations of the 1st Marine Division During the First Nine Months of the Korean War."

Books

The following books are the principal ones consulted in preparation of this history:

Acheson, Dean. *Present at the Creation.* New York: W. W. Norton and Company, 1969.

Appleman, Roy E. *South to the Naktong: North to the Yalu.* UNITED STATES ARMY IN THE KOREAN WAR. Washington, 1961.

Cagle, Malcolm W., and Frank A. Manson. *The Sea War in Korea.* Annapolis: U.S. Naval Institute, 1957.

Collins, J. Lawton. *War In Peacetime.* Boston, Houghton Mifflin, 1969.

Department of State. *The Conflict in Korea.* Washington, 1951.

Department of State. *In Quest of Peace and Security: Selected Documents on American Foreign Policy, 1941–1951.* Washington, 1951.

Department of State. *Korea, 1945 to 1948.* Washington, 1948.

Department of State. *Korea's Independence.* Washington, 1947.

Field, James A., Jr. *History of U.S. Naval Operations: Korea.* Washington, 1962.

Futrell, R. Frank. *The United States Air Force in Korea, 1950–1953.* New York: Duell, Sloan and Pearce, 1961.

Goodrich, Leland M. *Korea: A Study of U.S. Policy in the United Nations.* New York: Council on Foreign Relations, 1956.

Grajdanzev, Andrew J. *Modern Korea.* New York: John Day Company, 1944.

McCune, George M., with Arthur L. Gray, Jr. *Korea Today.* Cambridge: Harvard University Press, 1950.

Nelson, M. Frederick. *Korea and the Old Orders in Eastern Asia.* Baton Rouge: Louisiana State University Press, 1946.

Office of the Chief of Military History, Department of the Army. *Korea 1951–1953.* Washington, 1956.

Ridgway, Matthew B. *The Korean War.* Garden City: Doubleday & Company, Inc., 1967.

Sawyer, Maj. Robert K. *Military Advisors in Korea: KMAG in Peace and War.* Washington, 1963.

Truman, Harry S. *Memoirs: Years of Trial and Hope,* Vol. II. New York: Doubleday and Company, Inc., 1956.

Whiting, Allen S. *China Crosses the Yalu.* New York: The Macmillan Company, 1960.

List of Abbreviations

Abn	Airborne
ACofS	Assistant Chief of Staff
ADCOM	Advance Command and Liaison Group
Admin	Administration
AFFE	Army Forces Far East
AGO	Office of The Adjutant General
AMIK	American Mission in Korea
Asst	Assistant
ATIS	Allied Translator and Interpreter Section
CCF	Chinese Communist Forces
CG	Commanding general
CINCFE	Commander in Chief, Far East Command
CINCUNC	Commander in Chief, United Nations Command
Comd	Command
COMNAVFE	Commander, Naval Forces, Far East
DF	Disposition form
DIS	Daily Intelligence Summary
Div	Division
ECA	Economic Co-operation Administration
ERC	Enlisted Reserve Corps
EUSAK	Eighth U.S. Army in Korea
FEAF	Far East Air Forces
FEC	Far East Command
FMFPAC	Fleet Marine Force, Pacific
G–1	Personnel section of divisional or higher staff
G–2	Intelligence section of divisional or higher staff
G–3	Operations and training section of divisional or higher staff
G–4	Logistics section of divisional or higher staff
GO	General order
Interv	Interview
JANIS	Joint Army-Navy Intelligence Service
JCS	Joint Chiefs of Staff

JIC	Joint Intelligence Committee
JLC	Japan Logistical Command
Jnl	Journal
JSPOG	Joint Strategic Plans and Operations Group
JSSC	Joint Strategic Survey Committee
KMAG	United States Military Advisory Group to the Republic of Korea
LST	Landing ship, tank
Ltr	Letter
MARBO	Marianas-Bonins Command
MDAP	Mutual Defense Assistance Program
Memo	Memorandum
MOS	Military occupational speciality
MS	Manuscript
Msg	Message
MFR	Memorandum for record
Mono	Monograph
NATO	North Atlantic Treaty Organization
NAVFE	Naval Forces, Far East
NKA	North Korean Army
NPRJ	National Police Reserve of Japan
NSC	National Security Council
OCAFF	Office, Chief of Army Field Forces
OCMH	Office, Chief of Military History
Opn	Operation
ORC	Organized Reserve Corps
OSA	Office of the Secretary of the Army
OSI	Office of Strategic Intelligence
PHILCOM	Philippines Command
PMAG	Provisional Military Advisory Group
Rad	Radio
RCT	Regimental combat team
ROK	Republic of Korea
Rpt	Report
RYCOM	Ryukyus Command
SANACC	State-Army-Navy-Air Force Co-ordinating Committee
SCAP	Supreme Commander, Allied Powers
Sgd	Signed
SGS	Secretary of the General Staff
T/O	Table of Organization
TO&E	Table of Organization and Equipment
Telecon	Teleconference
Tng	Training

UNC	United Nations Command
UNCOK	United Nations Commission in Korea
UNCURK	United Nations Commission for the Unification and Rehabilitation of Korea
USAFIK	United States Army Forces in Korea
USAMGIK	United States Army Military Government in Korea
USMILAT	U.S. Military Attaché
ZI	Zone of Interior

Basic Military Map Symbols

Symbols within a rectangle indicate a military unit, within a triangle an observation post, and within a circle a supply point.

Military Units—Identification

Antiaircraft Artillery	
Armored Command	
Army Air Forces	
Artillery, except Antiaircraft and Coast Artillery	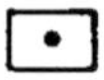
Cavalry, Horse	
Cavalry, Mechanized	
Chemical Warfare Service	
Coast Artillery	
Engineers	
Infantry	
Medical Corps	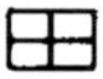
Ordnance Department	
Quartermaster Corps	
Signal Corps	
Tank Destroyer	
Transportation Corps	
Veterinary Corps	

Airborne units are designated by combining a gull wing symbol with the arm or service symbol:

Airborne Artillery	
Airborne Infantry	

Size Symbols

The following symbols placed either in boundary lines or above the rectangle, triangle, or circle inclosing the identifying arm or service symbol indicate the size of military organization:

Squad	•
Section	••
Platoon	•••
Company, troop, battery, Air Force flight	I
Battalion, cavalry squadron, or Air Force squadron	I I
Regiment or group; combat team (with abbreviation CT following identifying numeral)	I I I
Brigade, Combat Command of Armored Division, or Air Force Wing	X
Division or Command of an Air Force	XX
Corps or Air Force	XXX
Army	XXXX
Group of Armies	XXXXX

EXAMPLES

The letter or number to the left of the symbol indicates the unit designation; that to the right, the designation of the parent unit to which it belongs. Letters or numbers above or below boundary lines designate the units separated by the lines:

Company A, 137th Infantry

8th Field Artillery Battalion

Combat Command A, 1st Armored Division

Observation Post, 23d Infantry

Command Post, 5th Infantry Division

Boundary between 137th and 138th Infantry

Weapons

Machine gun

Gun

Gun battery

Howitzer or Mortar

Tank

Self-propelled gun

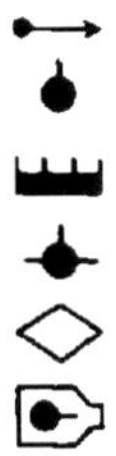

Index

U.S. GOVERNMENT PRINTING OFFICE: 1991 O 304-303

Made in United States
North Haven, CT
07 March 2025

66543435R00257